Universitext

Springer
New York
Berlin
Heidelberg
Hong Kong
London
Milan
Paris
Tokyo

Universitext

Editors (North America): S. Axler, F.W. Gehring, and K.A. Ribet

(continued after index)

Bruce van Brunt

The Calculus of Variations

With 24 Figures

Springer

Bruce van Brunt
Institute of Fundamental Sciences
Palmerston North Campus
Private Bag 11222
Massey University
Palmerston North 5301
New Zealand
b.vanbrunt@massey.ac.nz

Mathematics Subject Classification (2000): 34Bxx, 49-01, 70Hxx

Library of Congress Cataloging-in-Publication Data
van Brunt, B. (Bruce)
   The calculus of variations / Bruce van Brunt.
      p. cm. — (Universitext)
   Includes bibliographical references and index.

      1. Calculus of variations.   I. Title.
   QA315.V35 2003
      515′.64—dc21   2003050661
   ISBN 978-1-4419-2316-5
   e-ISBN 978-0-387-21697-3   Printed on acid-free paper.

www.springer-ny.com

Springer-Verlag   New York Berlin Heidelberg
A member of BertelsmannSpringer Science+Business Media GmbH

To Anne, Anastasia, and Alexander

Preface

The calculus of variations has a long history of interaction with other branches of mathematics such as geometry and differential equations, and with physics, particularly mechanics. More recently, the calculus of variations has found applications in other fields such as economics and electrical engineering. Much of the mathematics underlying control theory, for instance, can be regarded as part of the calculus of variations.

This book is an introduction to the calculus of variations for mathematicians and scientists. The reader interested primarily in mathematics will find results of interest in geometry and differential equations. I have paused at times to develop the proofs of some of these results, and discuss briefly various topics not normally found in an introductory book on this subject such as the existence and uniqueness of solutions to boundary-value problems, the inverse problem, and Morse theory. I have made "passive use" of functional analysis (in particular normed vector spaces) to place certain results in context and reassure the mathematician that a suitable framework is available for a more rigorous study. For the reader interested mainly in techniques and applications of the calculus of variations, I leavened the book with numerous examples mostly from physics. In addition, topics such as Hamilton's Principle, eigenvalue approximations, conservation laws, and nonholonomic constraints in mechanics are discussed. More importantly, the book is written on two levels. The technical details for many of the results can be skipped on the initial reading. The student can thus learn the main results in each chapter and return as needed to the proofs for a deeper understanding. Several key results in this subject have tractable analogues in finite-dimensional optimization. Where possible, the theory is motivated by first reviewing the theory for finite-dimensional problems.

The book can be used for a one-semester course, a shorter course, or independent study. The final chapter on the second variation has been written with these options in mind, so that the student can proceed directly from Chapter 3 to this topic. Throughout the book, asterisks have been used to flag material that is not central to a first course.

The target audience for this book is advanced undergraduate/ beginning graduate students in mathematics, physics, or engineering. The student is assumed to have some familiarity with linear ordinary differential equations, multivariable calculus, and elementary real analysis. Some of the more theoretical material from these topics that is used throughout the book such as the implicit function theorem and Picard's theorem for differential equations has been collected in Appendix A for the convenience of the reader.

Like many textbooks in mathematics, this book can trace its origins back to a set of lecture notes. The transformation from lecture notes to textbook, however, is nontrivial, and one is faced with myriad choices that, in part, reflect one's own interests and experiences teaching the subject. While writing this book I kept in mind three quotes spanning a few generations of mathematicians. The first is from the introduction to a volume of Spivak's multivolume treatise on differential geometry [64]:

> I feel somewhat like a man who has tried to cleanse the Augean stables
> with a Johnny-Mop.

It is tempting, when writing a textbook, to give some modicum of completeness. When faced with the enormity of literature on this subject, however, the task proves daunting, and it soon becomes clear that there is just too much material for a single volume. In the end, I could not face picking up the Johnny-Mop, and my solution to this dilemma was to be savage with my choice of topics. Keeping in mind that the goal is to produce a book that should serve as a text for a one-semester introductory course, there were many painful omissions. Firstly, I have tried to steer a reasonably consistent path by keeping the focus on the simplest type problems that illustrate a particular aspect of the theory. Secondly, I have opted in most cases for the "no frills" version of results if the "full feature" version would take us too far afield, or require a substantially more sophisticated mathematical background. Topics such as piecewise smooth extremals, fields of extremals, and numerical methods arguably belong in any introductory account. Nonetheless, I have omitted these topics in favour of other topics, such as a solution method for the Hamilton-Jacobi equation and Noether's theorem, that are accessible to the general mathematically literate undergraduate student but often postponed to a second course in the subject.

The second quote comes from the introduction to Titchmarsh's book on eigenfunction expansions [70]:

> I believe in the future of 'mathematics for physicists', but it seems
> desirable that a writer on this subject should understand both physics
> as well as mathematics.

The words of Titchmarsh remind me that, although I am a mathematician interested in the applications of mathematics, I am not a physicist, and it is best to leave detailed accounts of physical models in the hands of experts. This is not to say that the material presented here lies in some vacuum of pure mathematics, where we merely acknowledge that the material has found some applications. Indeed, the book is written with a definite slant towards "applied mathematics," but it focuses on no particular field of applied mathematics in any depth. Often it is the application not the mathematics that perplexes the student, and a study in depth of any particular field would require either the student to have the necessary prerequisites or the author to develop the subject. The former case restricts the potential audience; the latter case shifts away from the main topic. In any event, I have not tried to write a book on the calculus of variations with a particular emphasis on one of its many fields of applications. There are many splendid books that merge the calculus of variations with particular applications such as classical mechanics or control theory. Such texts can be read with profit in conjunction with this book.

The third quote comes from G.H. Hardy, who made the following comment about A.R. Forsyth's 656-page treatise [27] on the calculus of variations :[1]

> In this enormous volume, the author never succeeds in proving that the shortest distance between two points is a straight line.

Hardy did not mince words when it came to mathematics. The prospective author of any text on the calculus of variations should bear in mind that, although there are many mathematical avenues to explore and endless minutiæ to discuss, certain basic questions that can be answered by the calculus of variations in an elementary text should be answered. There are certain problems such as geodesics in the plane and the catenary that can be solved within our self-imposed regime of elementary theory. I do not hesitate to use these simple problems as examples. At the same time, I also hope to give the reader a glimpse of the power and elegance of a subject that has fascinated mathematicians for centuries.

I wish to acknowlege the help of my former students, whose input shaped the final form of this book. I wish also to thank Fiona Davies for helping me with the figures. Finally, I would like to acknowledge the help of my colleagues at the Institute of Fundamental Sciences, Massey University.

The earlier drafts of many chapters were written while travelling on various mountaineering expeditions throughout the South Island of New Zealand. The hospitality of Clive Marsh and Heather North is gratefully acknowledged along with that of Andy Backhouse and Zoe Hart. I should also like to acknowledge the New Zealand Alpine Club, in whose huts I wrote many early (and later) drafts during periods of bad weather. In particular, I would like to thank Graham and Eileen Jackson of Unwin Hut for providing a second home conducive to writing (and climbing).

Fox Glacier, New Zealand   *Bruce van Brunt*
February 2003

---

[1] F. Smithies reported this comment in an unpublished talk, "Hardy as I Knew Him," given to the *British Society for the History of Mathematics*, 19 December 1990.

Contents

1

Introduction

1.1 Introduction

The calculus of variations is concerned with finding extrema and, in this sense, it can be considered a branch of optimization. The problems and techniques in this branch, however, differ markedly from those involving the extrema of functions of several variables owing to the nature of the domain on the quantity to be optimized. A **functional** is a mapping from a set of functions to the real numbers. The calculus of variations deals with finding extrema for functionals as opposed to functions. The candidates in the competition for an extremum are thus functions as opposed to vectors in $\mathbb{R}^n$, and this gives the subject a distinct character. The functionals are generally defined by definite integrals; the sets of functions are often defined by boundary conditions and smoothness requirements, which arise in the formulation of the problem/model.

The calculus of variations is nearly as old as the calculus, and the two subjects were developed somewhat in parallel. In 1927 Forsyth [27] noted that the subject "attracted a rather fickle attention at more or less isolated intervals in its growth." In the eighteenth century, the Bernoulli brothers, Newton, Leibniz, Euler, Lagrange, and Legendre contributed to the subject, and their work was extended significantly in the next century by Jacobi and Weierstraß. Hilbert [38], in his renowned 1900 lecture to the International Congress of Mathematicians, outlined 23 (now famous) problems for mathematicians. His 23rd problem is entitled *Further development of the methods of the calculus of variations.* Immediately before describing the problem, he remarks:

> ...I should like to close with a general problem, namely with the indication of a branch of mathematics repeatedly mentioned in this lecture—which, in spite of the considerable advancement lately given it by Weierstraß, does not receive the general appreciation which in my opinion it is due—I mean the calculus of variations.

Hilbert's lecture perhaps struck a chord with mathematicians.[1] In the early twentieth century Hilbert, Noether, Tonelli, Lebesgue, and Hadamard among others made significant contributions to the field. Although by Forsyth's time the subject may have "attracted rather fickle attention," many of those who did pay attention are numbered among the leading mathematicians of the last three centuries. The reader is directed to Goldstine [36] for an in-depth account of the history of the subject up to the late nineteenth century.

The enduring interest in the calculus of variations is in part due to its applications. Of particular note is the relationship of the subject with classical mechanics, where it crosses the boundary from being merely a mathematical tool to encompassing a general philosophy. Variational principles abound in physics and particularly in mechanics. The application of these principles usually entails finding functions that minimize definite integrals (e.g., energy integrals) and hence the calculus of variations comes naturally to the fore. Hamilton's Principle in classical mechanics is a prominent example. An earlier example is Fermat's Principle of Minimum Time in geometrical optics. The development of the calculus of variations in the eighteenth and nineteenth centuries was motivated largely by problems in mechanics. Most textbooks on classical mechanics (old and new) discuss the calculus of variations in some depth. Conversely, many books on the calculus of variations discuss applications to classical mechanics in detail. In the introduction of Carathéodory's book [21] he states:

> I have never lost sight of the fact that the calculus of variations, as it is presented in Part II, should above all be a servant of mechanics.

Certainly there is an intimate relationship between mechanics and the calculus of variations, but this should not completely overshadow other fields where the calculus of variations also has applications. Aside from applications in traditional fields of continuum mechanics and electromagnetism, the calculus of variations has found applications in economics, urban planning, and a host of other "nontraditional fields." Indeed, the theory of optimal control is centred largely around the calculus of variations.

Finally it should be noted the calculus of variations does not exist in a mathematical vacuum or as a closed chapter of classical analysis. Historically, this field has always intersected with geometry and differential equations, and continues to do so. In 1974, Stampacchia [17], writing on Hilbert's 23rd problem, summed up the situation:

> One might infer that the interest in this branch of Analysis is weakening and that the Calculus of Variations is a Chapter of Classical Analysis. In fact this inference would be quite wrong since new problems like those in control theory are closely related to the problems of

---

[1] His nineteenth and twentieth problems were also devoted to the calculus of variations.

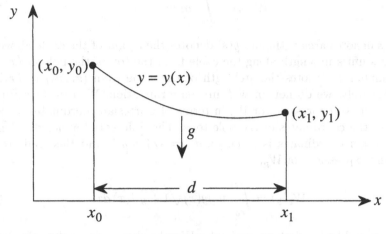

Fig. 1.1.

the Calculus of Variations while classical theories, like that of boundary value problems for partial differential equations, have been deeply affected by the development of the Calculus of Variations. Moreover, the natural development of the Calculus of Variations has produced new branches of mathematics which have assumed different aspects and appear quite different from the Calculus of Variations.

The field is far from dead and it continues to attract new researchers.

In the remainder of this chapter we discuss some typical problems in the calculus of variations that are easy to model (although perhaps not so easy to solve). These problems illustrate the above comments and give the reader a taste of the subject. We return to most of these examples later in the book as the mathematics to solve them develops.

1.2 The Catenary and Brachystochrone Problems

1.2.1 The Catenary

Consider a thin heavy uniform flexible cable suspended from the top of two poles of height y_0 and y_1 spaced a distance d apart (figure 1.1). At the base of each pole the cable is assumed to be coiled. The cable follows up the pole to the top, runs through a pulley, and then spans the distance d to the next pole. The problem is to determine the shape of the cable between the two poles.

The cable will assume the shape that makes the potential energy minimum. The potential energy associated with the vertical parts of the cable will be the same for any configuration of the cable and hence we may ignore this component. If m denotes the mass per unit length of the cable and g the gravitational constant, the potential energy of the cable between the poles is

$$W_p(y) = \int_0^L mgy(s)\,ds, \tag{1.1}$$

where s denotes arclength, and $y(s)$ denotes the height of the cable above the ground s units in length along the cable from the top of the pole at (x_0, y_0). The number L denotes the arclength of the cable from (x_0, y_0) to (x_1, y_1). Unfortunately, we do not know L in this formulation. We can, however, recast the above expression for W_p in terms of Cartesian coördinates since we do know the coördinates of the pole tops. The differential arclength element in Cartesian coördinates is given by $ds = \sqrt{1 + y'^2}$, and this leads to the following expression for W_p,

$$W_p(y) = \int_{x_0}^{x_1} mgy(x)\sqrt{1 + y'^2(x)}\,dx. \tag{1.2}$$

Note that unlike our first expression for W_p, the above one involves the derivative of y. We have implicitly assumed here that the solution curve can be represented by a function $y : [x_0, x_1] \to \mathbb{R}$ and that this function is continuous and at least piecewise differentiable. Given the nature of the problem these seem reasonable assumptions.

The cable will assume the shape that minimizes W_p. The constant factor mg in the expression for W_p can be ignored for the purposes of optimizing the potential energy. The essence of the problem is thus to determine a function y such that the quantity

$$J(y) = \int_{x_0}^{x_1} y\sqrt{1 + y'^2}\,dx \tag{1.3}$$

is minimum. The model requires that any candidate $\hat{y}$ for an extremum satisfies the boundary conditions

$$\hat{y}(x_0) = y_0, \quad \hat{y}(x_1) = y_1. \tag{1.4}$$

In addition, the candidates must also be continuous and at least piecewise differentiable in the interval $[x_0, x_1]$.

We find the extrema for J in Chapter 2, where we show that the shape of the cable can be described by a hyperbolic cosine function. The curve itself is called a **catenary**.[2]

The same functional J arises in a problem in geometry concerning a minimal surface of revolution, i.e., a surface of revolution having minimal surface area. Suppose that the x-axis corresponds to the axis of rotation. Any surface of revolution can be generated by a curve in the xy-plane (figure 1.2). The

---

[2] The name "catenary" is particularly descriptive. The name comes from the Latin word *catena* meaning chain. Catenary refers to the curve formed by a uniform chain hanging freely between two poles. Leibniz is credited with coining the term (ca. 1691).

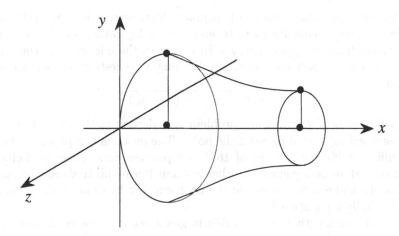

Fig. 1.2.

problem thus translates to finding the curve γ that generates the surface of revolution having the minimal surface area. As with the catenary problem, we make the assumption that γ can be described by a function $y : [x_0, x_1] \to \mathbb{R}$ that is continuous and piecewise differentiable in the interval $[x_0, x_1]$. Under these assumptions we have that the surface area of the corresponding surface of revolution is

$$A(y) = 2\pi \int_{x_0}^{x_1} |y(x)|\sqrt{1 + y'^2(x)}\, dx. \tag{1.5}$$

Here we need also make the assumption that $y(x) > 0$ for all $x \in [x_0, x_1]$.[3] The problem of finding the minimal surface thus reduces to finding the function y such that the quantity

$$J(y) = \int_{x_0}^{x_1} y\sqrt{1 + y'^2}\, dx$$

is minimum. The two problems thus produce the same functional to be minimized. The generating curve that produces the minimal surface of revolution is thus a catenary. The surface itself is called a **catenoid**.

---

[3] If $y = 0$ at some point $\tilde{x} \in (x_0, x_1)$ we can still generate a rotationally symmetric "object," but technically it would not be a surface. Near $(\tilde{x}, 0, 0)$ the "object" would resemble (i.e., be homeomorphic to) a double cone. The double cone fails the requirements to be a surface because any neighbourhood containing the common vertex is not homeomorphic to the plane.

Let us return to the original problem. A modification of the problem would be to first specify the length of the cable. Evidently, if L is the length of the cable we must require that

$$L \geq \sqrt{(x_1 - x_0)^2 + (y_1 - y_0)^2}$$

in order that the cable span the two poles. Moreover, it is intuitively clear that in the case of equality there is only one configuration possible viz., the line segment from (x_0, y_0) to (x_1, y_1). In this case, there is no optimization to be done as there is only one candidate. We may thus restrict our attention to the case

$$L > \sqrt{(x_1 - x_0)^2 + (y_1 - y_0)^2}.$$

Given a cable of length L, the problem is to determine the shape the cable assumes when supported between the poles. The problem was posed by Jacob Bernoulli in 1690. By the end of 1691 the problem was solved by Leibniz, Huygens, and Jacob's younger brother Johann Bernoulli. It should be noted that Galileo had earlier considered the problem, but he thought the catenary was essentially a parabola.[4]

Since the arclength L of the cable is given, we can use expression (1.1) to look for a minimum potential energy configuration. Instead, we start with expression (1.2). The modified problem is now to find the function $y : [x_0, x_1] \rightarrow \mathbb{R}$ such that W_p is minimized subject to the arclength constraint

$$L = \int_{x_0}^{x_1} \sqrt{1 + y'^2} \, dx, \tag{1.6}$$

and the boundary conditions

$$y(x_0) = y_0, \quad y(x_1) = y_1.$$

This problem is thus an example of a constrained variational problem. The constraint (1.6) can be regarded as an integral equation (with, it is hoped, nonunique solutions). Constraints such as (1.6) are called **isoperimetric**. We discuss problems having isoperimetric constraints in Chapter 4.

Suppose that we use expression (1.1), which *prima facie* seems simpler than expression (1.2). We know L, so that the limits of the integral are known, but the parameter s is special and corresponds to arclength. We must somehow build in the requirement that s is arclength if we are to use expression (1.1). In order to do this we must use a parametric representation of the curve $(x(s), y(s))$, $s \in [0, L]$. The arclength parameter for such a curve is characterized by the differential equation

$$x'^2(s) + y'^2(s) = 1. \tag{1.7}$$

---

[4] There is still some dispute regarding whether Galileo thought the catenary to be the parabola. See Giaquinta and Hildebrandt [32], p. 133 for more details.

The problem thus entails finding the functions $x(s)$, $y(s)$ that minimize W_p subject to the constraint (1.7) and the boundary conditions

$$x(0) = x_0, \quad x(L) = x_1$$
$$y(0) = y_0, \quad y(L) = y_1.$$

In general, a constraint of this kind is more difficult to deal with than an isoperimetric constraint.

1.2.2 Brachystochrones

The history of the calculus of variations essentially begins with a problem posed by Johann Bernoulli (1696) as a challenge to the mathematical community and in particular to his brother Jacob. (There was significant sibling rivalry between the two brothers.) The problem is important in the history of the calculus of variations because the method developed by Johann's pupil, Euler, to solve this problem provided a sufficiently general framework to solve other variational problems.

The problem that Johann posed was to find the shape of a wire along which a bead initially at rest slides under gravity from one end to the other in minimal time. The endpoints of the wire are specified and the motion of the bead is assumed frictionless. The curve corresponding to the shape of the wire is called a **brachystochrone**[5] or a curve of fastest descent.

The problem attracted the attention of a number of mathematical luminaries including Huygens, L'Hôpital, Leibniz, and Newton, in addition of course to the Bernoulli brothers, and later Euler and Lagrange. This problem was at the cutting edge of mathematics at the turn of the eighteenth century.

Jacob was up to the challenge and solved the problem. Meanwhile (and independently) Johann and Leibniz also arrived at correct solutions. Newton was late to the party because he learned about the problem some six months later than the others. Nonetheless, he solved the problem that same evening and sent his solution anonymously the next day to Johann. Newton's cover was blown instantly. Upon looking at the solution, Johann exclaimed "Ah! I recognize the paw of the lion."

To model Bernoulli's problem we use Cartesian coördinates with the positive y-axis oriented in the direction of the gravitational force (figure 1.3). Let (x_0, y_0) and (x_1, y_1) denote the coördinates of the initial and final positions of the bead, respectively. Here, we require that $x_0 < x_1$ and $y_0 < y_1$. The Bernoulli problem consists of determining, among the curves that have (x_0, y_0) and (x_1, y_1) as endpoints, the curve on which the bead slides down from (x_0, y_0) to (x_1, y_1) in minimum time. The problem makes sense only for continuous curves. We make the additional simplifying (but reasonable) assumptions that the curve can be represented by a function $y : [x_0, x_1] \to \mathbb{R}$

---

[5] The word comes from the Greek words *brakhistos* meaning "shortest" and *khronos* meaning time.

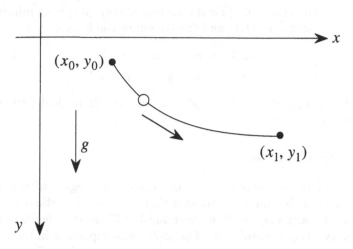

Fig. 1.3.

and that y is at least piecewise differentiable in the interval $[x_0, x_1]$. Now, the total time it takes the bead to slide down a curve is given by

$$T(y) = \int_0^L \frac{ds}{v(s)}, \tag{1.8}$$

where L denotes the arclength of the curve, s is the arclength parameter, and v is the velocity of the bead s units down the curve from (x_0, y_0). As with the catenary problem, we do not know the value of L, so we must seek an alternative formulation.

Our first job is to get an expression for the velocity in terms of the function y. We use the law of conservation of energy to achieve this. At any position $(x, y(x))$ on the curve, the sum of the potential and kinetic energies of the bead is a constant. Hence

$$\frac{1}{2}mv^2(x) + mgy(x) = c, \tag{1.9}$$

where m is the mass of the bead, v is the velocity of the bead at $(x, y(x))$, and c is a constant. Since the energy is constant along the curve, we know that

$$c = \frac{1}{2}mv^2(x_0) + mgy(x_0).$$

Solving equation (1.9) for v gives

$$v(x) = \sqrt{\frac{2c}{m} - 2gy(x)}.$$

Equation (1.8) thus implies that

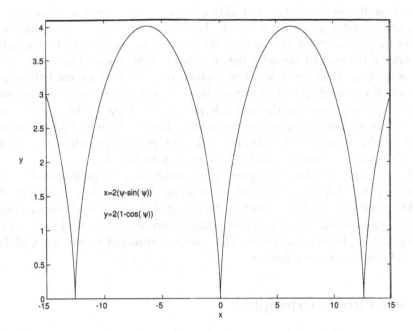

Fig. 1.4.

$$T(y) = \int_{x_0}^{x_1} \frac{\sqrt{1 + y'^2}}{\sqrt{\frac{2c}{m} - 2gy(x)}} \, dx. \tag{1.10}$$

We thus seek a function y such that T is minimum and

$$y(x_0) = y_0, \quad y(x_1) = y_1.$$

Note that for the purposes of optimization T can be replaced by the functional

$$J(y) = \int_{x_0}^{x_1} \frac{\sqrt{1 + w'^2}}{\sqrt{w}} \, dx, \tag{1.11}$$

and the relation

$$w(x) = \frac{1}{2g} \left(\frac{2c}{m} - 2gy(x) \right)$$

(the $\sqrt{2g}$ factor does not affect the extrema of J).

In Chapter 2 we find the extrema for J (and hence T), and show that the brachystochrone for this problem is a portion of a special type of curve called a **cycloid**. Figure 1.4 depicts a cycloid. You can visualize a cycloid in the safety of your own home by painting a white dot on a clean tyre and then rolling the tyre along a line. If you can follow the rolling dot, the curve traced out is a cycloid. Before the fabulous Bernoulli brothers came on the stage, Christiaan Huygens had already discovered a remarkable property of cycloids.

Christiaan discovered that a bead sliding down a cycloid generated by a circle of radius ρ under gravity reaches the bottom of the cycloid arch after the period $\pi\sqrt{\rho/g}$ *wherever* on the arch the bead starts from rest. This notable property of the cycloid earned it the appellation **isochrone**. The cycloid thus sports the names isochrone and brachystochrone.[6] Christiaan used the curve to good effect and designed what was then considered a remarkably accurate pendulum clock based on the laudable properties of the cycloid, which was used to govern the motion of the pendulum. The reader may find a diagram of the pendulum and further details on this interesting curve in an article by Tee [67] wherein several original references may be found.

Finally, we note that brachystochrone problems have proliferated in the three centuries following Bernoulli's challenge. Some models subjected the bead to a resisting medium whilst others changed the force field from a simple uniform gravitational field to more complicated scenarios. Research is still progressing on brachystochrones. The reader is directed to the work of Tee [67], [68], [69] for more references.

1.3 Hamilton's Principle

There are many fine books on classical (analytical) mechanics (e.g., [1], [6], [35], [48], [49], [59], and [73]) and we make no attempt here to give even a basic account of this seemingly vast subject. Nonetheless, it would be demeaning to the calculus of variations to ignore its rich heritage and fruitful interaction with classical mechanics. Moreover, many of our examples come from classical mechanics, so a few words from our sponsor seem in order.

Classical mechanics is teeming with variational principles of which Hamilton's Principle is perhaps the most important. [7] In this section we give a brief "no frills" statement of Hamilton's Principle as it applies to the motion of particles. The serious student of mechanics should consult one of the many specialized texts on this subject.

Let us first consider the motion of a single particle in $\mathbb{R}^3$. Let $\mathbf{r}(t) = (x(t), y(t), z(t))$ denote the position of the particle at time t. The **kinetic energy** of this particle is given by

$$T = \frac{1}{2}m\left(\dot{x}^2(t) + \dot{y}^2(t) + \dot{z}^2(t)\right),$$

where m is the mass of the particle and $\dot{\ }$ denotes d/dt. We assume that the forces on the particle can be derived from a single scalar function. Specifically, we assume there is a function V such that:

---

[6] It is also called a **tautochrone**, but we do not count this since the word is derived from the Greek word *tauto* meaning "same." The prefix iso comes from the Greek word *isos*, which also means "same."

[7] One need only scan through Lanczos' book [48] to find the "Principle of Virtual Work," "D'Alembert's Principle," "Gauss' Principle of Least Constraint," "Jacobi's Principle," and, of course, "Hamilton's Principle" among others.

1. V depends only on time and position; i.e., $V = V(t, x, y, z)$;
2. the force $\mathbf{f} = (f_1, f_2, f_3)$ acting on the particle has the components

$$f_1 = -\frac{\partial V}{\partial x}, \quad f_2 = -\frac{\partial V}{\partial y}, \quad f_3 = -\frac{\partial V}{\partial z}.$$

The function V is called the **potential energy**. Let

$$L = T - V.$$

The function L is called the **Lagrangian**. Suppose that the initial position of the particle $\mathbf{r}(t_0)$ and final position $\mathbf{r}(t_1)$ are specified. **Hamilton's Principle** states that the path of the particle $\mathbf{r}(t)$ in the time interval $[t_0, t_1]$ is such that the functional

$$J(\mathbf{r}) = \int_{t_0}^{t_1} L(t, \mathbf{r}, \dot{\mathbf{r}}) \, dt$$

is stationary, i.e., a local extremum or a "saddle point." (We define "stationary" more precisely in Section 2.2.) In the lingo of mechanics J is called **the action integral** or simply **the action**.

Problems in mechanics often involve several particles (or spatial coördinates); moreover, Cartesian coördinates are not always the best choice. Variational principles are thus usually given in terms of **generalized coördinates**. The letter q has been universally adopted to denote generalized position coördinates. The configuration of a system at time t is thus denoted by $\mathbf{q}(t) = (q_1(t), \ldots, q_n(t))$, where the q_k are position variables. If, for example, the system consists of three free particles in $\mathbb{R}^3$ then $n = 9$.

The kinetic energy T of a system is given by a quadratic form in the generalized velocities $\dot{q}_k$,

$$T(\mathbf{q}, \dot{\mathbf{q}}) = \frac{1}{2} \sum_{j,k=1}^{n} C_{j,k}(\mathbf{q}) \dot{q}_j \dot{q}_k.$$

Assuming the system has a potential energy function $V(t, \mathbf{q})$, the Lagrangian is given by

$$L(t, \mathbf{q}, \dot{\mathbf{q}}) = T(\mathbf{q}, \dot{\mathbf{q}}) - V(t, \mathbf{q}).$$

In this framework Hamilton's Principle takes the following form.

Theorem 1.3.1 (Hamilton's Principle) *The motion of a system of particles $\mathbf{q}(t)$ from a given initial configuration $\mathbf{q}(t_0)$ to a given final configuration $\mathbf{q}(t_1)$ in the time interval $[t_0, t_1]$ is such that the functional*

$$J(\mathbf{q}) = \int_{t_0}^{t_1} L(t, \mathbf{q}, \dot{\mathbf{q}}) \, dt$$

is stationary.

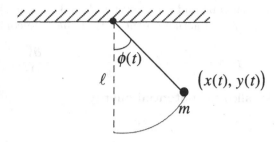

Fig. 1.5.

The dynamics of a system of particles is thus completely contained in the single scalar function L. We can derive the familiar equations of motion from Hamilton's Principle (cf. Section 3.2). The reader might rightfully question whether the motion predicted by Hamilton's Principle depends on the choice of coördinates. The variational approach would surely be of limited value were it sensitive to the observer's choice of coördinates. We show in Section 2.5 that Hamilton's Principle produces equations that are necessarily invariant with respect to coördinate choices.

Example 1.3.1:   Simple Pendulum
Consider a simple pendulum of mass m and length ℓ in the plane. Let $(x(t), y(t))$ denote the position of the mass at time t. Since $x^2 + y^2 = \ell^2$ we need in fact only one position variable. Rather than use x or y it is natural to use polar coördinates and characterize the position of the mass at time t by the angle $\phi(t)$ between the vertical and the string to which the mass is attached (figure 1.5). Now, the kinetic energy is

$$T = \frac{1}{2}m(\dot{x}^2(t) + \dot{y}^2(t)) = \frac{1}{2}m\ell^2\dot{\phi}^2(t),$$

and the potential energy is

$$V = mgh = mg\ell(1 - \cos\phi(t)),$$

where g is a gravitation constant. Thus,

$$L(\phi, \dot{\phi}) = \frac{1}{2}m\ell^2\dot{\phi}^2 - mg\ell(1 - \cos\phi),$$

and Hamilton's Principle implies that the motion from a given initial angle $\phi(t_0)$ to a fixed angle $\phi(t_1)$ is such that the functional

$$J(\phi) = \int_{t_0}^{t_1} \left(\frac{1}{2}m\ell^2\dot{\phi}^2 - mg\ell(1 - \cos\phi)\right) dt$$

is stationary.

Example 1.3.2:   Kepler problem

The Kepler problem models planetary motion. It is one of the most heavily studied problems in classical mechanics. Keeping with our no frills approach, we consider the simplest problem of a single planet orbiting around the sun, and ignore the rest of the solar system. Assuming the sun is fixed at the origin, the kinetic energy of the planet is

$$T = \frac{1}{2}m(\dot{x}^2(t) + \dot{y}^2(t)) = \frac{1}{2}m\left(\dot{r}^2(t) + r^2(t)\dot{\theta}^2(t)\right),$$

where r and θ denote polar coördinates and m is the mass of the planet. We can deduce the potential energy function V from the gravitational law of attraction

$$f = -\frac{GmM}{r^2},$$

where f is the force (acting in the radial direction), M is the mass of the sun, and G is the universal gravitation constant. Given that

$$f = -\frac{\partial V}{\partial r},$$

we have

$$V(r) = -\int f(r)\,dr = -\frac{GmM}{r};$$

hence,

$$L(r,\theta) = \frac{1}{2}m\left(\dot{r}^2 + r^2\dot{\theta}^2\right) + \frac{GmM}{r}.$$

Hamilton's Principle implies that the motion of the planet from an initial observation $(r(t_0), \theta(t_0))$ to a final observation $(r(t_1), \theta(t_1))$ is such that

$$J(r,\theta) = \int_{t_0}^{t_1} \left(\frac{1}{2}m\left(\dot{r}^2 + r^2\dot{\theta}^2\right) + \frac{GmM}{r}\right) dt$$

is stationary.

The reader may be wondering about the fate of the constant of integration in the last example. Any potential energy of the form $-GmM/r + const.$ will produce the requisite force f. In the pendulum problem we tacitly assumed that the potential energy was proportional to the height of the mass above the minimum possible height. In fact, for the purposes of describing the dynamics it does not matter; i.e., $V(t, \mathbf{q})$ and $V(t, \mathbf{q}) + c_1$ produce the same results for any constant c_1. We are optimizing J and the addition of a constant in the Lagrangian simply alters the functional $J(\mathbf{q})$ to $\tilde{J}(\mathbf{q}) = J(\mathbf{q}) + const.$ If one functional is stationary at $\mathbf{q}$ the other must also be stationary at $\mathbf{q}$.

In the lore of classical mechanics there is another variational principle that is sometimes called the "Principle of Least Action" or "Maupertuis'

Principle," which predates Hamilton's Principle. This principle is sometimes confused with Hamilton's and the situation is not mitigated by the fact that Hamilton's Principle is sometimes called the Principle of Least Action. [8] Maupertuis' Principle concerns systems that are **conservative**. In a conservative system we have that the total energy of the system at any time t along the path of motion is constant. In other words, $L + V = k$, where k is a constant. For this special case $L = 2T - k$, and Hamilton's Principle leads to Maupertuis' Principle that the functional

$$K(\mathbf{q}) = \int_{t_0}^{t_1} T(\mathbf{q}, \dot{\mathbf{q}}) \, dt$$

is stationary along a path of motion. Hence, Maupertuis' Principle is a special case of Hamilton's Principle. Most books on classical mechanics discuss these principles (along with others). Lanczos [48] gives a particularly complete and readable account that, in addition to mechanics, deals with the history and philosophy of these principles. The eminent scientist E. Mach [51] also writes at length about the history, significance, and philosophy underlying these principles. His perspective and sympathies are somewhat different from those of Lanczos. [9]

1.4 Some Variational Problems from Geometry

1.4.1 Dido's Problem

Dido was a Carthaginian queen (ca. 850 B.C.?) who came from a dysfunctional family. Her brother, Pygmalion, murdered her husband (who was also her uncle) and Dido, with the help of various gods, fled to the shores of North Africa with Pygmalion in pursuit. Upon landing in North Africa, legend has it that she struck a deal with a local chief to procure as much land as an oxhide could contain. She then selected an ox and cut its hide into very narrow strips, which she joined together to form a thread of oxhide more than two and a half miles long. Dido then used the oxhide thread and the North African sea coast to define the perimeter of her property. It is not clear what the immediate reaction of the chief was to this particular interpretation of the deal, but it is

----

[8] The translators of Landau and Lifshitz [49], p. 131, go so far as to draft a table to elucidate the different usages.

[9] Mach is not so generous with Maupertuis. In connexion with Maupertuis' Principle he writes, "It appears that Maupertuis reached this obscure expression by an unclear mingling of his ideas of *vis viva* and the principle of virtual velocities" (p. 365). In defense of Mach, we must note that Maupertuis suffered no lack of critics even in his own day. Voltaire wrote the satire *Histoire du docteur Akakia et du naif de Saint Malo* about Maupertuis. The situation at Frederick the Great's court regarding Maupertuis, König, and Voltaire is the stuff of soap operas (see Pars [59] p. 634).

clear that Dido sought to enclose the maximum area within her ox and the sea. The city of Carthage was then built within the perimeter defined by the thread and the sea coast. Dido called the place *Byrsa* meaning hide of bull.[10]

The problem that Dido faced on the shores of North Africa (aside from family difficulties) was to determine the optimal path along which to place the oxhide thread so as to provide Byrsa with the maximum amount of land. Dido did not have the luxury of waiting some 2500 years for the calculus of variations to develop and thus settled for an "intuitive solution."

Dido's problem entailed determining the curve γ of fixed length (the thread) such that the area enclosed by γ and a given curve σ (the North African shoreline) is maximum. Although this is perhaps the original version of Dido's problem, the term has been used to cover the more basic problem: among all closed curves in the plane of perimeter L determine the curve that encloses the maximum area. The problem did not escape the attention of ancient mathematicians, and as early as perhaps 200 B.C. the mathematician Zenodorus[11] is credited with a proof that the solution is a circle. Unfortunately, there were some technical loopholes in Zenodorus' proof (he compared the area of a circle with that of polygons having the same perimeter). The first complete proof of this result was given some 2000 years later by Karl Weierstraß in his Berlin lectures.

Prior to Weierstraß, Steiner (ca. 1841) proved that *if* there exists a "figure" γ whose area is never less than that of any other "figure" of the same perimeter, then γ is a circle. Not content with one proof, Steiner gave five proofs of this result. The proofs are based on simple geometric considerations (no calculus of variations). The operative word in the statement of his result, however, is "if." Steiner's contemporary, Dirichlet, pointed out that his proofs do not actually establish the existence of such a figure. Weierstraß and his followers resolved these subtle aspects of the problem. A lively account of Dido's problem and the first of Steiner's proofs can be found in Körner [45].

Some simple geometrical arguments can be used to show that if γ is a simple closed curve solution to Dido's problem then γ is convex (cf. Körner, *op. cit.*). This means that a chord joining any two points on γ lies within γ

---

[10] The reader will find various bits and pieces of Dido's history scattered in Latin works by authors such as Justin and Virgil. One account of the hide story comes from the *Aeneid*, Bk. I, vs. 367. The story gets even better once Aeneas arrives on the scene. Finally, good ideas never die. It is said that the Anglo-Saxon chieftains Hengist and Horsa (ca. 449 A.D.) acquired their land by circling it with oxhide strips [37]. Beware of real estate transactions that involve an ox.

[11] The proof may have been known even earlier, but Zenodorus in any event is the author of the proof that appears in the commentary of Theon to Ptolemy's *Almagest*. Zenodorus quotes Archimedes (who died in 212 B.C.) and is quoted by Pappus (ca. 340 A.D.). Aside from these rough dates we do not know exactly when Zenodorus lived. At any rate, the solution was of little comfort to Dido's heirs as the Romans obliterated Carthage/ Byrsa in the third Punic war just after 200 B.C. and sowed salt on the scorched ground so that nothing would grow.

and the area enclosed by γ. The convexity of γ is then used to show that Dido's problem can be distilled down to the problem of finding a function $y : [x_0, x_1] \to \mathbb{R}$ such that

$$A(y) = \int_{x_0}^{x_1} y(x)\, dx$$

is maximum subject to the constraint that the arclength of the curve γ^+ described by y is $L/2$. If we assume that y is at least piecewise differentiable then this amounts to the condition

$$\frac{L}{2} = \int_{x_0}^{x_1} \sqrt{1 + y'^2}\, dx.$$

The problem with this formulation is that we do not know the limits of the integral. The geometrical character of the problem indicates that we do not need to know both x_0 and x_1 (we could always normalize the construction so that $x_0 = 0 < x_1$), but we do need to know $x_1 - x_0$. This problem is effectively the opposite of the problem we had with the first formulation of the catenary. Since we know arclength, a natural formulation to use would be one in terms of arclength.

Suppose that γ^+ is described parametrically by $(x(s), y(s))$, $s \in [0, L/2]$, where s is arclength. Suppose further that x and y are at least piecewise differentiable. Green's theorem in the plane can then be used to show that the area of the set enclosed by γ^+ and the x-axis is

$$A(y) = \frac{1}{2} \int_0^{L/2} y(s) \sqrt{1 - y'^2(s)}\, ds, \qquad (1.12)$$

where we have used the relation $x'^2(s) + y'^2(s) = 1$. The basic Dido problem is thus to determine a positive function $y : [0, L/2] \to \mathbb{R}$ such that A is maximum.

1.4.2 Geodesics

Let Σ be a surface, and let p_0, p_1 be two distinct points on Σ. The geodesic problem concerns finding the curve(s) on Σ with endpoints p_0, p_1 for which the arclength is minimum. A curve having this property is called a **geodesic**. The theory of geodesics is one of the most developed subjects in differential geometry. The general theory is complicated analytically by the situation that simple, common surfaces such as the sphere require more than one vector function to describe them completely. In the language of geometry, the sphere is a manifold that requires at least two charts. We have encountered and side-stepped the analogous problem for curves, and we do so here in the interest of simplicity. We focus on the local problem and refer the reader to any general text on differential geometry such as Stoker [66] or Willmore [75] for a more precise and in-depth treatment of geodesics.[12]

Suppose that Σ is described by the position vector function $\mathbf{r} : \sigma \to \mathbb{R}^3$, where σ is a nonempty connected open subset of $\mathbb{R}^2$, and for $(u, v) \in \sigma$,

$$\mathbf{r}(u, v) = (x(u, v), y(u, v), z(u, v)).$$

We assume that $\mathbf{r}$ is a smooth function on σ; i.e., x, y, and z are smooth functions of (u, v), and that

$$\left| \frac{\partial \mathbf{r}}{\partial u} \wedge \frac{\partial \mathbf{r}}{\partial v} \right| \neq \mathbf{0}, \tag{1.13}$$

so that $\mathbf{r}$ is a one-to-one mapping of σ onto Σ. If γ is a curve on Σ, then there is a curve $\hat{\gamma}$ in σ that maps to γ under $\mathbf{r}$. Any curve on Σ may thus be regarded as a curve in σ. Suppose that the points p_0 and p_1 correspond to $\mathbf{r}_0 = \mathbf{r}(u_0, v_0)$ and $\mathbf{r}_1 = \mathbf{r}(u_1, v_1)$, respectively. Any curve γ from $\mathbf{r}_0$ to $\mathbf{r}_1$ maps to a curve $\hat{\gamma}$ from (u_0, v_0) to (u_1, v_1).

For the geodesic problem we restrict our attention to smooth simple curves (no self-intersections) on Σ from $\mathbf{r}_0$ to $\mathbf{r}_1$. Let Γ denote the set of all such curves. Thus, if $\gamma \in \Gamma$, then there exists a parametrization of γ of the form

$$\mathbf{R}(t) = \mathbf{r}(u(t), v(t)), \quad t \in [t_0, t_1], \tag{1.14}$$

where $\mathbf{R}(t_0) = \mathbf{r}_0$, $\mathbf{R}(t_1) = \mathbf{r}_1$, and u and v are smooth functions on the interval $[t_0, t_1]$ such that

$$u'^2(t) + v'^2(t) \neq 0 \tag{1.15}$$

for all $t \in [t_0, t_1]$. In the parameter space σ, the last condition ensures that the curve $\hat{\gamma}$ is also a smooth curve and has a well-defined unit tangent vector. The differential of arclength along γ is given by

$$\begin{aligned} ds^2 &= \left| \mathbf{R}'(t) \right|^2 dt^2 \\ &= \left| \frac{\partial \mathbf{r}}{\partial u} u'(t) + \frac{\partial \mathbf{r}}{\partial v} v'(t) \right|^2 dt^2 \\ &= \left(E u'^2 + 2F u' v' + G v'^2 \right) dt^2, \end{aligned}$$

where

$$E = \left| \frac{\partial \mathbf{r}}{\partial u} \right|^2, \quad F = \frac{\partial \mathbf{r}}{\partial u} \cdot \frac{\partial \mathbf{r}}{\partial v}, \quad G = \left| \frac{\partial \mathbf{r}}{\partial v} \right|^2.$$

The functions E, F, and G are called components of the **first fundamental form** or **metric tensor**. Note that these components depend only on u and v. Note also that the identity

$$\left| \frac{\partial \mathbf{r}}{\partial u} \wedge \frac{\partial \mathbf{r}}{\partial v} \right| = EG - F^2$$

---

[12] A more specialized discussion can be found in Postnikov [62].

and condition (1.13) indicate that the quadratic form

$$I = Eu'^2 + 2Fu'v' + Gv'^2$$

is positive definite.

The arclength of γ is given by

$$L(\gamma) = \int_{t_0}^{t_1} \sqrt{Eu'^2 + 2Fu'v' + Gv'^2}\, dt.$$

The geodesic problem is thus to find the functions u and v (i.e., the curve $\hat{\gamma}$) such that L is a minimum and

$$u(t_0) = u_0, \quad v(t_0) = v_0$$
$$u(t_1) = u_1, \quad v(t_1) = v_1.$$

Example 1.4.1:   Geodesics on a Sphere

Let Σ be an octant of the unit sphere. The surface Σ can be described parametrically by

$$\mathbf{r}(u, v) = (\sin u \cos v, \sin u \sin v, \cos u)$$

for $\sigma = \{(u, v) : 0 < u < \pi/2, 0 < v < \pi/2\}$. Now,

$$E = \left| \frac{\partial \mathbf{r}}{\partial u} \right|^2 = |\,(\cos u \cos v, \cos u \sin v, -\sin u)\,|^2$$
$$= 1,$$

$$F = \frac{\partial \mathbf{r}}{\partial u} \cdot \frac{\partial \mathbf{r}}{\partial v}$$
$$= (\cos u \cos v, \cos u \sin v, -\sin u) \cdot (-\sin u \sin v, \sin u \cos v, 0)$$
$$= 0,$$

$$G = \left| \frac{\partial \mathbf{r}}{\partial v} \right|^2 = |\,(-\sin u \sin v, \sin u \cos v, 0)\,|^2$$
$$= \sin^2 u.$$

The arclength integral is thus

$$L(\gamma) = \int_{t_0}^{t_1} \sqrt{u'^2 + v'^2 \sin^2 u}\, dt.$$

A feature of the basic geodesic problem described above is that it does not involve the function $\mathbf{r}$ directly. The arclength of a curve depends only on the three scalar functions $E, F,$ and G. Geodesics are part of the **intrinsic geometry** of the surface, i.e., the geometry defined by the metric tensor. The metric tensor does not define a surface uniquely even modulo translations and rotations. There are any number of distinct surfaces in $\mathbb{R}^3$ that have the same metric tensor. For example, a plane, a cone, and a cylinder all have the same metric tensor. If a cylinder is "unrolled" and "flattened" to form a portion of the plane, then a geodesic on the cylinder would become a geodesic on the plane.

One direction for a generalization of the above problem is to focus on the space $\sigma \subseteq \mathbb{R}^2$ and *define* the components of the metric tensor. For notational simplicity, let $u = u^1$, $v = u^2$, and $\mathbf{u} = (u, v)$. We can choose scalar functions $g_{jk} : \sigma \to \mathbb{R}$ $j, k = 1, 2$ and define the arclength element ds by

$$ds^2 = g_{11}(du^1)^2 + g_{12}du^1du^2 + g_{21}du^2du^1 + g_{22}(du^2)^2$$
$$= g_{jk}du^jdu^k,$$

where the last expression uses the Einstein summation convention: summation of repeated indices when one is a superscript and the other is a subscript. Of course we must place some restrictions on the g_{jk} in order to ensure that our arclength element is positive and that the length of a curve does not depend on the choice of coördinates $\mathbf{u}$. We can take care of these concerns by requiring that the g_{jk} produce a quadratic form that is positive definite and that the g_{jk} form a second order covariant tensor. To mimic the earlier case we also impose the symmetry condition

$$g_{jk} = g_{kj},$$

so that

$$ds^2 = g_{11}(du^1)^2 + 2g_{12}du^1du^2 + g_{22}(du^2)^2. \tag{1.16}$$

In terms of the former notation, $E = g_{11}$, $F = g_{12} = g_{21}$, and $G = g_{22}$. For this case, the positive definite requirement amounts to the condition

$$g_{11}g_{22} - g_{12}^2 > 0$$

with $g_{11} > 0$. The condition that the g_{jk} form a second-order covariant tensor means that under a smooth coördinate transformation from $\mathbf{u} = (u^1, u^2)$ to $\hat{\mathbf{u}} = (\hat{u}^1, \hat{u}^2)$, the components $g_{jk}(\mathbf{u})$ transform to $\hat{g}_{lm}(\hat{\mathbf{u}})$ according to the relation

$$\hat{g}_{lm} = g_{jk}\frac{\partial u^j}{\partial \hat{u}^l}\frac{\partial u^k}{\partial \hat{u}^m}.$$

The set σ equipped with such a tensor can be viewed as defining a geometrical object in itself (as the surface Σ was). It is a special case of what is called a **Riemannian manifold**. Let $\mathcal{M}$ denote this geometrical object. A curve $\hat{\gamma}$ in σ generates a curve γ in $\mathcal{M}$, and the arclength is given by

$$L(\gamma) = \int_{t_0}^{t_1} \sqrt{g_{jk}u^{j'}u^{k'}}\, dt,$$

where $(u^1(t), u^2(t))$, $t \in [t_0, t_1]$ is a parametrization of $\hat{\gamma}$. The condition that the g_{jk} form a second-order covariant tensor ensures that $L(\gamma)$ is invariant with respect to changes in the curvilinear coördinates $\mathbf{u}$ used to represent $\mathcal{M}$. Note also that $L(\gamma)$ is invariant with respect to orientation-preserving parametrizations of $\hat{\gamma}$.

The advantage of the above abstraction is that it can be readily modified to accommodate higher dimensions. Suppose that $\sigma \subseteq \mathbb{R}^n$ and $\mathbf{u} = (u^1, \ldots, u^n)$. We can define an n-dimensional (Riemannian) manifold $\mathcal{M}$ by introducing a metric tensor with components g_{jk} such that:

1. the quadratic form $g_{jk}du^j du^k$ is positive definite;
2. $g_{jk} = g_{kj}$ for $j, k = 1, 2, \ldots, n$;
3. under any smooth transformation $\mathbf{u} = \mathbf{u}(\hat{\mathbf{u}})$ the g_{jk} transform to $\hat{g}_{lm}$ according to the relation

$$\hat{g}_{lm} = g_{jk} \frac{\partial u^j}{\partial \hat{u}^l} \frac{\partial u^k}{\partial \hat{u}^m}.$$

A curve γ on $\mathcal{M}$ is generated by a curve $\hat{\gamma}$ in $\sigma \subseteq \mathbb{R}^n$. Suppose that $\mathbf{u}(t) = (u^1(t), \ldots, u^n(t))$, $t \in [t_0, t_1]$ is a parametrization of $\hat{\gamma}$. The arclength of γ is then defined as

$$L(\gamma) = \int_{t_0}^{t_1} \sqrt{g_{jk} u^{j\prime} u^{k\prime}} \, dt.$$

A generalization of the geodesic problem is thus to find the curve(s) $\hat{\gamma}$ in σ with specified endpoints $\mathbf{u}_0 = \mathbf{u}(t_0)$, $\mathbf{u}_1 = \mathbf{u}(t_1)$ such that $L(\gamma)$ is a minimum.

Geodesics are of interest not only in differential geometry, but also in mathematical physics and other subjects. It turns out that many problems can be interpreted as geodesic problems on a suitably defined manifold.[13] In this regard, the geodesic problem is even more important because it provides a unifying framework for many problems.

1.4.3 Minimal Surfaces

We have already encountered a special minimal surface problem in our discussion of the catenary. The rotational symmetry of the problem reduced the problem to that of finding a function y of a single variable x, the graph of which generates the surface of revolution having minimal surface area. Locally, any surface can be represented in "graphical" form,

$$\mathbf{r}(x, y) = (x, y, z(x, y)), \tag{1.17}$$

where $\mathbf{r}$ is the position function in $\mathbb{R}^3$. Unless some symmetry condition is imposed, a surface parametrization requires two independent variables. Thus the problem of finding a surface with minimal surface area involves two independent variables in contrast to the problems discussed earlier.

---

[13] In the theory of relativity, where differential geometry is widely used, the condition that the metric tensor be positive definite is relaxed to positive semidefinite.

Given a simple closed space curve γ, the basic minimal surface problem entails finding, among all smooth simply connected surfaces with γ as a boundary, the surface having minimal surface area. Suppose that the curve γ can be represented parametrically by $(x(t), y(t), z(t))$ for $t \in [t_0, t_1]$, and for simplicity suppose that the projection of γ on the xy-plane is also a simple closed curve; i.e., the curve $\hat{\gamma}$ described by $(x(t), y(t))$ for $t \in [t_0, t_1]$ is a simple closed curve in the xy-plane. Let Ω denote the region in the xy-plane enclosed by $\hat{\gamma}$. Suppose further that we restrict the class of surfaces under consideration to those that can be represented in the form (1.17), where z is a smooth function for $(x, y) \in \Omega$. The differential area element is given by

$$dA = \sqrt{1 + \left(\frac{\partial z}{\partial x}\right)^2 + \left(\frac{\partial z}{\partial y}\right)^2}\, dx\, dy,$$

and the surface area is thus

$$A(z) = \int\int_\Omega \sqrt{1 + \left(\frac{\partial z}{\partial x}\right)^2 + \left(\frac{\partial z}{\partial y}\right)^2}\, dx\, dy.$$

The (simplified) minimal surface problem thus concerns determining a smooth function $z : \Omega \to \mathbb{R}$ such that $z(x(t), y(t)) = z(t)$ for $t \in [t_0, t_1]$, and $A(z)$ is a minimum. There is a substantial body of information about minimal surfaces. The reader can find an overview of the subject in Osserman [58].

1.5 Optimal Harvest Strategy

Our final example in this chapter concerns a problem in economics dealing with finding a harvest strategy that maximizes profit. Here, we follow the example given by Wan [71], p. 6 and use a fishery to illustrate the model.

Let $y(t)$ denote the total tonnage of fish at time t in a region Ω of the ocean, and let y_c denote the carrying capacity of the region Ω for the fish. The growth of the fish population without any harvesting is typically modelled by a first-order differential equation

$$y'(t) = f(t, y). \tag{1.18}$$

If y is small compared to y_c, then f is often approximated by a linear function in y; i.e., $f(t, y) = ky + g(t)$, where k is a constant. More complicated models are available for a wider range of $y(t)$ such as logistic growth

$$f(t, y) = ky(t)\left(1 - \frac{y(t)}{y_c}\right).$$

The ordinary differential equation (1.18) is accompanied by an initial condition

$$y(0) = y_0 \qquad\qquad (1.19)$$

that reflects the initial fish population.

Suppose now that the fish are harvested at a rate $w(t)$. Equation (1.18) for the population growth can then be modified to the relation

$$y'(t) = f(t, y) - w(t). \qquad\qquad (1.20)$$

Given the function f, the problem is to determine the function w so that the profit in a given time interval T is maximum.

It is reasonable to expect that the cost of harvesting the fish depends on the season, the fish population, and the harvest rate. Let $c(t, y, w)$ denote the cost to harvest a unit of fish biomass. Suppose that the fish commands a price p per unit fish biomass and that the price is perhaps season dependent, but not dependent on the volume of fish on the market. The profit gained by harvesting the fish in a small time increment is $(p(t) - c(t, y, w))w(t)\, dt$. Given a fixed period T with which to plan the strategy, the total profit is thus

$$P(y, w) = \int_0^T (p(t) - c(t, y, w))w(t)\, dt.$$

The problem is to identify the function w so that P is maximum.

The above problem is an example of a constrained variational problem. The functional P is optimized subject to the constraint defined by the differential equation (1.20) (a nonholonomic constraint) and initial condition (1.19). We can convert the problem into an unconstrained one by simply eliminating w from the integrand defining P using equation (1.20). The problem then becomes the determination of a function y that maximizes the total profit. This approach is not necessarily desirable because we want to keep track of w, the only physical quantity we can regulate.

A feature of this problem that distinguishes it from earlier problems is the absence of a boundary condition for the fish population at time T. Although we are given the initial fish population, it is not necessarily desirable to specify the final fish population after time T. As Wan points out, the condition $y(T) = 0$, for example, is not always the best strategy: "green issues" aside, it may cost far more to harvest the last few fish than they are worth. This simple model thus provides an example of a variational problem with only one endpoint fixed in contrast to the catenary and brachystochrone.

In passing we note that economic models such as this one are generally framed in terms of "present value." A pound sterling invested earns interest, and this should be incorporated into the overall profit. If the interest is compounded continuously at a rate r, then a pound invested yields e^{rt} pounds after time t. Another way of looking at this is to view a pound of income at time t as worth e^{-rt} pounds now. Considerations of this sort lead to profit functionals of the form

$$P(y, w) = \int_0^T e^{-rt}(p(t) - c(t, y, w))w(t)\, dt.$$

2

The First Variation

In this chapter we develop a necessary condition for a function to yield an extremum for a functional. The centrepiece of the chapter is a second-order differential equation, the Euler-Lagrange equation, which plays a rôle analogous to the gradient of a function. We first motivate the analysis by reviewing necessary conditions for functions to have local extrema. The Euler-Lagrange equations are derived in Section 2.2 and some special cases where the differential equation can be simplified are discussed in Section 2.3. The remaining three sections are devoted to more qualitative topics concerning degenerate cases, invariance, and existence of solutions. We postpone a discussion of sufficient conditions until Chapter 10.

2.1 The Finite-Dimensional Case

The theory underlying the necessary conditions for extrema in the calculus of variations is motivated by that for functions of n independent variables. Problems in the calculus of variations are inherently infinite-dimensional. The character of the analytical tools needed to solve infinite-dimensional problems differs from that required for finite-dimensional problems, but many of the underlying ideas have tractable analogues in finite dimensions. In this section we review a necessary condition for a function of n independent variables to have a local extremum.

2.1.1 Functions of One Variable

Let f be a real-valued function defined on the interval $I \subseteq \mathbb{R}$. The function $f : I \to \mathbb{R}$ is said to have a **local maximum** at $x \in I$ if there exists a number $\epsilon > 0$ such that for any $\hat{x} \in (x - \epsilon, x + \epsilon) \subset I$, $f(\hat{x}) \leq f(x)$. The function $f : I \to \mathbb{R}$ is said to have a **local minimum** at $x \in I$ if $-f$ has a local maximum at x. A function may have several local extrema in a given interval.

It may be that a function attains a maximum or minimum value for the entire interval. The function $f : I \to \mathbb{R}$ has a **global maximum** on I at $x \in I$ if $f(\hat{x}) \leq f(x)$ for *all* $\hat{x} \in I$. The function f is said to have a **global minimum** on I at $x \in I$ if $-f$ has a global maximum at x. Note that if I has boundary points then f may have a global maximum on the boundary. If f is differentiable on I then the presence of local maxima or minima on I is characterized by the first derivative.

Theorem 2.1.1 *Let f be a real-valued function differentiable on the open interval I. If f has a local extremum at $x \in I$ then $f'(x) = 0$.*

Proof: The proof of this result is essentially the same for a local maximum or minimum. Suppose that x is a local maximum. Then there is a number $\epsilon > 0$ such that for any $\hat{x} \in (x - \epsilon, x + \epsilon) \subset I$ the inequality $f(x) \geq f(\hat{x})$ is satisfied. Now the derivative of f at x is given by

$$f'(x) = \lim_{\hat{x} \to x} (f(\hat{x}) - f(x))/(\hat{x} - x).$$

The numerator of this limit is never positive since $f(x)$ is a maximum, but the denominator is positive when $\hat{x} > x$ and negative when $\hat{x} < x$. Since the function f is differentiable at x the right- and left-sided limits exist and are equal. The only way this can be true is if $f'(x) = 0$. □

It is illuminating to examine the situation for smooth functions. We use the generic term "smooth" to indicate that the function has as many continuous derivatives as are necessary to perform whatever operations are required. Suppose that f is smooth in the interval $(x - \epsilon, x + \epsilon)$, where $\epsilon > 0$. Let $\hat{x} - x = \epsilon \eta$. Taylor's theorem indicates that, for ϵ sufficiently small, f can be represented by

$$f(\hat{x}) = f(x) + \epsilon \eta f'(x) + \frac{\epsilon^2}{2!} \eta^2 f''(x) + O(\epsilon^3). \tag{2.1}$$

If $f'(x) \neq 0$ and ϵ is small, the sign of $f(\hat{x}) - f(x)$ is determined by $\eta f'(x)$. Suppose that $f'(x) \neq 0$. If f has a local extremum at x then the sign of $f(\hat{x}) - f(x)$ cannot change in $(x - \epsilon, x + \epsilon)$, so that $\eta f'(x)$ must have the same sign for all η. But it is clear that η can be positive or negative and hence $\eta f'(x)$ can be positive or negative. We must therefore have that $f'(x) = 0$. If $f'(x) = 0$, then the above expansion indicates that the sign of the difference is that of the quadratic term, i.e., the sign of $f''(x)$. If this derivative is negative then $f(x)$ is a local maximum; if it is positive then $f(x)$ is a local minimum. It may be that $f''(x) = 0$. In this case the sign of the difference depends on the cubic term, which contains a factor η^3. Like the linear term, however, this factor can be either positive or negative depending on the choice of η. Thus, if $f'''(x) \neq 0$, $f(x)$ cannot be a local extremum. We can continue in this manner as long as f has the requisite derivatives in $(x - \epsilon, x + \epsilon)$.

For a differentiable function it is easy to see graphically why the condition $f'(x) = 0$ is necessary for a local extremum. The Taylor expansion for a smooth function indicates that at any point x at which the first derivative vanishes an $O(\epsilon)$ change in the independent variable produces an $O(\epsilon^2)$ change in the function value as $\epsilon \to 0$. For this reason points such as x are called **stationary points**. The functions $f_n(x) = x^n$, where $n \in \mathbb{N}$, $x \in \mathbb{R}$ provide simple paradigms for the various possibilities

Example 2.1.1:   Let $f(x) = 3x^2 - x^3$. The function f is smooth for $x \in \mathbb{R}$ and therefore if any local extrema exist they must satisfy the equation $6x - 3x^2 = 0$. This equation is satisfied if $x = 0$ or $x = 2$. The second derivative is $6 - 6x$, so that $f''(0) = 6$ and consequently $f(0)$ is a local minimum. On the other hand, $f''(2) = -6$ and thus $f(2)$ is a local maximum.

Example 2.1.2:   Let

$$f(x) = \begin{cases} x^2 \sin^2(1/x), & \text{if } x \neq 0 \\ 0, & \text{if } x = 0. \end{cases}$$

This function is differentiable for all $x \in \mathbb{R}$. Now $f'(0) = 0$, and thus $x = 0$ is a stationary point but the derivative is not continuous there and so $f''(0)$ does not exist. We can deduce that f has a local minimum at $x = 0$ because $f(x) \geq 0$ for all $x \in \mathbb{R}$.

Example 2.1.3:   Let $f(x) = |x|$. This function is differentiable for all $x \in \mathbb{R} - \{0\}$. The derivative is given by $f'(x) = -1$ for $x < 0$, and $f'(x) = 1$ for $x > 0$. Thus f cannot have a local extremum in $\mathbb{R} \setminus \{0\}$. Nonetheless it is clear that $f(0) = 0$ is a local (and global) minimum for f in $\mathbb{R}$.

Example 2.1.4:   Let $f(x) = e^x$. This function is smooth for all $x \in \mathbb{R}$ and its derivative never vanishes; consequently, f does not have any local extrema.

The relationship between local and global extrema is limited. Certainly if f has a global extremum at some *interior* point x of an interval then $f(x)$ is also a local extremum. If, in addition, f is differentiable in I, then it must also satisfy the condition $f'(x) = 0$. But it may be (as often is the case) that a global extremum is attained at one of the boundary points of I, in which case even if f is differentiable nothing regarding the value of the derivative can be asserted.

2.1.2 Functions of Several Variables

The definitions for local and global extrema in n dimensions are formally the same as for the one-variable case. Let $\Omega \subseteq \mathbb{R}^n$ be a region and suppose that $f : \Omega \to \mathbb{R}$. For $\epsilon > 0$ and $\mathbf{x} = (x_1, x_2, \ldots, x_n)$, let

$$B(\mathbf{x}; \epsilon) = \{\hat{\mathbf{x}} \in \mathbb{R}^n : |\hat{x}_1 - x_1|^2 + |\hat{x}_2 - x_2|^2 + \cdots |\hat{x}_n - x_n|^2 < \epsilon^2\}.$$

The function $f : \Omega \to \mathbb{R}$ has a **global maximum (global minimum)** on Ω at $\mathbf{x} \in \Omega$ if $f(\hat{\mathbf{x}}) \leq f(\mathbf{x})$ $(f(\hat{\mathbf{x}}) \geq f(\mathbf{x}))$ for all $\hat{\mathbf{x}} \in \Omega$. The function f has a **local maximum (local minimum)** at $\mathbf{x} \in \Omega$ if there exists a number $\epsilon > 0$ such that for any $\hat{\mathbf{x}} \in B(\mathbf{x}; \epsilon) \subset \Omega$, $f(\hat{\mathbf{x}}) \leq f(\mathbf{x})$ $(f(\hat{\mathbf{x}}) \geq f(\mathbf{x}))$. As with the one-variable case if Ω has boundary points f may have a global maximum/minimum on the boundary.

Necessary conditions for a smooth function of two independent variables to have local extrema can be derived from considerations similar to those used in the single-variable case. Suppose that $f : \Omega \to \mathbb{R}$ is a smooth function on the region $\Omega \subseteq \mathbb{R}^2$, and that f has a local extremum at $\mathbf{x} = (x_1, x_2) \in \Omega$. Then there exists an $\epsilon > 0$ such that $f(\hat{\mathbf{x}}) - f(\mathbf{x})$ does not change sign for all $\hat{\mathbf{x}} \in B(\mathbf{x}; \epsilon)$. Let $\hat{\mathbf{x}} = \mathbf{x} + \epsilon\eta$, where $\eta = (\eta_1, \eta_2) \in \mathbb{R}^2$. For ϵ small, Taylor's theorem implies

$$f(\hat{\mathbf{x}}) = f(\mathbf{x}) + \epsilon\left\{\eta_1 \frac{\partial f(\mathbf{x})}{\partial x_1} + \eta_2 \frac{\partial f(\mathbf{x})}{\partial x_2}\right\}$$
$$+ \frac{\epsilon^2}{2!}\left\{\eta_1^2 \frac{\partial^2 f(\mathbf{x})}{\partial x_1^2} + 2\eta_1\eta_2 \frac{\partial^2 f(\mathbf{x})}{\partial x_1 \partial x_2} + \eta_2^2 \frac{\partial^2 f(\mathbf{x})}{\partial x_2^2}\right\}$$
$$+ O(\epsilon^3),$$

and the sign of $f(\hat{\mathbf{x}}) - f(\mathbf{x})$ is given by the linear term in the Taylor expansion, unless this term is zero. But, if $\mathbf{x} + \epsilon\eta \in B(\mathbf{x}; \epsilon)$, then $\mathbf{x} - \epsilon\eta \in B(\mathbf{x}; \epsilon)$ and these points yield different signs for the linear term unless it is zero. If $\mathbf{x}$ is a local extremum we must therefore have that

$$(\eta_1, \eta_2) \cdot \left(\frac{\partial f}{\partial x_1}, \frac{\partial f}{\partial x_2}\right) = 0, \tag{2.2}$$

for *all* $\eta \in \mathbb{R}^2$. In particular, equation (2.2) must hold for the special choices $\mathbf{e}_1 = (1, 0)$ and $\mathbf{e}_2 = (0, 1)$. The former choice implies that $\partial f / \partial x_1 = 0$ and the latter choice implies that $\partial f / \partial x_2 = 0$. We thus have that if f has a local extremum at $\mathbf{x}$ then

$$\nabla f(\mathbf{x}) = \mathbf{0}. \tag{2.3}$$

Geometrically, equation (2.2) implies that the tangent plane to the graph of f is horizontal at a local extremum. Points $\mathbf{x}$ at which $\nabla f(\mathbf{x}) = \mathbf{0}$ are called **stationary points**. If $\mathbf{x}$ is a stationary point and $\hat{\mathbf{x}} = \mathbf{x} + \epsilon\eta$, then $f(\hat{\mathbf{x}}) - f(\mathbf{x})$ is $O(\epsilon^2)$ as $\epsilon \to 0$, in contrast to the generic case where an $O(\epsilon)$ change in the independent variables produces an $O(\epsilon)$ change in the difference.

Example 2.1.5:   Let $f(x_1, x_2) = x_1^2 - x_2^2 + x_1^3$. The stationary points for f are given by $\nabla f(x_1, x_2) = (2x_1 + 3x_1^2, -2x_2) = \mathbf{0}$. This equation has two solutions $(0, 0)$ and $(-2/3, 0)$. It can be shown that $(0, 0)$ produces neither a local minimum nor a local maximum for f (it is a saddle point). In contrast, at $(-2/3, 0)$ it can be shown that f has a local maximum.

Example 2.1.6:   The monkey saddle[1] is a surface described by $f(x_1, x_2) = x_2^3 - 3x_1^2 x_2$. If $\mathbf{x}$ is a stationary point for f then the equations

$$-6x_1 x_2 = 0,$$
$$3x_2^2 - 3x_1^2 = 0,$$

must be satisfied and this means that $x_1 = x_2 = 0$. The function f does not have a local extremum at this point. Note that even the second derivatives at this point are zero.

The extension of the above arguments to functions of n independent variables is straightforward. Let $f : \Omega \to \mathbb{R}$ be a smooth function on the region $\Omega \subset \mathbb{R}^n$, and suppose that f has a local extremum at $\mathbf{x} \in \Omega$. Then, for $\epsilon > 0$ sufficiently small, the sign of $f(\hat{\mathbf{x}}) - f(\mathbf{x})$ does not change for all $\hat{\mathbf{x}} \in B(\mathbf{x}; \epsilon)$. Let $\hat{\mathbf{x}} = \mathbf{x} + \epsilon\eta$, where $\eta = (\eta_1, \eta_2, \ldots, \eta_n)$. For ϵ is sufficiently small, Taylor's theorem implies

$$f(\hat{\mathbf{x}}) = f(\mathbf{x}) + \epsilon\eta \cdot \nabla f(\mathbf{x}) + O(\epsilon^2),$$

and the sign of $f(\hat{\mathbf{x}}) - f(\mathbf{x})$ is determined by the linear term in the Taylor expansion, provided this term is not zero. But the linear term must be zero since $\mathbf{x} + \epsilon\eta$ and $\mathbf{x} - \epsilon\eta$ are both in $B(\mathbf{x}; \epsilon)$; hence,

$$\eta \cdot \nabla f(\mathbf{x}) = 0 \tag{2.4}$$

for all $\eta \in \mathbb{R}^n$. The special choices $\mathbf{e}_1 = (1, 0, \ldots, 0)$, $\mathbf{e}_2 = (0, 1, \ldots, 0), \ldots, \mathbf{e}_n = (0, 0, \ldots, 1)$ for η yield the n conditions $\partial f / \partial x_k = 0$ at $\mathbf{x}$ for $k = 1, 2, \ldots, n$. In summary we have the following result.

Theorem 2.1.2 *Let $f : \Omega \to \mathbb{R}$ be a smooth function on the region $\Omega \subseteq \mathbb{R}^n$. If f has a local extremum at a point $\mathbf{x} \in \Omega$ then*

$$\nabla f(\mathbf{x}) = \mathbf{0}. \tag{2.5}$$

---

[1] The graph of this surface near $\mathbf{x} = \mathbf{0}$ has three valleys and three hills. A monkey requires a saddle with two depressions for its legs and a third for its tail.

2.2 The Euler-Lagrange Equation

Local extrema for a functional can be defined in a manner analogous to that
used for functions of n variables. The transition from finite to infinite dimensional domains, however, carries with it some complications. For instance,
there may be several vector spaces for which the problem is well defined, and
once a function space is chosen, there may be several suitable norms available. The vector space $C^n[x_0, x_1]$, for example, can be equipped with any of
the $\| \cdot \|_{k,\infty}$ norms, $k = 1, 2, \ldots, n$ or even any L^p norm. [2] Unlike the finite-dimensional case, different norms need not be equivalent and thus may lead to
different extrema. Functions "close" in one norm need not be close in another
norm. In applications, the choice of a vector space and norm form an integral
part of the mathematical model.

   Let $J : X \to \mathbb{R}$ be a functional defined on the function space $(X, \| \cdot \|)$
and let $S \subseteq X$. The functional J is said to have a **local maximum** in S at
$y \in S$ if there exists an $\epsilon > 0$ such that $J(\hat{y}) - J(y) \leq 0$ for all $\hat{y} \in S$ such
that $\|\hat{y} - y\| < \epsilon$. The functional J is said to have a **local minimum** in S at
$y \in S$ if y is a local maximum in S for $-J$. In this chapter, the set S is a set
of functions satisfying certain boundary conditions.

   Functions $\hat{y} \in S$ in an ϵ-neighbourhood of a function $y \in S$ can be represented in a convenient way as a perturbation of y. Specifically, if $\hat{y} \in S$ and
$\|\hat{y} - y\| < \epsilon$, then there is some $\eta \in X$ such that

$$\hat{y} = y + \epsilon \eta.$$

All the functions in an ϵ-neighbourhood of y can be generated from a suitable
set H_ϵ of functions η. Certainly any such η must be an element of X, but η
must also be such that $y + \epsilon \eta \in S$. The set H_ϵ is thus defined by

$$H_\epsilon = \{\eta \in X : y + \epsilon \eta \in S \text{ and } \|\eta\| < 1\}.$$

Since the inequalities defining the extrema must be valid when ϵ is replaced
by any number $\hat{\epsilon}$ such that $0 < \hat{\epsilon} < \epsilon$, it is clear that ϵ can always be made
arbitrarily small when convenient. The auxiliary set H_ϵ can thus be replaced
by the set

$$H = \{\eta \in X : y + \epsilon \eta \in S\},$$

for the purposes of analysis.

   At this stage we specialize to a particular class of problem called the **fixed
endpoint variational problem**, [3] and work with the vector space $C^2[x_0, x_1]$
that consists of functions on $[x_0, x_1]$ that have continuous second derivatives.
Let $J : C^2[x_0, x_1] \to \mathbb{R}$ be a functional of the form

---

[2] See Appendix B.1.

[3] More accurately, it is called the nonparametric fixed endpoint problem in the
plane.

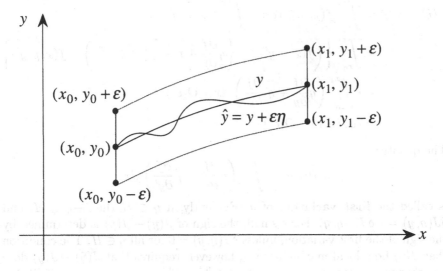

Fig. 2.1.

$$J(y) = \int_{x_0}^{x_1} f(x, y, y') \, dx,$$

where f is a function assumed to have at least second-order continuous partial derivatives with respect to x, y, and y'. Given two values $y_0, y_1 \in \mathbb{R}$, the fixed endpoint variational problem consists of determining the functions $y \in C^2[x_0, x_1]$ such that $y(x_0) = y_0$, $y(x_1) = y_1$, and J has a local extremum in S at $y \in S$. Here,

$$S = \{y \in C^2[x_0, x_1] : y(x_0) = y_0 \text{ and } y(x_1) = y_1\},$$

and

$$H = \{\eta \in C^2[x_0, x_1] : \eta(x_0) = \eta(x_1) = 0\}$$

(cf. figure 2.1).

Suppose that J has a local extremum in S at y. For definiteness, let us assume that J has a local maximum at y. Then there is an $\epsilon > 0$ such that $J(\hat{y}) - J(y) \leq 0$ for all $\hat{y} \in S$ such that $\|\hat{y} - y\| < \epsilon$. For any $\hat{y} \in S$ there is an $\eta \in H$ such that $\hat{y} = y + \epsilon \eta$, and for ϵ small Taylor's theorem implies that

$$f(x, \hat{y}, \hat{y}') = f(x, y + \epsilon \eta, y' + \epsilon \eta')$$
$$= f(x, y, y') + \epsilon \left\{ \eta \frac{\partial f}{\partial y} + \eta' \frac{\partial f}{\partial y'} \right\} + O(\epsilon^2).$$

Here, we regard f as a function of the three independent variables x, y, and y', and the partial derivatives in the above expression are all evaluated at the point (x, y, y'). Now,

$$J(\hat{y}) - J(y) = \int_{x_0}^{x_1} f(x, \hat{y}, \hat{y}') \, dx - \int_{x_0}^{x_1} f(x, y, y') \, dx$$

$$= \int_{x_0}^{x_1} \left\{ \left(f(x, y, y') + \epsilon \left\{ \eta \frac{\partial f}{\partial y} + \eta' \frac{\partial f}{\partial y'} \right\} + O(\epsilon^2) \right) - f(x, y, y') \right\} dx$$

$$= \epsilon \int_{x_0}^{x_1} \left(\eta \frac{\partial f}{\partial y} + \eta' \frac{\partial f}{\partial y'} \right) dx + O(\epsilon^2)$$

$$= \epsilon \delta J(\eta, y) + O(\epsilon^2).$$

The quantity

$$\delta J(\eta, y) = \int_{x_0}^{x_1} \left(\eta \frac{\partial f}{\partial y} + \eta' \frac{\partial f}{\partial y'} \right) dx$$

is called the **first variation** of J. Evidently, if $\eta \in H$ then $-\eta \in H$, and $\delta J(\eta, y) = -\delta J(-\eta, y)$. For ϵ small, the sign of $J(\hat{y}) - J(y)$ is determined by the sign of the first variation, unless $\delta J(\eta, y) = 0$ for all $\eta \in H$. The condition that $J(y)$ be a local maximum in S, however, requires that $J(\hat{y}) - J(y)$ does not change sign for any $\hat{y} \in S$ such that $\|\hat{y} - y\| < \epsilon$; consequently, if $J(y)$ is a local maximum then

$$\delta J(\eta, y) = \int_{x_0}^{x_1} \left(\eta \frac{\partial f}{\partial y} + \eta' \frac{\partial f}{\partial y'} \right) dx = 0, \qquad (2.6)$$

for *all* $\eta \in H$. A similar chain of arguments can be used to show that equation (2.6) must be satisfied for all $\eta \in H$ if J has a local minimum in S at y.

So far we have shown that if J has a local extremum in S at y then equation (2.6) must be satisfied for all $\eta \in H$. As in the finite-dimensional case, the converse is not true: satisfaction of equation (2.6) does not necessarily mean that y produces a local extremum for J. If y satisfies equation (2.6) for all $\eta \in H$, we say that J is **stationary** at y, and following common convention, y is called an **extremal** for J even though it may not produce a local extremum for J.

Equation (2.6) is the infinite-dimensional analogue of the equation (2.5). Recall that the condition $\nabla f = \mathbf{0}$ is derived from the fact that $\eta \cdot \nabla f = 0$ must hold for all $\eta \in \mathbb{R}^n$. By a suitable choice of vectors in $\mathbb{R}^n$ it was shown that each component of ∇f must vanish separately. A similar strategy can be used to divorce the necessary condition (2.6) from the arbitrary function η. It is not yet clear, however, which special choices of functions in H will accomplish this. Moreover the integrand in equation (2.6) contains not only η but also η' to complicate matters.

The η' term in equation (2.6) can be eliminated using integration by parts. In detail,

$$\int_{x_0}^{x_1} \eta' \frac{\partial f}{\partial y'} \, dx = \eta \frac{\partial f}{\partial y'} \Big|_{x_0}^{x_1} - \int_{x_0}^{x_1} \eta \frac{d}{dx} \left(\frac{\partial f}{\partial y'} \right) dx$$

$$= - \int_{x_0}^{x_1} \eta \frac{d}{dx} \left(\frac{\partial f}{\partial y'} \right) dx,$$

where we have used the conditions $\eta(x_0) = 0$ and $\eta(x_1) = 0$. Equation (2.6) can thus be written

$$\int_{x_0}^{x_1} \eta \left\{ \frac{\partial f}{\partial y} - \frac{d}{dx} \left(\frac{\partial f}{\partial y'} \right) \right\} dx = 0. \tag{2.7}$$

Now,

$$\frac{\partial f}{\partial y} - \frac{d}{dx} \left(\frac{\partial f}{\partial y'} \right) = \frac{\partial f}{\partial y} - \frac{\partial^2 f}{\partial x \partial y'} - \frac{\partial^2 f}{\partial y \partial y'} y' - \frac{\partial^2 f}{\partial y' \partial y'} y'',$$

and given that f has at least two continuous derivatives, we see that for any fixed $y \in C^2[x_0, x_1]$ the function $E : [x_0, x_1] \to \mathbb{R}$ defined by

$$E(x) = \frac{\partial f}{\partial y} - \frac{d}{dx} \left(\frac{\partial f}{\partial y'} \right)$$

is continuous on the interval $[x_0, x_1]$. Here, for a given function y the partial derivatives defining E are evaluated at the point $(x, y(x), y'(x))$. In fact, E can be regarded as an element in the Hilbert space $L^2[x_0, x_1]^4$ and since any $\eta \in H$ is also in $L^2[x_0, x_1]$ we can draw a closer analogy with the finite-dimensional case by noting that equation (2.7) is equivalent to the inner product condition

$$\langle \eta, E \rangle = \int_{x_0}^{x_1} \eta(x) E(x) \, dx = 0 \tag{2.8}$$

for all $\eta \in H$. As with the finite-dimensional case, we can show that the above condition leads to $E = 0$ by considering a special subset of H. First we establish two technical results.

Lemma 2.2.1 *Let α and β be two real numbers such that $\alpha < \beta$. Then there exists a function $\nu \in C^2(\mathbb{R})$ such that $\nu(x) > 0$ for all $x \in (\alpha, \beta)$ and $\nu(x) = 0$ for all $x \in \mathbb{R} - (\alpha, \beta)$.*

Proof: Let

$$\nu(x) = \begin{cases} (x - \alpha)^3 (\beta - x)^3, & \text{if } x \in (\alpha, \beta) \\ 0, & \text{otherwise.} \end{cases}$$

The function ν clearly has all the properties claimed in the lemma except perhaps continuous derivatives at $x = \alpha$ and $x = \beta$. Now,

$$\lim_{x \to \alpha^+} \frac{\nu(x) - \nu(\alpha)}{x - \alpha} = \lim_{x \to \alpha^+} \frac{(x - \alpha)^3 (\beta - x)^3 - 0}{x - \alpha}$$
$$= \lim_{x \to \alpha^+} (x - \alpha)^2 (\beta - x)^3 = 0,$$

and

---

[4] Hilbert spaces are discussed in Appendix B.2. Any function continuous on the interval $[x_0, x_1]$ is in this space. There are a lot "rougher" functions in this space as well.

$$\lim_{x \to \alpha^-} \frac{\nu(x) - \nu(\alpha)}{x - \alpha} = \lim_{x \to \alpha^-} \frac{0 - 0}{x - \alpha} = 0,$$

so that $\nu' = 0$. Similarly,

$$\lim_{x \to \alpha^+} \frac{\nu'(x) - \nu'(\alpha)}{x - \alpha} = \lim_{x \to \alpha^+} \frac{3(x - \alpha)^2 (\beta - x)^2 (\beta + \alpha - 2x) - 0}{x - \alpha}$$
$$= \lim_{x \to \alpha^+} 3(x - \alpha)(\beta - x)^2 (\beta + \alpha - 2x) = 0,$$

and

$$\lim_{x \to \alpha^-} \frac{\nu'(x) - \nu'(\alpha)}{x - \alpha} = \lim_{x \to \alpha^-} \frac{0 - 0}{x - \alpha} = 0,$$

so that $\nu''(\alpha) = 0$. Similar arguments can be used to show that $\nu''(\beta) = 0$. The second derivative is thus

$$\nu''(x) = \begin{cases} 6(x - \alpha)(\beta - x) \left\{ (x - \alpha)^2 + (\beta - x)^2 \right. \\ \left. -3(x - \alpha)(\beta - x) \right\}, & \text{if } x \in (\alpha, \beta) \\ \\ 0, & \text{otherwise,} \end{cases}$$

and it is clear that

$$\lim_{x \to \alpha} \nu''(x) = \nu''(\alpha) = 0$$

and

$$\lim_{x \to \beta} \nu''(x) = \nu''(\beta) = 0;$$

hence, $\nu \in C^2(\mathbb{R})$. □

Lemma 2.2.2 *Suppose that* $\langle \eta, g \rangle = 0$ *for all* $\eta \in H$. *If* $g : [x_0, x_1] \to \mathbb{R}$ *is a continuous function then* $g = 0$ *on the interval* $[x_0, x_1]$.

Proof: Suppose that $g \neq 0$ for some $c \in [x_0, x_1]$. Without loss of generality it can be assumed that $g(c) > 0$, and by continuity that $c \in (x_0, x_1)$. Since g is continuous on $[x_0, x_1]$ there are numbers α, β such that $x_0 < \alpha < c < \beta < x_1$ and $g(x) > 0$ for $x \in (\alpha, \beta)$. Lemma 2.2.1 implies that there exists a function $\nu \in C^2[x_0, x_1]$ such that $\nu(x) > 0$ for all $x \in (\alpha, \beta)$ and $\nu(x) = 0$ for all $x \in [x_0, x_1] - (\alpha, \beta)$. Therefore, $\nu \in H$ and

$$\langle \nu, g \rangle = \int_{x_0}^{x_1} \nu(x) g(x) \, dx = \int_{\alpha}^{\beta} \nu(x) g(x) \, dx > 0,$$

which contradicts the assumption that $\langle \eta, g \rangle = 0$ for all $\eta \in H$. Thus $g = 0$ on (x_0, x_1) and by continuity $g = 0$ on $[x_0, x_1]$. □

The above result indicates that if y is an extremal, then $E = 0$ for all $x \in [x_0, x_1]$. Formally, this result is summarized in the next theorem.

Theorem 2.2.3 *Let $J : C^2[x_0, x_1] \to \mathbb{R}$ be a functional of the form*

$$J(y) = \int_{x_0}^{x_1} f(x, y, y') \, dx,$$

where f has continuous partial derivatives of second order with respect to x, y, and y', and $x_0 < x_1$. Let

$$S = \{y \in C^2[x_0, x_1] : y(x_0) = y_0 \text{ and } y(x_1) = y_1\},$$

where y_0 and y_1 are given real numbers. If $y \in S$ is an extremal for J, then

$$\frac{d}{dx}\left(\frac{\partial f}{\partial y'}\right) - \frac{\partial f}{\partial y} = 0 \tag{2.9}$$

for all $x \in [x_0, x_1]$.

Equation (2.9) is a second-order (generally nonlinear) ordinary differential equation that any (smooth) extremal y must satisfy. This differential equation is called the **Euler-Lagrange equation**. The boundary values associated with this equation for the fixed endpoint problem are

$$y(x_0) = y_0, \quad y(x_1) = y_1. \tag{2.10}$$

The Euler-Lagrange equation is the infinite-dimensional analogue of the equation (2.5). In the transition from finite to infinite dimensions, an algebraic condition for the determination of points $\mathbf{x} \in \mathbb{R}^n$ which might lead to local extrema is replaced by a boundary-value problem involving a second-order differential equation.

Example 2.2.1:   Geodesics in the Plane
Let $(x_0, y_0) = (0, 0)$ and $(x_1, y_1) = (1, 1)$. The arclength of a curve described by $y(x)$, $x \in [0, 1]$ is given by

$$J(y) = \int_0^1 \sqrt{1 + y'^2} \, dx.$$

The geodesic problem in the plane entails determining the function y such that the arclength is minimum. We limit our investigation to functions in $C^2[0, 1]$ such that

$$y(0) = 0, \quad y(1) = 1.$$

If y is an extremal for J then the Euler-Lagrange equation must be satisfied; hence,

$$\frac{d}{dx}\left(\frac{\partial f}{\partial y'}\right) - \frac{\partial f}{\partial y} = \frac{d}{dx}\left(\frac{y'}{\sqrt{1 + y'^2}}\right) - 0 = 0;$$

i.e.,

$$\frac{y'}{\sqrt{1+y'^2}} = const.$$

The last equation is equivalent to the condition that $y' = c_1$, where c_1 is a constant. Consequently, an extremal for J must be of the form

$$y(x) = c_1 x + c_2,$$

where c_2 is another constant of integration. Since $y(0) = 0$, we see that $c_2 = 0$, and since $y(1) = 1$, we see that $c_1 = 1$. Thus, the only extremal y is given by $y(x) = x$, which describes the line segment from $(0,0)$ to $(1,1)$ in the plane (as expected). We have not shown that this extremal is in fact a minimum. (This is shown in Example 10.7.1.)

Example 2.2.2:   Let $(x_0, y_0) = (0,0)$, $(x_1, y_1) = (1,1)$, and consider the functional defined by

$$J(y) = \int_0^1 (y'^2 - y^2 + 2xy)\, dx.$$

The Euler-Lagrange equation for this functional is

$$\frac{d}{dx}(2y') - (-2y + 2x) = 0;$$

i.e.,

$$y'' + y = x.$$

The homogeneous solution is $y_h(x) = c_1 \cos(x) + c_2 \sin(x)$, where c_1 and c_2 are constants, and the particular solution is $y_p(x) = x$. The general solution to the Euler-Lagrange equation is thus given by

$$y(x) = c_1 \cos(x) + c_2 \sin(x) + x.$$

The condition $y(0) = 0$ implies that $c_1 = 0$, and the condition $y(1) = 1$ implies that $c_2 = -1/\sin(1)$. The only extremal for this functional is thus given by

$$y(x) = x - \frac{\sin(x)}{\sin(1)}.$$

Example 2.2.3:   Let k denote some positive constant and let J be the functional defined by

$$J(y) = \int_0^\pi (y'^2 - ky^2)\, dx,$$

with endpoint conditions $y(0) = 0$ and $y(\pi) = 0$. If y is an extremal for J then

$$\frac{d}{dx}(2y') + 2ky = 0;$$

i.e.,

$$y'' + ky = 0.$$

The general solution to the Euler-Lagrange equation is

$$y(x) = c_1\cos(\sqrt{k}x) + c_2\sin(\sqrt{k}x).$$

Now $y(0) = 0$ implies that $c_1 = 0$, and $y(\pi) = 0$ implies that $c_2\sin(\sqrt{k}\pi) = 0$. If $\sqrt{k}$ is not an integer, then $c_2 = 0$, and the only extremal is $y = 0$. If $\sqrt{k}$ is an integer, then $\sin(\sqrt{k}\pi) = 0$ and c_2 can be any number. In the latter case we have an infinite number of extremals of the form $y(x) = c_2\sin(\sqrt{k}x)$.

Exercises 2.2:

1. **Alternative Proof of Condition** (2.6): Let $y \in S$ and $\eta \in H$ be fixed functions. Then the quantity $J(y + \epsilon\eta)$ can be regarded as a function of the single real variable ϵ. Show that the equation $dJ/d\epsilon = 0$ at $\epsilon = 0$ leads to condition (2.6) under the same hypotheses for f.

2. **The First Variation:** Let $J : S \to \Omega$ and $K : S \to \Omega$, be functionals defined by

$$J(y) = \int_{x_0}^{x_1} f(x, y, y')\,dx, \quad K(y) = \int_{x_0}^{x_1} g(x, y, y')\,dx,$$

where f and g are smooth functions of the indicated arguments and $\Omega \subseteq \mathbb{R}$.

(a) Show that for any real numbers A and B,

$$\delta(AJ + BK)(\eta, y) = A\delta J(\eta, y) + B\delta K(\eta, y) \qquad (2.11)$$

(i.e., δ is a linear operator), and

$$\delta(JK)(\eta, y) = K(\eta, y)\delta J(\eta, y) + J(\eta, y)\delta K(\eta, y) \qquad (2.12)$$

(a product rule).

(b) Suppose that $G : \Omega \times \Omega \to \mathbb{R}$ is differentiable on $\Omega \times \Omega$. Show that

$$\delta G(J, K)(\eta, y) = \frac{\partial G}{\partial J}\delta J(\eta, y) + \frac{\partial G}{\partial K}\delta K(\eta, y) \qquad (2.13)$$

(a "chain rule" for the δ operator).

3. Let n be any positive integer. Extend Lemma 2.2.1 by showing that there exists a $\nu \in C^n(\mathbb{R})$ such that $\nu(x) > 0$ for all $x \in (\alpha, \beta)$ and $\nu = 0$ for all $x \in \mathbb{R} \setminus (\alpha, \beta)$.

4. Let J be the functional defined by

$$J(y) = \int_0^1 \left(y'^2 + y^2 + 4ye^x \right) dx,$$

with boundary conditions $y(0) = 0$ and $y(1) = 1$. Find the extremal(s) in $C^2[0,1]$ for J.

5. Consider the functional defined by

$$J(y) = \int_{-1}^1 x^4 y'^2 \, dx.$$

(a) Show that no extremals in $C^2[-1,1]$ exist which satisfy the boundary conditions $y(-1) = -1$, $y(1) = 1$.

(b) Without resorting to the Euler-Lagrange equation, prove that J cannot have a local minimum in the set

$$S = \{ y \in C^2[-1,1] : y(-1) = -1 \text{ and } y(1) = 1 \}.$$

2.3 Some Special Cases

The Euler-Lagrange equation is a second-order nonlinear differential equation, and such equations are usually difficult to simplify let alone solve. There are, however, certain cases when this differential equation can be simplified. We examine two such cases in this section. We suppose throughout that the functional satisfies the conditions of Theorem 2.2.3.

2.3.1 Case I: No Explicit y Dependence

Suppose that the functional is of the form

$$J(y) = \int_{x_0}^{x_1} f(x, y') \, dx,$$

where the variable y does not appear explicitly in the integrand. Evidently, the Euler-Lagrange equation reduces to

$$\frac{\partial f}{\partial y'} = c_1, \tag{2.14}$$

where c_1 is a constant of integration. Now $\partial f / \partial y'$ is a known function of x and y', so that equation (2.14) is a first-order differential equation for y. In principle, equation (2.14) is solvable for y', provided $\partial^2 f / \partial y'^2 \neq 0$, [5] so that equation (2.14) could be recast in the form

---

[5] One can invoke a variant of the implicit function theorem (Appendix A.2).

$$y' = g(x, c_1),$$

for some function g and then integrated. In practice, however, solving equation (2.14) for y' can prove formidable if not impossible, and there may be several solutions available. Nonetheless, the absence of y in the integrand simplifies the problem of solving a second-order differential equation to solving an implicit equation and quadratures.

Example 2.3.1: Let

$$J(y) = \int_{x_0}^{x_1} e^x \sqrt{1 + y'^2}\, dx.$$

The Euler-Lagrange equation for this functional leads to the equation

$$\frac{\partial f}{\partial y'} = \frac{e^x y'}{\sqrt{1 + y'^2}} = c_1, \tag{2.15}$$

where c_1 is a constant of integration. Note that $|y'/\sqrt{1 + y'^2}| \leq 1$ so that $|c_1| \leq e^{x_0}$. Equation (2.15) can be solved for y' to get

$$y' = \frac{c_1}{\sqrt{e^{2x} - c_1^2}},$$

and integrating this expression with respect to x yields

$$y(x) = \sec^{-1}\left(\frac{e^x}{c_1}\right) + c_2,$$

where c_2 is another constant.

Example 2.3.2: Geodesics on a Sphere
In Example 1.4.1, let $u = \theta$ and $v = \phi$. Suppose that we choose $t = u$, so that we regard ϕ as a function of θ. The arclength functional for the sphere is then

$$J(\phi) = \int_{\theta_0}^{\theta_1} \sqrt{1 + \phi'^2 \sin^2 \theta}\, d\theta, \tag{2.16}$$

where ϕ' denotes $d\phi/d\theta$. The integrand does not contain ϕ explicitly, and therefore the Euler-Lagrange equation gives

$$\frac{\phi' \sin^2 \theta}{\sqrt{1 + \phi'^2 \sin^2 \theta}} = c_1, \tag{2.17}$$

where c_1 is a constant. Now, $\phi'^2 \sin^4 \theta \leq \phi'^2 \sin^2 \theta \leq 1 + \phi'^2 \sin^2 \theta$, and therefore $-1 \leq c_1 \leq 1$. Hence, we can replace c_1 by the constant $\sin \alpha$. Equation (2.17) implies

$$\phi' = \frac{\sin \alpha}{\sin \theta \sqrt{\sin^2 \theta - \sin^2 \alpha}};$$

thus,

$$\phi = \int_{\theta_0}^{\theta} \frac{\sin \alpha}{\sin \xi \sqrt{\sin^2 \xi - \sin^2 \alpha}} \, d\xi + \beta,$$

where $\beta = \phi(\theta_0)$. The above equation yields the implicit relation

$$\cos(\phi + \beta) = \frac{\tan \alpha}{\tan \theta}, \tag{2.18}$$

or in Cartesian coördinates,

$$x \cos \beta - y \sin \beta = z \tan \alpha. \tag{2.19}$$

Equation (2.19) is the equation of a plane through the centre of the sphere. The geodesic corresponds to the intersection of this plane with the sphere; hence, it must be an arc of great circle.

2.3.2 Case II: No Explicit x Dependence

Another simplification is available when the integrand does not contain the independent variable x explicitly.

Theorem 2.3.1 *Let J be a functional of the form*

$$J(y) = \int_{x_0}^{x_1} f(y, y') \, dx,$$

and define the function H by

$$H(y, y') = y' \frac{\partial f}{\partial y'} - f.$$

Then H is constant along any extremal y.

Proof: Suppose that y is an extremal for J. Now,

$$\begin{aligned}
\frac{d}{dx} H(y, y') &= \frac{d}{dx} \left(y' \frac{\partial f}{\partial y'} - f \right) \\
&= y'' \frac{\partial f}{\partial y'} + y' \frac{d}{dx} \frac{\partial f}{\partial y'} - \left(y' \frac{\partial f}{\partial y} + y'' \frac{\partial f}{\partial y'} \right) \\
&= y' \left(\frac{d}{dx} \frac{\partial f}{\partial y'} - \frac{\partial f}{\partial y} \right),
\end{aligned}$$

and since y is an extremal, the Euler-Lagrange equation (2.9) is satisfied; hence,

$$\frac{d}{dx}H(y,y') = 0.$$

Consequently, H must be constant along an extremal.   □

Note that the function H depends only on y and y', and thus the equation

$$H(y,y') = const. \tag{2.20}$$

is a *first*-order differential equation for the extremal y.

Example 2.3.3:   Catenary
The catenary problem (Section 1.2) has a functional of the form

$$J(y) = \int_{x_0}^{x_1} y\sqrt{1+y'^2}\,dx.$$

The above integrand does not contain x explicitly and therefore

$$H(y,y') = y'\frac{\partial f}{\partial y'} - f$$

$$= y'\frac{yy'}{\sqrt{1+y'^2}} - y\sqrt{1+y'^2}$$

is constant along an extremal. Any extremal y must consequently satisfy the first-order differential equation

$$\frac{y^2}{1+y'^2} = c_1^2, \tag{2.21}$$

where c_1 is a constant. If $c_1 = 0$, then the only solution to equation (2.21) is $y = 0$. Suppose that $c_1 \neq 0$; then equation (2.21) can be replaced by

$$y' = \sqrt{\frac{y^2}{c_1^2} - 1}. \tag{2.22}$$

We integrate equation (2.22) for x as a function of y, viz.,

$$x = \int \frac{dy}{\sqrt{\frac{y^2}{c_1^2} - 1}}$$

$$= c_1 \ln\left(\frac{y + \sqrt{y^2 - c_1^2}}{c_1}\right) + c_2,$$

where c_2 is a constant of integration. Now,

$$c_1 e^{(x-c_2)/c_1} = y + \sqrt{y^2 - c_1^2},$$

and

$$c_1 e^{-(x-c_2)/c_1} = \frac{c_1^2}{y + \sqrt{y^2 - c_1^2}};$$

therefore,

$$c_1 \left(e^{(x-c_2)/c_1} + e^{-(x-c_2)/c_1} \right) = y + \sqrt{y^2 - c_1^2} + \frac{c_1^2}{y + \sqrt{y^2 - c_1^2}}$$

$$= 2y.$$

The extremals are thus given by

$$y(x) = c_1 \cosh\left(\frac{x - c_2}{c_1}\right).$$

Example 2.3.4:   Brachystochrone

The brachystochrone problem (Section 1.2) has a functional of the form

$$J(y) = \int_{x_0}^{x_1} \sqrt{\frac{1 + y'^2}{y}}\, dx.$$

The integrand does not depend on x explicitly; thus,

$$H(y, y') = y' \frac{\partial f}{\partial y'} - f$$

$$= \frac{y'^2}{\sqrt{y(1 + y'^2)}} - \sqrt{\frac{1 + y'^2}{y}}$$

$$= -\frac{1}{\sqrt{y(1 + y'^2)}}$$

is constant along an extremal. If y is an extremal for J then it must satisfy the first-order differential equation

$$y(1 + y'^2) = c_1, \tag{2.23}$$

where c_1 is a constant. Equation (2.23) can be solved parametrically. Let $y' = \tan \psi$; then $1 + y'^2 = \sec^2 \psi$ and

$$y = \frac{c_1}{\sec^2 \psi} = c_1 \cos^2 \psi = \kappa_1 (1 + \cos(2\psi)), \tag{2.24}$$

where $\kappa_1 = c_1/2$. Now,

$$dy = -4\kappa_1 \cos \psi \sin \psi \, d\psi$$

and

$$dx = \cot \psi \, dy = -4\kappa_1 \cos^2 \psi \, d\psi$$
$$= -2\kappa_1 (1 + \cos(2\psi)) \, d\psi.$$

Therefore,

$$x = \kappa_2 - \kappa_1 (2\psi + \sin(2\psi)), \tag{2.25}$$

where κ_2 is an integration constant. Equations (2.24) and (2.25) provide a parametric solution to the problem. The solution curve is a well-known class of plane curves called **cycloids** (Section 1.2).

The simplification when f does not depend on y explicitly is more or less obvious from the Euler-Lagrange equation; the simplification when x is absent in f is less obvious. In particular, what leads one to consider a function such as H in the first place? Equation (2.20) is an example of a **conservation law**: along any extremal, the quantity H is conserved. In problems concerning classical mechanics, H often represents the total energy of the system. One can thus be led to consider a function such as H from the physics of whatever the functional is modelling if a conservation law is known. Mathematically, this approach is not very satisfactory. One immediately questions whether other conservation laws exist and if there are any other special cases for the integrand leading to conservation laws. In fact, there are ways to deduce conservation laws mathematically. Noether's theorem provides a general framework in which to derive conservation laws. We discuss this theorem in Chapter 9.

Exercises 2.3:

1. Find the general solution to the Euler-Lagrange equation corresponding to the functional
$$J(y) = \int_{x_0}^{x_1} f(x)\sqrt{1 + y'^2} \, dx,$$
where $x_0 > 0$, and investigate the special cases: (i) $f(x) = \sqrt{x}$, (ii) $f(x) = x$.

2. Find the extremals for the functional defined by
$$\int_{x_0}^{x_1} \frac{y'^2}{x^3} \, dx,$$
where $x_0 > 0$.

3. Let
$$J(y) = \int_{2}^{3} y^2 (1 - y')^2 \, dx.$$
Find a smooth extremal for J satisfying the boundary conditions $y(2) = 1$ and $y(3) = \sqrt{3}$.

2.4 A Degenerate Case

In the examples so far, the integrand of the functional depends on y' in some nonlinear way. If the integrand is linear in y', the problem becomes degenerate in a sense that is explained in this section.

Suppose that J is a functional of the form

$$J(y) = \int_{x_0}^{x_1} (A(x, y)y' + B(x, y))\ dx,$$

where A and B are smooth functions of x and y. The Euler-Lagrange equation for this functional is

$$\frac{d}{dx}A(x, y) - \left(y'\frac{\partial A}{\partial y} + \frac{\partial B}{\partial y}\right) = 0.$$

But

$$\frac{d}{dx}A(x, y) = \frac{\partial A}{\partial x} + y'\frac{\partial A}{\partial y},$$

so that the Euler-Lagrange equation reduces to

$$\frac{\partial A}{\partial x} - \frac{\partial B}{\partial y} = 0. \tag{2.26}$$

Note that equation (2.26) is not even a differential equation for y: it is an implicit equation for y that may or may not have solutions depending on the given functions A and B. Moreover, equation (2.26) contains no arbitrary constants so that arbitrary boundary conditions cannot be imposed on any solutions.

It may be that equation (2.26) is satisfied for all x and y; i.e., $A_x = B_y$ is an identity. In this case equation (2.26) places no restriction on y, but it does imply the existence of a function $\phi(x, y)$ such that $\phi_y = A$ and $\phi_x = B$. In this case the integrand can be written as

$$f = \frac{\partial \phi}{\partial x} + y'\frac{\partial \phi}{\partial y} = \frac{d\phi}{dx};$$

that is, $f\ dx = d\phi$ (an exact differential).[6] Consequently,

$$J(y) = \int_{x_0}^{x_1} d\phi = \phi(x_1, y(x_1)) - \phi(x_0, y(x_0)),$$

so that J depends only on ϕ and the endpoints $(x_0, y(x_0))$ and $(x_1, y(x_1))$. The value of J is thus independent of y, so that the integral is path independent.

---

[6] Equation (2.26) is a well-known integrability condition (cf. [44], p. 529).

Example 2.4.1: Let $f(x, y, y') = (x^2 + 3y^2)y' + 2xy$. Here, $A_x = 2x = B_y$, so that the value of the functional defined by

$$J(y) = \int_{x_0}^{x_1} \left((x^2 + 3y^2)y' + 2xy \right) \, dx$$

is independent of the choice of y. A function ϕ can be found by integrating the equations $B = \phi_x$ and $A = \phi_y$. For example $\phi_x = B = 2xy$; hence,

$$\phi = x^2 y + C(y),$$

where C is some function of y to be determined. Now

$$\phi_y = x^2 + C'(y) = A = x^2 + 3y^2,$$

and so

$$\phi = x^2 y + y^3 + k,$$

where k is an arbitrary constant. Thus,

$$\begin{aligned} J(y) &= \phi(x_1, y(x_1)) - \phi(x_0, y(x_0)) \\ &= x_1^2 y_1 + y_1^3 - (x_0^2 y_0 + y_0^3). \end{aligned}$$

(Note that the arbitrary constant k vanishes from the final answer.)

In summary, variational problems with integrands of the form $A(x, y)y' + B(x, y)$ are degenerate in that either y is determined implicitly and can satisfy only very special sets of boundary data, or the value of the corresponding functional does not depend on the choice of y. In the latter case the determination of local extrema is vacuous.

An immediate concern is that there may be other forms of integrands that lead to path independent functionals. These functionals are characterized by the property that the Euler-Lagrange equation reduces to an identity valid for all x and y in the space under consideration. The next theorem shows that in fact the integrand must be linear in y' for such an identity to be valid.

Theorem 2.4.1 *Suppose that the functional J satisfies the conditions of Theorem 2.2.3 and that the Euler-Lagrange equation (2.9) reduces to an identity. Then, the integrand must be linear in y', and the value of the functional is independent of y.*

Proof: If the Euler-Lagrange equation is an identity, then

$$\frac{\partial f}{\partial y} - \frac{\partial^2 f}{\partial x \partial y'} - \frac{\partial^2 f}{\partial y \partial y'} y' - \frac{\partial^2 f}{\partial y'^2} y'' = 0 \tag{2.27}$$

for all $x \in [x_0, x_1]$ and $y \in S$. Now, y'' appears only in the last term on the left-hand side of the equation, and since equation (2.27) must hold for all

$y \in S$ we must have that $\partial^2 f / \partial y'^2 = 0$. The integrand must therefore be of the form

$$f(x, y, y') = A(x, y)y' + B(x, y),$$

where $A_x = B_y$ for all $x \in [x_0, x_1]$ and $y \in S$. $\square$

2.5 Invariance of the Euler-Lagrange Equation

The principles in physics that lead to variational formulations do not depend on coördinate systems. Geometrical problems such as the determination of geodesics are likewise "coördinate free" in character. The path of a particle, for instance, does not depend on the coördinate system the observer uses to describe it; a geodesic does not depend on a particular parametrization of the surface. These types of problems can be framed in terms of maximizing functionals and ultimately lead to solutions to an Euler-Lagrange equation. On physical (and geometrical) grounds one thus expects the Euler-Lagrange equation to also be invariant with respect to coördinate transformations. In this section we take an informal but practical look at the invariance of the Euler-Lagrange equation.

A coördinate transformation

$$x = x(u, v), \quad y = y(u, v), \tag{2.28}$$

is called **smooth** if the functions x and y have continuous partial derivatives with respect to u and v. A smooth transformation is called **nonsingular** if the Jacobian

$$\frac{\partial(x, y)}{\partial(u, v)} = \det \begin{pmatrix} x_u & y_u \\ x_v & y_v \end{pmatrix}$$

satisfies the condition

$$\frac{\partial(x, y)}{\partial(u, v)} \neq 0. \tag{2.29}$$

Here we use the notation $x_u = \partial x / \partial u$ etc. for succinctness. Note that condition (2.29) implies that the transformation is invertible: to every pair (x, y) there corresponds a unique pair (u, v) satisfying equation (2.28).[7] We assume that the coördinate transformation defined by equation (2.28) is smooth and nonsingular.

Let J be a functional of the form

$$J(y) = \int_{x_0}^{x_1} f(x, y, y') \, dx, \tag{2.30}$$

and let S be defined by

---

[7] This result follows from the implicit function theorem; see Theorem A.2.2.

$$S = \{y \in C^2[x_0, x_1] : y(x_0) = y_0 \text{ and } y(x_1) = y_1\},$$

where y_0 and y_1 are given numbers. Suppose now that we write the functional in terms of the (u, v) coördinates and, for definiteness, let us regard v as a function of u. Then,

$$\frac{dy}{dx} = \frac{dy/du}{dx/du} = \frac{y_u + y_v \dot{v}}{x_u + x_v \dot{v}},$$

and

$$dx = \frac{dx}{du} du = (x_u + x_v \dot{v}) du,$$

where $\dot{v}$ denotes dv/du. The functional defined by equation (2.30) thus becomes

$$\begin{aligned}
J(y) &= \int_{x_0}^{x_1} f(x, y, y') \, dx \\
&= \int_{u_0}^{u_1} f(x(u, v), y(u, v), \frac{y_u + y_v \dot{v}}{x_u + x_v \dot{v}})(x_u + x_v \dot{v}) \, du \\
&= \int_{u_0}^{u_1} F(u, v, \dot{v}) \, du.
\end{aligned}$$

Here, the numbers u_0, u_1 and the new boundary values $v(u_0) = v_0$, $v(u_1) = v_1$ are the unique solutions to the equations

$$\begin{aligned}
x_0 = x(u_0, v_0), \quad x_1 = x(u_1, v_1), \\
y_0 = y(u_0, v_0), \quad y_1 = y(u_1, v_1).
\end{aligned}$$

For clarity, let

$$K(v) = \int_{u_0}^{u_1} F(u, v, \dot{v}) \, du, \tag{2.31}$$

and let T be the set defined by

$$T = \{v \in C^2[u_0, u_1] : v(u_0) = v_0 \text{ and } v(u_1) = v_1\}.$$

Given a curve in the xy-plane described by a function $y = y(x)$, the transformation (2.28) defines the curve in the uv-plane described by some function $v = v(u)$. The essence of the invariant question is: if $v \in T$ is an extremal for K, is $y \in S$ and extremal for J (and vice versa)? The next theorem resolves this question.

Theorem 2.5.1 *Let $y \in S$ and $v \in T$ be two functions that satisfy the smooth nonsingular transformation (2.28). Then y is an extremal for J if and only if v is an extremal for K.*

Proof: Suppose that $v \in T$ is an extremal for K. Then, v satisfies the Euler-Lagrange equation

$$\frac{d}{du} \frac{\partial F}{\partial \dot{v}} - \frac{\partial F}{\partial v} = 0. \tag{2.32}$$

Now,

$$F(u, v, \dot{v}) = f(x(u,v), y(u,v), \frac{y_u + y_v \dot{v}}{x_u + x_v \dot{v}})(x_u + x_v \dot{v}),$$

so that

$$\frac{\partial F}{\partial \dot{v}} = \frac{\partial f}{\partial y'}(x_u + x_v \dot{v})\frac{\partial}{\partial \dot{v}}\left(\frac{y_u + y_v \dot{v}}{x_u + x_v \dot{v}}\right)$$

$$+ x_v f,$$

and

$$\frac{\partial F}{\partial v} = \left(\frac{\partial f}{\partial x}x_v + \frac{\partial f}{\partial y}y_v + \frac{\partial f}{\partial y'}\frac{\partial}{\partial v}\left(\frac{y_u + y_v \dot{v}}{x_u + x_v \dot{v}}\right)\right)(x_u + x_v \dot{v})$$

$$+ f\frac{\partial}{\partial v}(x_u + x_v \dot{v}).$$

A straightforward but tedious calculation shows that

$$\frac{d}{du}\frac{\partial F}{\partial \dot{v}} - \frac{\partial F}{\partial v} = \frac{\partial(x,y)}{\partial(u,v)}\left(\frac{d}{dx}\frac{\partial f}{\partial y'} - \frac{\partial f}{\partial y}\right). \tag{2.33}$$

Since the transformation is nonsingular, the Jacobian is nonzero; hence, if v is an extremal for K then equations (2.32) and (2.33) imply that

$$\frac{d}{dx}\frac{\partial f}{\partial y'} - \frac{\partial f}{\partial y} = 0,$$

so that y must be an extremal for J. Equation (2.33) also implies the converse.
□

It is philosophically reassuring that the path of a particle is independent of the observer's choice of coördinates. There is also a practical implication: coördinate transformations can be made in the functional before the Euler-Lagrange equation is formulated. An example suffices to illustrate the value of this observation.

Example 2.5.1:   Let J be the functional defined by

$$J(y) = \int_{x_0}^{x_1} \sqrt{x^2 + y^2}\sqrt{1 + y'^2}\, dx.$$

The integrand contains both x and y so that there are no conspicuous first integrals of the Euler-Lagrange equation

$$\frac{d}{dx}\left(\sqrt{\frac{x^2 + y^2}{1 + y'^2}}y'\right) - y\sqrt{\frac{1 + y'^2}{x^2 + y^2}} = 0. \tag{2.34}$$

On the other hand, the presence of the term $\sqrt{x^2 + y^2}$ suggests the use of polar coördinates. Let

$$x = x(\phi, r) = r \cos \phi,$$
$$y = y(\phi, r) = r \sin \phi.$$

This transformation is evidently smooth, and since

$$\frac{\partial(x, y)}{\partial(\phi, r)} = \det \begin{pmatrix} x_\phi & y_\phi \\ x_r & y_r \end{pmatrix}$$

$$= \det \begin{pmatrix} -r \sin \phi & r \cos \phi \\ \cos \phi & \sin \phi \end{pmatrix} = -r,$$

the transformation is nonsingular, provided $r \neq 0$. Now, suppose that r is regarded as a function of ϕ, then

$$y' = \frac{y_\phi + y_r \dot{r}}{x_\phi + x_r \dot{r}} = \frac{r \cos \phi + \sin \phi \dot{r}}{-r \sin \phi + \cos \phi \dot{r}},$$

so that

$$\sqrt{1 + y'^2}\, dx = \sqrt{r^2 + \dot{r}^2}\, d\phi.$$

The functional J thus becomes

$$K(r) = \int_{\phi_0}^{\phi_1} r\sqrt{r^2 + \dot{r}^2}\, d\phi = \int_{\phi_0}^{\phi_1} F(r, \dot{r})\, d\phi. \qquad (2.35)$$

The integrand does not depend on ϕ explicitly, and therefore the corresponding Euler-Lagrange equation has a first integral

$$H(r, \dot{r}) = \dot{r}\frac{\partial F}{\partial \dot{r}} - F$$

$$= \frac{r\dot{r}^2}{\sqrt{r^2 + \dot{r}^2}} - r\sqrt{r^2 + \dot{r}^2}$$

$$= const.;$$

i.e.,

$$\dot{r} = r\sqrt{c_1^2 r^4 - 1}, \qquad (2.36)$$

where c_1 is a nonzero constant. Equation (2.36) can be integrated to solve for ϕ as a function of r,

$$\int \frac{dr}{r\sqrt{c_1^2 r^4 - 1}} = -\frac{1}{2}\sin^{-1}\left(\frac{1}{c_1 r^2}\right) = \phi + c_2,$$

where c_2 is a constant. Thus, for $\kappa_1 = 1/c_1$, and $\kappa_2 = -2c_2$, the function $r(\phi)$ is given implicitly by

$$\frac{\kappa_1}{r^2} = \sin(-2\phi + \kappa_2)$$
$$= -\sin(2\phi)\cos\kappa_2 + \cos(2\phi)\sin\kappa_2$$
$$= -2\sin\phi\cos\phi\cos\kappa_2 + \left(2\cos^2\phi - 1\right)\sin\kappa_2.$$

In terms of the original Cartesian coördinate system, the above expression is equivalent to

$$\kappa_1 = x^2\sin\kappa_2 - 2xy\cos\kappa_2 - y^2\sin\kappa_2. \tag{2.37}$$

Exercises 2.5:

1. **Change of Variable:** Let $\psi : [t_0, t_1] \to \mathbb{R}$ be a smooth function on the interval $[t_0, t_1]$ such that $\psi'(t) > 0$ for all $t \in [t_0, t_1]$ and let $\psi(t_0) = x_0$, $\psi(t_1) = x_1$. Using the transformation $x = \psi(t)$, the functional J defined by

$$J(y) = \int_{x_0}^{x_1} f(x, y, y')\, dx$$

can be transformed to the functional K defined by

$$K(Y) = \int_{t_0}^{t_1} F(t, Y, \dot{Y})\, dt,$$

where, for $Y(t) = y(\psi(t))$, $\dot{Y}$ denotes dY/dt and

$$F(t, Y, \dot{Y}) = f(\psi(t), Y, \dot{Y})\psi'(t).$$

Prove by direct calculation that

$$\frac{d}{dt}\frac{\partial F}{\partial \dot{Y}} - \frac{\partial F}{\partial Y} = \psi'\left(\frac{d}{dx}\frac{\partial f}{\partial y'} - \frac{\partial f}{\partial y}\right),$$

and hence that y is an extremal for J if and only if Y is an extremal for K.

2. Let J be the functional defined by

$$J(r) = \int_{\pi/2}^{\pi} \sqrt{r^2 + \dot{r}^2}\, d\phi.$$

Find an extremal for J satisfying the boundary conditions $r(\pi/2) = 1$ and $r(\pi) = -1$.

3. Let J be a functional of the form

$$J(y) = \int_{x_0}^{x_1} g(x^2 + y^2)\sqrt{1 + y'^2}\, dx,$$

where g is some function of $x^2 + y^2$. Use the polar coördinate transformation to find the general form of the extremals in terms of g, r, and ϕ.

2.6 Existence of Solutions to the Boundary-Value Problem*

In this section we discuss briefly and informally the question of existence and uniqueness of solutions to the boundary-value problem associated with finding extremals. Generally questions of this nature are difficult to answer even for specific cases owing to two features. Firstly, the Euler-Lagrange equation is usually a nonlinear differential equation and thus difficult if not impossible to solve analytically. Secondly, boundary-value problems are global in character: the solutions must be defined on the entire interval $[x_0, x_1]$. In contrast to initial-value problems, which are local in character,[8] there are few general results analogous to Picard's theorem[9] available. Our discussion is limited primarily to examples that illustrate some of the pathologies of boundary-value problems. An example of a general existence result for certain boundary-value problems is given at the end of this section.

Under the conditions of Theorem 2.2.3, the determination of extremals for a functional J of the form

$$J(y) = \int_{x_0}^{x_1} f(x, y, y') \, dx, \qquad (2.38)$$

with $x_0 < x_1$ and given boundary values y_0, y_1, entails finding solutions to the Euler-Lagrange equation

$$\frac{d}{dx} \frac{\partial f}{\partial y'} - \frac{\partial f}{\partial y} = 0 \qquad (2.39)$$

subject to the conditions

$$y(x_0) = y_0, \quad y(x_1) = y_1. \qquad (2.40)$$

In this context a solution to the boundary-value problem is a function y such that:

(a) $y \in C^2[x_0, x_1]$;
(b) y satisfies the Euler-Lagrange equation (2.39) for all $x \in [x_0, x_1]$; and
(c) y satisfies the boundary conditions (2.40).

The definition of a solution can certainly be relaxed to include "rougher" functions such as piecewise smooth functions, but we do not pursue this generalization and limit our discussion to smooth solutions.

Much of the discussion in the earlier sections of this chapter focused on determining the general solution $y(x, c_1, c_2)$ to equation (2.39). Even if the

---

[8] Initial-value problems entail solving a differential equation subject to conditions of the form $y(x_0) = y_0$, $y'(x_0) = y_0'$. The conditions are defined at the same point x_0 and the solution need exist only in a small neighbourhood of x_0.
[9] See Appendix A.3.

two-parameter family of functions that comprises the general solution can be found, however, there is no guarantee that constants c_1 and c_2 can be found such that

$$y(x_0, c_1, c_2) = y_0, \quad y(x_1, c_1, c_2) = y_1, \tag{2.41}$$

for a given choice of points (x_0, y_0) and (x_1, y_1). In fact, there is no a priori reason why $y(x, c_1, c_2)$ need even be in the space $C^2[x_0, x_1]$ for any particular choice of constants. It may be that no solution exists to the boundary-value problem even though a general solution can be found to the Euler-Lagrange equation. At the other extreme, equations (2.41) may have an infinite number of solutions for c_1 and/or c_2, and in this case the boundary-value problem would have an infinite number of solutions. Examples 2.2.1, 2.2.3, and Exercises 2.2-5 illustrate some of the possible scenarios.

Example 2.2.1:   The general solution for geodesics in the plane is

$$y(x, c_1, c_2) = c_1 x + c_2,$$

and given any set of points (x_0, y_0), (x_1, y_1) (such that $x_0 \neq x_1$) it is clear that the function

$$y(x) = \frac{y_1 - y_0}{x_1 - x_0} x + \frac{y_0 x_1 - y_1 x_0}{x_1 - x_0}$$

is the unique solution to the boundary-value problem.

Example 2.2.3:   The general solution to this problem is

$$y(x, c_1, c_2) = c_1 \cos(\sqrt{k}x) + c_2 \sin(\sqrt{k}x),$$

and if $\sqrt{k}$ is not an integer, then $c_1 = c_2 = 0$ is the only solution to equations (2.41). If $\sqrt{k}$ is an integer, however, then any function of the form

$$y(x) = c_2 \sin(\sqrt{k}x)$$

is a solution to the boundary-value problem. In the above expression c_2 is an arbitrary number and hence there are an infinite number of solutions to the boundary-value problem.

Exercises 2.2-5:   The general solution to the Euler-Lagrange equation is of the form

$$y(x, c_1, c_2) = \frac{c_1}{x^3} + c_2.$$

If y is a solution to the boundary-value problem, then it must be in the space $C^2[-1, 1]$, and since $0 \in [-1, 1]$ this means that $c_1 = 0$. Evidently, there is no constant c_2 such that

$$y(-1, 0, c_2) = -1$$

and

$$y(1, 0, c_2) = 1;$$

consequently, this boundary-value problem has no solutions.

A more involved but illuminating example is afforded by the catenary.[10]

Example 2.6.1:   **Catenary** Recall that the functional J defined by

$$J(y) = \int_{x_0}^{x_1} y\sqrt{1 + y'^2}\, dx$$

with boundary conditions $y(x_0) = y_0 \geq 0$ and $y(x_1) = y_1 \geq 0$ models the shape of a uniform flexible cable suspended from a pole of height y_0 to another pole of height y_1, where the poles are a distance of $x_1 - x_0$ apart. The cable is assumed to be coiled at the base of each pole so that there is no restriction regarding the arclength of cable between the poles.

There are three important parameters in the model: the heights y_0 and y_1, and the separation distance $x_1 - x_0$. We can always normalize the problem by assuming that the separation distance is one unit, say $x_0 = 0$ and $x_1 = 1$. We can then work with the parameters y_0 and y_1.

Recall from Example 2.3.3 that the general solution to the boundary-value problem is

$$y(x) = c_1 \cosh\left(\frac{x - c_2}{c_1}\right),$$

where c_1 and c_2 are constants. It is required that

$$y_0 = c_1 \cosh\left(\frac{-c_2}{c_1}\right),$$

$$y_1 = c_1 \cosh\left(\frac{1 - c_2}{c_1}\right),$$

but these are transcendental equations for the constants c_1 and c_2, and it is not clear whether solutions exist for all $y_0, y_1 > 0$. Let $\kappa_1 = c_1$ and $\kappa_2 = -c_2/c_1$; then the above equations can be recast as

$$y_0 = \kappa_1 \cosh(\kappa_2),$$

$$y_1 = \kappa_1 \cosh\left(\frac{1}{\kappa_1} + \kappa_2\right).$$

At this stage let us specialize (and simplify) the problem by assuming that $y_0 = 1$. Although this does not capture all the possibilities, it does display the basic pathologies. We thus look at the availability of solutions for various values of the remaining parameter $y_1 > 0$. Under the assumption that $y_0 = 1$, the above equations imply that

$$\begin{aligned}
y_1 &= \kappa_1 \cosh(\cosh(\kappa_2) + \kappa_1) \\
&= \frac{\cosh(\cosh(\kappa_2) + \kappa_1)}{\cosh(\kappa_2)} \\
&= F(\kappa_2).
\end{aligned} \tag{2.42}$$

---

[10] Carathéodory [21], p. 297 discusses this problem in detail.

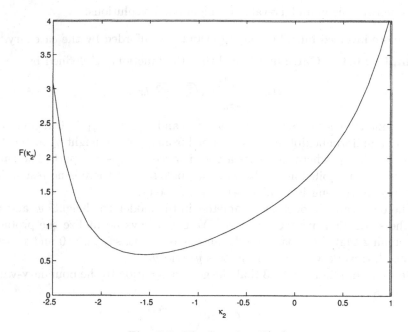

Fig. 2.2. The function $F(\kappa_2)$

Now $\cosh x > 0$ for all $x \in \mathbb{R}$ and so $F(\kappa_2) > 0$ for all $\kappa_2 \in \mathbb{R}$. Moreover, F is a smooth function of κ_2 and $F(\kappa_2) \to \infty$ as $\kappa_2 \to \pm\infty$; consequently, F must have a *positive* minimum at some point $\kappa^* \in \mathbb{R}$. In fact, it can be shown that F has precisely one local minimum at $\kappa^* \approx -1.56$ at which $F(\kappa^*) \approx 0.59$. A graph of F is given in figure 2.2.

Evidently y_1 must be positive for a solution (no physical surprise here), but the above calculation indicates that we need $y_1 \geq F(\kappa^*) \approx 0.59$. This means if $y_1 < F(\kappa^*)$ then there is no solution to the boundary value problem (in the space $C^2[0,1]$). A quick study of the curve of F reveals the following cases:

(a) if $y_1 < F(\kappa^*)$, then there are no solutions;
(b) if $y_1 = F(\kappa^*)$, then there is precisely one solution; and
(c) if $y_1 > F(\kappa^*)$, then there are precisely two solutions.

Case (c) requires further comment: physically we do not expect two possible solutions to the problem. We must remember, however, that the above analysis predicts two extremals in this case but implies nothing regarding the nature of these extremals. In fact, only one of the extremals corresponds to a local minimum for J. We show this in Example 10.6.3.

The catenary is a revealing example of possible problems with solutions to boundary-value problems. As with the earlier examples, however, questions regarding the existence and uniqueness of solutions were resolved only because the general solution was available explicitly. Typically, the Euler-Lagrange equation cannot be solved and we do not have the luxury of knowing the general solution before we investigate these questions. Even if we cannot solve the Euler-Lagrange equation analytically, qualitative properties such as existence and uniqueness of solutions to the boundary-value problem are nonetheless important. These properties test the veracity of the model especially when experiment shows that a solution must exist. Moreover, the investigation of solution existence and uniqueness highlights any special parameter regions where no solution or multiple solutions may exist. Generally this type of investigation provides useful information in preparation for a more efficient numerical approach to the problem.

There is, unfortunately, a paucity of *general* results concerning boundary-value problems involving nonlinear second-order differential equations, and the results that are available are often fettered with numerous special (and usually complicated) conditions. It is well beyond the scope of this book to give even a brief overview of the various results/techniques used to address existence/uniqueness questions for boundary-value problems. Instead, we leave the reader with an "older" but useful result due to Bernstein [7], which we do not prove.

Theorem 2.6.1 (Bernstein) *Consider the boundary-value problem that consists of solving the equation*

$$y'' = F(x, y, y'),   \qquad (2.43)$$

subject to the boundary conditions

$$y(x_0) = y_0, \quad y(x_1) = y_1,   \qquad (2.44)$$

where y_0 and y_1 are given real numbers and $x_0 \neq x_1$. Suppose that on the set $\Omega = [x_0, x_1] \times \mathbb{R} \times \mathbb{R}$ the function F is continuous and has continuous partial derivatives with respect to y and y'. Suppose further that there exists a positive constant μ such that

$$\frac{\partial F(x, y, y')}{\partial y} > \mu   \qquad (2.45)$$

for all $(x, y, y') \in \Omega$ and that there exist nonnegative functions $A, B : [x_0, x_1] \times \mathbb{R} \to \mathbb{R}$ bounded in any compact subset of $[x_0, x_1] \times \mathbb{R}$ such that

$$|F(x, y, y')| \leq A(x, y)y'^2 + B(x, y).   \qquad (2.46)$$

Then, there exists precisely one function y such that equations (2.43) and (2.44) are satisfied.

Remarks:

(a) We need continuity in a set such as Ω since y is unknown and hence its range as well as that of y' is unknown.

(b) Although the theorem does not state it explicitly, a solution requires y to be at least twice differentiable. This means that y and y' are continuous functions on the interval $[x_0, x_1]$. Since F is continuous, equation (2.43) implies that y'' must also be continuous on the interval $[x_0, x_1]$; i.e., $y \in C^2[x_0, x_1]$.

In closing, we stress that existence and uniqueness results for the boundary-value problem do not necessarily transfer to the original variational problem, which is generally concerned with finding local extrema. The catenary is an example of this situation. The basic question concerns the existence of a local extremum for a given functional not merely an extremal. Some results concerning this question can be found in Carathéodory *loc. cit.* and Ewing [26].

3

Some Generalizations

3.1 Functionals Containing Higher-Order Derivatives

The arguments leading to the Euler-Lagrange equation in Section 2.2 can be extended to functionals involving higher-order derivatives. Naturally, the function spaces must be further restricted to account for the higher-order derivatives. Consider a functional of the form

$$J(y) = \int_{x_0}^{x_1} f(x, y, y', y'') \, dx,$$

along with boundary conditions of the form $y(x_0) = y_0$, $y'(x_0) = y_0'$, $y(x_1) = y_1$, and $y'(x_1) = y_1'$. Here we assume that f has continuous partial derivatives of the third order with respect to x, y, y', and y'', and that $y \in C^4[x_0, x_1]$. The set S is thus

$$S = \{y \in C^4[x_0, x_1] : y(x_0) = y_0, y'(x_0) = y_0', y(x_1) = y_1, y'(x_1) = y_1'\},$$

and the set H is defined by

$$H = \{\eta \in C^4[x_0, x_1] : \eta(x_0) = \eta'(x_0) = \eta(x_1) = \eta'(x_1) = 0\}.$$

Suppose that J has a local extremum in S at $y \in S$. Proceeding as in Section 2.2, let $\hat{y} = y + \epsilon\eta$ and consider the difference $J(\hat{y}) - J(y)$. Taylor's theorem implies that

$$f(x, \hat{y}, \hat{y}', \hat{y}'') = f(x, y + \epsilon\eta, y' + \epsilon\eta', y'' + \epsilon\eta'')$$
$$= f(x, y, y', y'') + \epsilon\left(\eta\frac{\partial f}{\partial y} + \eta'\frac{\partial f}{\partial y'} + \eta''\frac{\partial f}{\partial y''}\right) + O(\epsilon^2),$$

and consequently,

$$J(\hat{y}) - J(y) = \epsilon \int_{x_0}^{x_1} \left(\eta\frac{\partial f}{\partial y} + \eta'\frac{\partial f}{\partial y'} + \eta''\frac{\partial f}{\partial y''}\right) dx + O(\epsilon^2).$$

The first variation for this functional is therefore

$$\delta J(\eta, y) = \int_{x_0}^{x_1} \left(\eta \frac{\partial f}{\partial y} + \eta' \frac{\partial f}{\partial y'} + \eta'' \frac{\partial f}{\partial y''} \right) dx.$$

If J has a local extremum at y then using the same arguments as employed in Section 2.2 we see that

$$\delta J(\eta, y) = 0, \tag{3.1}$$

for *all* $\eta \in H$. As with the earlier case we can integrate by parts to eliminate the derivatives of η. The presence of η'' in the first variation indicates that we must integrate by parts twice. Specifically,

$$\int_{x_0}^{x_1} \eta'' \frac{\partial f}{\partial y''} dx = \eta' \frac{\partial f}{\partial y''} \Big|_{x_0}^{x_1} - \int_{x_0}^{x_1} \eta' \frac{d}{dx} \left(\frac{\partial f}{\partial y''} \right) dx$$

$$= -\eta \frac{d}{dx} \left(\frac{\partial f}{\partial y''} \right) \Big|_{x_0}^{x_1} + \int_{x_0}^{x_1} \eta \frac{d^2}{dx^2} \left(\frac{\partial f}{\partial y''} \right) dx$$

$$= \int_{x_0}^{x_1} \eta \frac{d^2}{dx^2} \left(\frac{\partial f}{\partial y''} \right) dx,$$

where we have used the boundary conditions $\eta(x_0) = 0$, $\eta'(x_0) = 0$, $\eta(x_1) = 0$, and $\eta'(x_1) = 0$. Now,

$$\int_{x_0}^{x_1} \eta' \frac{\partial f}{\partial y'} dx = \eta \frac{\partial f}{\partial y'} \Big|_{x_0}^{x_1} - \int_{x_0}^{x_1} \eta \frac{d}{dx} \left(\frac{\partial f}{\partial y'} \right) dx$$

$$= -\int_{x_0}^{x_1} \eta \frac{d}{dx} \left(\frac{\partial f}{\partial y'} \right) dx,$$

using the boundary conditions $\eta(x_0) = 0$ and $\eta(x_1) = 0$, and so condition (3.1) reduces to the equation

$$\int_{x_0}^{x_1} \eta \left\{ \frac{\partial f}{\partial y} - \frac{d}{dx} \left(\frac{\partial f}{\partial y'} \right) + \frac{d^2}{dx^2} \left(\frac{\partial f}{\partial y''} \right) \right\} dx = 0, \tag{3.2}$$

which must hold for all $\eta \in H$. The integrand f by assumption has continuous third order partial derivatives so that for any $y \in C^4[x_0, x_1]$ the term

$$E(x) = \frac{\partial f}{\partial y} - \frac{d}{dx} \left(\frac{\partial f}{\partial y'} \right) + \frac{d^2}{dx^2} \left(\frac{\partial f}{\partial y''} \right)$$

must be continuous on the interval $[x_0, x_1]$. A suitable modification of Lemma 2.2.2 (cf. Exercises 2.2-3) can be used to show that y must satisfy the fourth-order Euler-Lagrange differential equation

$$\frac{d^2}{dx^2} \left(\frac{\partial f}{\partial y''} \right) - \frac{d}{dx} \left(\frac{\partial f}{\partial y'} \right) + \frac{\partial f}{\partial y} = 0. \tag{3.3}$$

The above equation is a necessary condition for a function $y \in S$ to be an extremal for the functional J.

Example 3.1.1:   Let

$$J(y) = \int_0^1 \left((y'')^2 - 2\rho y \right) \, dx,$$

where ρ is a constant, and suppose it is required that $y(0) = y'(0) = 0$ and $y(1) = y'(1) = 1$. The Euler-Lagrange equation for this functional is

$$y^{(iv)}(x) = \rho,$$

which has the general solution

$$y(x) = \frac{1}{4!}\rho x^4 + c_1 x^3 + c_2 x^2 + c_3 x + c_4,$$

where the c_ks are constants. The conditions $y(0) = 0$ and $y'(0) = 0$ imply that $c_4 = c_3 = 0$. The conditions $y(1) = 1$ and $y'(1) = 1$ imply that $\rho/4! + c_1 + c_2 = 1$ and $\rho/3! + 3c_1 + 2c_2 = 1$; hence, $c_1 = -1 - \rho/12$ and $c_2 = 2 + \rho/24$. The extremal is thus given by

$$y(x) = \frac{\rho}{24} - \left(1 + \frac{\rho}{12}\right) x^3 + \left(2 + \frac{\rho}{24}\right) x^2.$$

Results such as Theorem 2.3.1 have analogues for functionals containing higher-order derivatives. For the second-order case, if the integrand does not contain y explicitly then it is plain that a first integral for the Euler-Lagrange equation can be obtained, viz.,

$$\frac{d}{dx}\left(\frac{\partial f}{\partial y''}\right) - \frac{\partial f}{\partial y'} = const.$$

If the integrand does not contain the variable x explicitly then it is left as an exercise to show that along any extremal

$$H(y, y', y'') = y''\frac{\partial f}{\partial y''} - y'\left(\frac{d}{dx}\frac{\partial f}{\partial y''} - \frac{\partial f}{\partial y'}\right) - f = const. \tag{3.4}$$

Example 3.1.2:   Let

$$J(y) = \int_{x_0}^{x_1} \frac{(1 + y'^2)^2}{y''} \, dx.$$

The integrand defining J does not contain y explicitly; therefore, any extremal y satisfies the differential equation

$$\frac{d}{dx}\left(\frac{\partial f}{\partial y''}\right) - \frac{\partial f}{\partial y'} = c_1,$$

where c_1 is some constant. The integrand also does not contain x explicitly, and so for any extremal

$$H(y, y', y'') = y'' \frac{\partial f}{\partial y''} - y' \left(\frac{d}{dx} \frac{\partial f}{\partial y''} - \frac{\partial f}{\partial y'} \right) - f$$

$$= y'' \frac{\partial f}{\partial y''} - y' c_1 - f$$

$$= -2 \frac{(1 + y'^2)^2}{y''} - y' c_1 = c_2,$$

where c_2 is another constant. The above expression can be recast in the form

$$y'' \frac{k_1 y' + k_2}{(1 + y'^2)^2} = 1, \qquad (3.5)$$

where k_1 and k_2 are constants. The two simplifications thus enable us to reduce the fourth-order Euler-Lagrange equation (3.3) to a second-order differential equation. We can solve equation (3.5) parametrically: let

$$y' = \tan \psi; \qquad (3.6)$$

then $y'' = \psi' \sec^2 \psi$ and equation (3.5) becomes

$$\left(k_1 \cos \psi \sin \psi + k_2 \cos^2 \psi \right) \psi' = 1.$$

Integrating both sides of the above equation yields

$$x = k_3 + \frac{k_1}{4} (1 - \cos(2\psi)) + \frac{k_2}{2} \left(\psi + \frac{1}{2} \sin(2\psi) \right),$$

where k_3 is an integration constant. Simplifying the above expression and using $k_1 = 4\kappa_1$, $k_2 = 4\kappa_2$, $\kappa_3 = k_3 + k_1/4$ gives

$$x = \kappa_3 + 2\kappa_2 \psi + \kappa_2 \sin(2\psi) - \kappa_1 \cos(2\psi). \qquad (3.7)$$

Equations (3.6) and (3.7) imply that

$$dy = \tan \psi \, dx$$

$$= (2\kappa_2(1 + \cos(2\psi)) + 2\kappa_1 \sin(2\psi)) \tan \psi \, d\psi$$

$$= (2\kappa_1 + 2\kappa_2 \sin(2\psi) - 2\kappa_1 \cos(2\psi)) \, d\psi;$$

hence,

$$y = \kappa_4 + 2\kappa_1 \psi - \kappa_2 \cos(2\psi) + \kappa_1 \sin(2\psi), \qquad (3.8)$$

where κ_4 is another integration constant. The solution is thus given parametrically by equations (3.7) and (3.8).

The methods used for integrands containing second-order derivatives can be extended to integrands containing derivatives of the nth order. We leave as an exercise the proof that the Euler-Lagrange equation for a functional of the form

$$J(y) = \int_{x_0}^{x_1} f(x, y, y', \ldots, y^{(n)}) \, dx$$

is

$$(-1)^n \frac{d^n}{dx^n} \left(\frac{\partial f}{\partial y^n} \right) + (-1)^{n-1} \frac{d^{n-1}}{dx^{n-1}} \left(\frac{\partial f}{\partial y^{n-1}} \right) + \cdots + \frac{\partial f}{\partial y} = 0. \qquad (3.9)$$

Exercises 3.1:

1. Find the general solution for the extremals to the functional J defined by

$$J(y) = \int_{x_0}^{x_1} \left((y'')^2 - y^2 + 2yx^3 \right) \, dx.$$

2. **Conservation Law:** Suppose the integrand f defining the functional J does not depend on x explicitly. Prove that equation (3.4) is satisfied along any extremal.

3. For the functional J defined by

$$J(y) = \int_0^1 y' \sqrt{1 + (y'')^2} \, dx,$$

find an extremal satisfying the conditions $y(0) = 0$, $y'(0) = 0$, $y(1) = 1$, and $y'(1) = 2$.

4. **Degenerate Case:** Let J be a functional of the form

$$J(y) = \int_{x_0}^{x_1} \left(A(x, y, y')y'' + B(x, y, y') \right) \, dx,$$

where A and B are smooth functions of x, y, and y'. Prove that the Euler-Lagrange equation for this functional is a differential equation of at most second order and that consequently any solutions can satisfy at most two arbitrary boundary conditions.

5. Let J and K be functionals defined by

$$J(y) = \int_{x_0}^{x_1} f(x, y, y', y'') \, dx$$

$$K(y) = \int_{x_0}^{x_1} F(x, y, y', y'') \, dx,$$

where, for some smooth function G,

$$F(x, y, y', y'') = f(x, y, y', y'') + \frac{d}{dx} G(x, y, y').$$

Prove that any extremals for J must also be extremals for K.

6. Let J be a functional of the form

$$J(y) = \int_{x_0}^{x_1} f(x, y, y', \ldots, y^{(n)}) \, dx,$$

where $y^{(n)}$ denotes the nth derivative of y.

(a) Formulate the fixed endpoint variational problem for this functional and prove that any smooth extremal must satisfy the Euler-Lagrange equation (3.9). Note any assumptions on the function f and the function space.

(b) If f is of the form $A(x, y, y', \ldots, y^{(n-1)})y^{(n)} + B(x, y, y', \ldots, y^{(n-1)})$ what is the maximum order the Euler-Lagrange equation can be?

3.2 Several Dependent Variables

Variational problems typically involve functionals that depend on several dependent variables. In classical mechanics, for example, even the motion of a single particle in space requires three dependent variables $(x(t), y(t), z(t))$ to describe the position of the particle at time t. In this section we derive the Euler-Lagrange equations for functionals that depend on several dependent variables and one independent variable.

Let $\mathbf{C}^2[t_0, t_1]$ denote the set of functions $\mathbf{q} : [t_0, t_1] \to \mathbb{R}^n$ such that for $\mathbf{q} = (q_1, q_2, \ldots, q_n)$ we have $q_k \in C^2[t_0, t_1]$ for $k = 1, 2, \ldots, n$. The set $\mathbf{C}^2[t_0, t_1]$ is a vector space and a norm such as

$$\|\mathbf{q}\| = \max_{k=1,2,\ldots,n} \sup_{t \in [t_0, t_1]} |q_k(t)|$$

can be defined on this space. As with the single dependent variable case, the choice of norm really depends on the application.

Consider a functional of the form

$$J(\mathbf{q}) = \int_{t_0}^{t_1} L(t, \mathbf{q}, \dot{\mathbf{q}}) \, dt, \tag{3.10}$$

where $\dot{}$ denotes differentiation with respect to t, and L is a function having continuous partial derivatives of second order with respect to t, q_k, and $\dot{q}_k$, for $k = 1, 2, \ldots, n$. Given two vectors $\mathbf{q}_0, \mathbf{q}_1 \in \mathbb{R}^n$, the fixed endpoint problem consists of determining the local extrema for J subject to the conditions $\mathbf{q}(t_0) = \mathbf{q}_0$ and $\mathbf{q}(t_1) = \mathbf{q}_1$. Here,

$$S = \{\mathbf{q} \in \mathbf{C}^2[t_0, t_1] : \mathbf{q}(t_0) = \mathbf{q}_0 \text{ and } \mathbf{q}(t_1) = \mathbf{q}_1\}.$$

Again we can represent a "nearby" function $\hat{\mathbf{q}}$ as a perturbation,

$$\hat{\mathbf{q}} = \mathbf{q} + \epsilon\eta,$$

where $\eta = (\eta_1, \eta_2, \ldots, \eta_n)$. For this case,

$$H = \{\eta \in \mathbf{C}^2[t_0, t_1] : \eta(t_0) = \eta(t_1) = 0\}.$$

For ϵ small Taylor's theorem implies that

$$L(t, \hat{\mathbf{q}}, \dot{\hat{\mathbf{q}}}) = L(t, \mathbf{q} + \epsilon\eta, \dot{\mathbf{q}} + \epsilon\dot{\eta})$$

$$= L(t, \mathbf{q}, \dot{\mathbf{q}}) + \epsilon \sum_{k=1}^{n} \left(\eta_k \frac{\partial L}{\partial q_k} + \dot{\eta}_k \frac{\partial L}{\partial \dot{q}_k} \right) + O(\epsilon^2);$$

consequently,

$$J(\hat{\mathbf{q}}) - J(\mathbf{q}) = \int_{t_0}^{t_1} L(t, \hat{\mathbf{q}}, \dot{\hat{\mathbf{q}}}) \, dt - \int_{t_0}^{t_1} L(t, \mathbf{q}, \dot{\mathbf{q}}) \, dt$$

$$= \epsilon \int_{t_0}^{t_1} \sum_{k=1}^{n} \left(\eta_k \frac{\partial L}{\partial q_k} + \dot{\eta}_k \frac{\partial L}{\partial \dot{q}_k} \right) \, dt + O(\epsilon^2).$$

The first variation for this functional is thus

$$\delta J(\eta, \mathbf{q}) = \int_{t_0}^{t_1} \sum_{k=1}^{n} \left(\eta_k \frac{\partial L}{\partial q_k} + \dot{\eta}_k \frac{\partial L}{\partial \dot{q}_k} \right) \, dt.$$

If J has a local extremum at $\mathbf{q}$ then arguments similar to those used in Section 2.2 show that a necessary condition for $\mathbf{q}$ to be an extremal is that

$$\delta J(\eta, \mathbf{q}) = 0 \tag{3.11}$$

for all $\eta \in H$.

Condition (3.11) is more complicated than its analogue (2.6) owing to the presence of n arbitrary functions and their derivatives, but judicious choices of functions $\eta \in H$ can be made to make the problem more tractable. Consider the set of functions H_1 defined by $H_1 = \{(\eta_1, 0, \ldots, 0) \in H\}$. Condition (3.11) must hold for all $\eta \in H_1$, and for any $\eta \in H_1$ this condition reduces to

$$\int_{t_0}^{t_1} \left(\eta_1 \frac{\partial L}{\partial q_1} + \dot{\eta}_1 \frac{\partial L}{\partial \dot{q}_1} \right) \, dt = 0. \tag{3.12}$$

We know from Section 2.2 that this condition leads to the Euler-Lagrange equation

$$\frac{d}{dt} \frac{\partial L}{\partial \dot{q}_1} - \frac{\partial L}{\partial q_1} = 0,$$

as a necessary condition for an extremal. Evidently we can modify the above approach by selecting appropriate subsets of H to argue that if J has a local extremum at $\mathbf{q}$ then

$$\frac{d}{dt}\frac{\partial L}{\partial \dot{q}_1} - \frac{\partial L}{\partial q_1} = 0,$$

$$\frac{d}{dt}\frac{\partial L}{\partial \dot{q}_2} - \frac{\partial L}{\partial q_2} = 0,$$

$$\vdots$$

$$\frac{d}{dt}\frac{\partial L}{\partial \dot{q}_n} - \frac{\partial L}{\partial q_n} = 0.$$

The above condition is a system of n second-order differential equations for the n unknown functions $q_1, \ldots, q_n$. Note that if $\mathbf{q}$ satisfies this system then condition (3.11) is satisfied for any $\eta \in H$. In summary, we have the following result.

Theorem 3.2.1 *Let* $J : \mathbf{C}^2[t_0, t_1] \to \mathbb{R}$ *be a function of the form*

$$J(\mathbf{q}) = \int_{t_0}^{t_1} L(t, \mathbf{q}, \dot{\mathbf{q}}) \, dt,$$

where $\mathbf{q} = (q_1, q_2, \ldots, q_n)$, *and* L *has continuous second-order partial derivatives with respect to* t, q_k, *and* $\dot{q}_k$, $k = 1, 2, \ldots, n$. *Let*

$$S = \{\mathbf{q} \in \mathbf{C}^2[t_0, t_1] : \mathbf{q}(t_0) = \mathbf{q}_0 \text{ and } \mathbf{q}(t_1) = \mathbf{q}_1\},$$

where $\mathbf{q}_0, \mathbf{q}_1 \in \mathbb{R}^n$ *are given vectors. If* $\mathbf{q}$ *is an extremal for* J *in* S *then*

$$\frac{d}{dt}\frac{\partial L}{\partial \dot{q}_k} - \frac{\partial L}{\partial q_k} = 0 \qquad (3.13)$$

for $k = 1, 2, \ldots, n$.

Example 3.2.1:   Let

$$J(\mathbf{q}) = \int_0^1 \left(\dot{q}_1^2 + (\dot{q}_2 - 1)^2 + q_1^2 + q_1 q_2 \right) dt,$$

with $\mathbf{q}(0) = \mathbf{q}_0$, $\mathbf{q}(1) = \mathbf{q}_1$. The Euler-Lagrange equations for this functional correspond to the system

$$\ddot{q}_1 - q_1 - \frac{1}{2}q_2 = 0, \qquad (3.14)$$

$$\ddot{q}_2 - \frac{1}{2}q_1 = 0. \qquad (3.15)$$

Equation (3.15) can be used to eliminate q_1 from equation (3.14) to give the fourth-order equation

$$2q_2^{(iv)} - 2\ddot{q}_2 - \frac{1}{2}q_2 = 0. \qquad (3.16)$$

The characteristic equation for this linear differential equation is

$$2\mu^4 - 2\mu^2 - \frac{1}{2} = 0,$$

which has roots

$$\mu_1, \mu_2 = \pm\sqrt{\frac{1}{2} + \frac{1}{\sqrt{2}}} \in \mathbb{R}$$

$$\mu_3, \mu_4 = \pm\sqrt{\frac{1}{2} - \frac{1}{\sqrt{2}}} = \pm im, \quad m \in \mathbb{R}.$$

The general solution to equation (3.16) is therefore

$$q_2(t) = c_1 e^{\mu_1 t} + c_2 e^{\mu_2 t} + c_3 \cos(mt) + c_4 \sin(mt),$$

where the c_ks are determined by the boundary conditions $\mathbf{q}(0) = \mathbf{q}_0$, $\mathbf{q}(1) = \mathbf{q}_1$. The function $q_1(t)$ can be readily deduced from $q_2(t)$ by use of equation (3.15).

The special cases detailed in Section 2.3 can also be extended to several dependent variables. In particular, if L does not depend on t explicitly it can be shown that

$$H = \sum_{k=1}^{n} \dot{q}_k \frac{\partial L}{\partial \dot{q}_k} - L = const. \tag{3.17}$$

along any extremal.

Example 3.2.2:   The familiar equations of motion for a particle can be derived from Hamilton's Principle (Section 1.3). Let $\mathbf{q}(t) = (q_1(t), q_2(t), q_3(t))$ denote the Cartesian coördinates of a free particle of mass m at time t. The kinetic energy of this particle is

$$T(\mathbf{q}, \dot{\mathbf{q}}) = \frac{1}{2} m (\dot{q}_1^2 + \dot{q}_2^2 + \dot{q}_3^2).$$

Let $V(t, \mathbf{q})$ denote the potential energy. The Lagrangian is

$$L(t, \mathbf{q}, \dot{\mathbf{q}}) = T(\mathbf{q}, \dot{\mathbf{q}}) - V(t, \mathbf{q})$$
$$= \frac{1}{2} m (\dot{q}_1^2 + \dot{q}_2^2 + \dot{q}_3^2) - V(t, \mathbf{q}),$$

and Hamilton's Principle implies that the path of the motion for the particle from $\mathbf{q}(t_0)$ to $\mathbf{q}(t_1)$ is such that $\mathbf{q}$ is an extremal for

$$J(\mathbf{q}) = \int_{t_0}^{t_1} L(t, \mathbf{q}, \dot{\mathbf{q}}) \, dt.$$

The Euler-Lagrange equations (3.13) give immediately the Lagrange equations of motion,

$$\frac{d}{dt}\frac{\partial L}{\partial \dot{q}_k} = \frac{\partial L}{\partial q_k},$$

which, in turn, lead to the relations

$$m\ddot{q}_k = -\frac{\partial V}{\partial q_k},$$

for $k = 1, 2, 3$. Recall from Section 1.3 that the kth component of force, f_k on the particle is given by

$$f_k = -\frac{\partial V}{\partial q_k}.$$

Hence, the Euler-Lagrange equations imply **Newton's equation**

$$\mathbf{f} = m\mathbf{a}.$$

where $\mathbf{a} = \ddot{\mathbf{q}}$ is the acceleration and $\mathbf{f} = (f_1, f_2, f_3)$ is the force on the particle.

For this example note that if the potential energy V does not depend on time explicitly then neither does L. In this case, we have the conservation law (3.17), which gives

$$H = \frac{1}{2}m(\dot{q}_1^2 + \dot{q}_2^2 + \dot{q}_3^2) + V(\mathbf{q}) = const.;$$

i.e., the total energy of the particle is conserved along an extremal.

Exercises 3.2:

1. Let

$$L(t, \mathbf{q}, \dot{\mathbf{q}}) = \frac{1}{2}\left(\dot{q}_1^2 + \dot{q}_2^2\right) - gq_2,$$

where g is a constant.
(a) Find the extremals for the functional J defined by

$$J(\mathbf{q}) = \int_{t_0}^{t_1} L(t, \mathbf{q}, \dot{\mathbf{q}})\, dt.$$

(b) Verify that equation (3.17) is satisfied.
2. Prove equation (3.17).
3. Let

$$L(t, \mathbf{q}, \dot{\mathbf{q}}) = \sqrt{\dot{q}_1^2 + q_2^2 \dot{q}_2^2} - kq_2,$$

where k is a constant. Find the extremals for the functional J defined by

$$J(\mathbf{q}) = \int_{t_0}^{t_1} L(t, \mathbf{q}, \dot{\mathbf{q}})\, dt.$$

4. Let

$$\Xi = \Xi(t, \mathbf{q})$$

be any smooth function and let

$$\Theta(t, \mathbf{q}, \dot{\mathbf{q}}) = \frac{\partial \Xi}{\partial t} + \sum_{k=1}^{n} \frac{\partial \Xi}{\partial q_k} \dot{q}_k.$$

(a) Prove that the Euler-Lagrange equations (3.13) for the functional

$$A(y) = \int_{t_0}^{t_1} \Theta(t, \mathbf{q}, \dot{\mathbf{q}}) \, dt$$

are satisfied for any smooth function y. (This is the degenerate case.)

(b) Let

$$J(\mathbf{q}) = \int_{t_0}^{t_1} L(t, \mathbf{q}, \dot{\mathbf{q}}) \, dt$$

and

$$K(\mathbf{q}) = \int_{t_0}^{t_1} \left(L(t, \mathbf{q}, \dot{\mathbf{q}}) + \Theta(t, \mathbf{q}, \dot{\mathbf{q}}) \right) \, dt,$$

where L satisfies the conditions of Theorem 3.2.1. Prove that $\mathbf{q}$ is an extremal for J if and only if it is an extremal for K.

3.3 Two Independent Variables*

This book is concerned primarily with functionals whose integrands contain a single independent variable. We pause here, however, to discuss briefly the first variation for functionals defined by multiple integrals. We focus on the simplest case when the integrand contains two independent variables.

Let Ω be a simply connected bounded region in $\mathbb{R}^2$ with boundary $\partial \Omega$ and closure $\bar{\Omega} = \partial \Omega \cup \Omega$. Let $C^2(\bar{\Omega})$ denote the space of all functions $u : \bar{\Omega} \to \mathbb{R}$ such that u has continuous derivatives of second order. Consider a functional $J : C^2(\bar{\Omega}) \to \mathbb{R}$ of the form

$$J(u) = \int \int_{\Omega} f(x, y, u, p, q) \, dx \, dy, \tag{3.18}$$

where $p = u_x$, $q = u_y$, and f is a smooth function of x, y, u, p, and q. An analogue of the fixed-endpoint variational problem is to find a function $u \in C^2(\bar{\Omega})$ such that J is an extremum subject to a boundary condition of the form

$$u(x, y) = u_0(x, y), \quad (x, y) \in \partial \Omega, \tag{3.19}$$

where $u_0 : \partial \Omega \to \mathbb{R}$ is a given function.

We can approach this problem in the same manner as the single independent variable case. Suppose that u is an extremal for J subject to the boundary condition (3.19), and let

$$\hat{u}(x,y) = u(x,y) + \epsilon\eta(x,y).$$

Here, ϵ is a small parameter and $\eta \in C^2(\bar{\Omega})$. In addition, it is required that $\hat{u}$ satisfy the boundary condition (3.19) and hence

$$\eta(x,y) = 0 \tag{3.20}$$

for all $(x,y) \in \partial\Omega$. The function η is otherwise arbitrary.

Taylor's theorem implies, for ϵ small, that

$$f(x,y,\hat{u},\hat{p},\hat{q}) = f(x,y,u+\epsilon\eta,p+\epsilon\eta_x,q+\epsilon\eta_y)$$
$$= f(x,y,u,p,q) + \epsilon\left\{\eta\frac{\partial f}{\partial u} + \eta_x\frac{\partial f}{\partial p} + \eta_y\frac{\partial f}{\partial q}\right\}$$
$$+ O(\epsilon^2),$$

where $\hat{p} = \hat{u}_x = p + \epsilon\eta_x$ and $\hat{q} = \hat{u}_y = q + \epsilon\eta_y$; hence,

$$J(\hat{u}) - J(u) = \epsilon\int\int_\Omega\left\{\eta\frac{\partial f}{\partial u} + \eta_x\frac{\partial f}{\partial p} + \eta_y\frac{\partial f}{\partial q}\right\}dx\,dy$$
$$+ O(\epsilon^2).$$

If J has an extremum at u, then the arguments of Section 2.2 can be modified to show that the terms of order ϵ must vanish; thus,

$$\int\int_\Omega\left\{\eta\frac{\partial f}{\partial u} + \eta_x\frac{\partial f}{\partial p} + \eta_y\frac{\partial f}{\partial q}\right\}dx\,dy = 0, \tag{3.21}$$

for all $\eta \in C^2(\bar{\Omega})$ satisfying condition (3.20). Lemma 2.2.2 can be generalized to accommodate multiple integrals. As with the fixed-endpoint problem, however, we need to eliminate the derivatives of the arbitrary function from condition (3.21).

Green's theorem states that

$$\int\int_\Omega\left(\frac{\partial\phi}{\partial x} + \frac{\partial\psi}{\partial y}\right)dx\,dy = \int_{\partial\Omega}\phi\,dy - \psi\,dx,$$

for any functions $\phi,\psi : \bar{\Omega} \to \mathbb{R}$ such that ϕ, ψ, ϕ_x, and ψ_y are continuous. Let

$$\phi = \eta\frac{\partial f}{\partial p}, \quad \psi = \eta\frac{\partial f}{\partial q}.$$

Since η and f are smooth functions, we can apply Green's theorem to get

$$\int\int_{\Omega}\left\{\eta_x\frac{\partial f}{\partial p}+\eta\frac{\partial}{\partial x}\left(\frac{\partial f}{\partial p}\right)+\eta_y\frac{\partial f}{\partial q}+\eta\frac{\partial}{\partial y}\left(\frac{\partial f}{\partial q}\right)\right\}dx\,dy$$
$$=\int_{\partial\Omega}\eta\frac{\partial f}{\partial p}\,dy+\eta\frac{\partial f}{\partial q}\,dx.$$

Here, $\frac{\partial}{\partial x}$ denotes partial differentiation holding (only) y fixed, and $\frac{\partial}{\partial y}$ denotes partial differentiation holding x fixed. Condition (3.20) implies that the boundary integral is zero; therefore,

$$\int\int_{\Omega}\left\{\eta_x\frac{\partial f}{\partial p}+\eta_y\frac{\partial f}{\partial q}\right\}dx\,dy=-\int\int_{\Omega}\eta\left\{\frac{\partial}{\partial x}\left(\frac{\partial f}{\partial p}\right)+\frac{\partial}{\partial y}\left(\frac{\partial f}{\partial q}\right)\right\}dx\,dy.$$

Condition (3.21) thus implies

$$\int\int_{\Omega}\eta\left\{\frac{\partial}{\partial x}\left(\frac{\partial f}{\partial p}\right)+\frac{\partial}{\partial y}\left(\frac{\partial f}{\partial q}\right)-\frac{\partial f}{\partial u}\right\}dx\,dy=0. \tag{3.22}$$

Equation (3.22) must be satisfied for arbitrary η, and the coefficient of η in the integrand is a continuous function since $u\in C^2(\bar{\Omega})$ and f is smooth. We can thus invoke a generalization of Lemma 2.2.2 to get the necessary condition

$$\frac{\partial}{\partial x}\left(\frac{\partial f}{\partial p}\right)+\frac{\partial}{\partial y}\left(\frac{\partial f}{\partial q}\right)-\frac{\partial f}{\partial u}=0. \tag{3.23}$$

Equation (3.23) is a second-order partial differential equation for the unknown function u, which must also satisfy the boundary condition (3.19). This differential equation is the analogue of equation (2.9); it is also called the **Euler-Lagrange equation**.

Example 3.3.1: Let Ω be the disc defined by $x^2+y^2<1$, and let

$$J(u)=\int\int_{\Omega}(p^2+q^2)\,dx\,dy. \tag{3.24}$$

For boundary conditions, suppose that

$$u_0(x,y)=2x^2-1, \tag{3.25}$$

for all $(x,y)\in\partial\Omega=\{(x,y):x^2+y^2=1\}$. The Euler-Lagrange equation for this functional is

$$\frac{\partial^2 u}{\partial x^2}+\frac{\partial^2 u}{\partial y^2}=0 \tag{3.26}$$

(Laplace's equation). If J has an extremum at $u\in C^2(\bar{\Omega})$, then u must be a solution to the partial differential equation (3.26) and satisfy the boundary condition (3.25). The reader can verify that the function $u(x,y)=x^2-y^2$ is a solution to this simple problem.

Example 3.3.2:   Let $\mathbf{r} : \Omega \to \mathbb{R}^3$ be a function of the form

$$\mathbf{r}(x, y) = (x, y, u(x, y)).\tag{3.27}$$

Then $\mathbf{r}$ describes a surface $\Sigma \subset \mathbb{R}^3$. The surface area of Σ is given by

$$J(u) = \int\!\!\int_{\Omega} \sqrt{1 + p^2 + q^2}\, dx\, dy.\tag{3.28}$$

Suppose we consider the minimal surface problem (Section 1.4), which consists of finding a minimum for J subject to boundary conditions of the form (3.19). Geometrically, the problem entails finding a surface that can be described parametrically in the form (3.27) such that the surface contains the (closed) space curve γ described by $\mathbf{r}_0 : \partial\Omega \to \mathbb{R}^3$, where

$$\mathbf{r}_0(x, y) = (x, y, u_0(x, y)),$$

and the surface area is minimum compared to other smooth surfaces containing the space curve γ. The Euler-Lagrange equation for this problem reduces to

$$(1 + p^2)t - 2pqs + (1 + q^2)r = 0,\tag{3.29}$$

where

$$r = \frac{\partial^2 u}{\partial x^2}, \quad s = \frac{\partial^2 u}{\partial x \partial y}, \quad t = \frac{\partial^2 u}{\partial y^2}.$$

The **mean curvature** of a surface described parametrically in the form (3.27) is given by

$$\mathcal{H} = \frac{(1 + p^2)t - 2pqs + (1 + q^2)r}{2(1 + p^2 + q^2)^{3/2}},$$

so that solutions to the minimal surface problem are characterized geometrically by the condition

$$\mathcal{H} = 0.$$

If J has an extremum at u, then u must satisfy an equation of the form

$$Ar + 2Bs + Ct + D = 0,\tag{3.30}$$

where A, B, C, and D are functions of the variables x, y, u, p, q. The Euler-Lagrange equation is thus a quasilinear second-order partial differential equation for the extremal u. Boundary-value problems involving such equations can be exceedingly difficult to solve and basic questions concerning the existence and uniqueness of solutions for specific problems can be difficult to answer. The boundary conditions for these problems play a central part in the solution method, and there are concerns here that do not manifest themselves strongly in the one-variable case such as whether the problem is well-posed.

A well-posed boundary-value problem has a unique solution, and the solution is stable with respect to small perturbations of the boundary conditions.

We eschew a general discussion on well-posed boundary-value problems. Suffice it to say that the matter is complicated especially for quasilinear (and fully nonlinear) partial differential equations. The reader is referred to standard works such as Garabedian [30] and John [42] for a fuller introductory account.

In some cases, it is possible to classify the Euler-Lagrange equation, and then general results concerning the class of equation can be exploited. The differential equation (3.30) is called:

(a) **hyperbolic**, if $AC - B^2 < 0$;
(b) **parabolic**, if $AC - B^2 = 0$;
(c) **elliptic**, if $AC - B^2 > 0$.

The classification is based on the existence of a special class of curves called **characteristics** on the integral surface. Roughly speaking, a characteristic is a curve on the integral surface along which the differential equation and the initial/boundary data do not determine all the second-order derivatives uniquely. Hyperbolic equations have integral surfaces with two real families of characteristics. Parabolic equations have integral surfaces with only one characteristic. Elliptic equations have integral surfaces with no real characteristics. The presence of characteristics influences strongly the type of problem for which the differential equation is well-posed. The type of boundary-value problem considered in this section is called a **Dirichlet problem**. It is well known that Dirichlet problems involving hyperbolic partial differential equations are ill-posed. In contrast, Dirichlet problems are generally well-posed for elliptic partial differential equations.

In general, the coefficients A, B, and C depend on the variables x, y, u, p, q, so that an Euler-Lagrange equation need not fit into any of the categories mentioned. The signs and magnitudes of these coefficients can change, an equation may be hyperbolic at some points in Ω and elliptic at other points. More importantly, the coefficients depend on the solution itself. The classification really depends on the equation, the domain, and the solution. Nonetheless, there are cases where the equation can be classified without knowing solutions. If the coefficients are all constants, for example, then the classification depends purely on these constants. Laplace's equation (3.26) is clearly elliptic; the wave equation,

$$r - t = 0,$$

is clearly hyperbolic. The reader can also verify that equation (3.29) is elliptic. Gilbarg and Trudinger [34] discuss the Dirichlet problem for quasilinear elliptic partial differential equations of this type in some depth.

3.4 The Inverse Problem*

The variational formulation of a boundary-value problem has some advantages. For example, in Chapter 5 we show how one can exploit the isoperimetric problem to approximate eigenvalues for Sturm-Liouville problems. In Chapter 8, we show how variational problems lead to Hamilton's equations and the Hamilton-Jacobi equation, which may be solvable through separation of variables. In addition, Noether's theorem (Chapter 9) provides a systematic algorithm for finding conservation laws for variational problems. [1] These and other features (e.g., Rayleigh-Ritz numerical methods) make it attractive to identify a given differential equation as the Euler-Lagrange equation of some functional.

Given a differential equation

$$y'' - F(x, y, y') = 0, \tag{3.31}$$

the **inverse problem** is to determine a function $f(x, y, y')$ such that y is a solution to (3.31) if and only if y is a solution to the Euler-Lagrange equation

$$\frac{d}{dx}\frac{\partial f}{\partial y'} - \frac{\partial f}{\partial y} = 0. \tag{3.32}$$

In this section we discuss briefly and informally some qualitative aspects of the inverse problem.

Let us first consider the general second-order linear differential equation

$$y'' + Py' + Qy - G = 0, \tag{3.33}$$

where P, Q, and G are functions of x. We know from the theory of differential equations that such equations can be put in an equivalent self-adjoint form

$$(py')' + qy - g = 0, \tag{3.34}$$

where

$$p = \exp\left(\int_{x_0}^{x} P(\xi)\, d\xi\right), \quad q = Qp, \quad g = Gp.$$

A quick comparison with equation (3.32) shows that equation (3.34) is equivalent to the Euler-Lagrange equation for

$$f(x, y, y') = \frac{1}{2}\left(py'^2 - qy^2 + 2gy\right).$$

In this manner, we see that the general linear equation (3.33) can always be transformed into an Euler-Lagrange equation. We discuss this relationship for Sturm-Liouville problems in more detail in Section 5.1.

---

[1] In fact, there are versions of Noether's theorem that do not require a variational formulation. Anco and Bluman [2], [3] describe the algorithm.

We now turn to the general nonlinear equation (3.31). Now, the Euler-Lagrange equation is

$$y'' f_{y'y'} + y' f_{yy'} + f_{xy'} - f_y = 0,$$

and y'' can be eliminated from the above equation using (3.31) to give

$$F f_{y'y'} + y' f_{yy'} + f_{xy'} - f_y = 0. \tag{3.35}$$

Equation (3.35) can be regarded as a second-order partial differential equation for the function f. From a practical standpoint, the above equation is of limited value owing to the paucity of methods for solving such equations. Fortunately, it is possible to transform equation (3.35) into a first-order partial differential equation for $\Phi = f_{y'y'}$. Differentiating both sides of equation (3.35) with respect to y' gives

$$F_{y'} f_{y'y'} + F f_{y'y'y'} + y' f_{yy'y'} + f_{xy'y'} = 0;$$

i.e.,

$$F \Phi_{y'} + y' \Phi_y + \Phi_x + F_{y'} \Phi = 0. \tag{3.36}$$

There is a general method for solving first-order partial differential equations, the method of characteristics, that entails solving a system of four ordinary differential equations. We do not go into this method here, but simply note that it can be used to show that solutions to equation (3.36) exist,[2] and hence the general second-order nonlinear equation (3.31) does have a variational formulation.

The inverse problem for systems of second-order differential equations poses a more formidable problem. Fortunately, there is a result that helps characterize systems that have variational formulations. Let

$$\mathbf{A}(t, \mathbf{q}, \dot{\mathbf{q}}, \ddot{\mathbf{q}}) = (A_1(t, \mathbf{q}, \dot{\mathbf{q}}, \ddot{\mathbf{q}}), \dots, A_n(t, \mathbf{q}, \dot{\mathbf{q}}, \ddot{\mathbf{q}})) = \mathbf{0}$$

denote a system of n second-order differential equations for $\mathbf{q} = (q_1, \dots, q_n)$, and let

$$E_j(L) = \frac{d}{dt} \frac{\partial L}{\partial \dot{q}_j} - \frac{\partial L}{\partial q_j}.$$

A necessary and sufficient condition that there exists an $L(t, \mathbf{q}, \dot{\mathbf{q}})$ such that

$$A_j(t, \mathbf{q}, \dot{\mathbf{q}}, \ddot{\mathbf{q}}) = E_j(L), \tag{3.37}$$

for $j = 1, \dots, n$, is that $\mathbf{A}$ satisfies the following integrability conditions,[3]

---

[2] We could also appeal to results such as the Cauchy-Kowalevski theorem ([30]) if F is analytic in x, y, and y'.

[3] These conditions correspond to the requirement that the Fréchet derivative of $\mathbf{A}$ be self-adjoint (cf. [57], p. 355).

$$\frac{\partial A_k}{\partial \ddot{q}_j} = \frac{\partial A_j}{\partial \ddot{q}_k}$$

$$\frac{\partial A_k}{\partial \dot{q}_j} = -\frac{\partial A_j}{\partial \dot{q}_k} + 2\frac{d}{dx}\left(\frac{\partial A_j}{\partial \ddot{q}_k}\right) \tag{3.38}$$

$$\frac{\partial A_k}{\partial q_j} = \frac{\partial A_j}{\partial q_k} - \frac{d}{dx}\left(\frac{\partial A_j}{\partial \dot{q}_k}\right) + \frac{d^2}{dx^2}\left(\frac{\partial A_j}{\partial \ddot{q}_k}\right),$$

for $j, k = 1, \ldots, n$. Relations (3.38) are called the **Helmholtz conditions**. If the A_j satisfy the Helmholtz conditions, then it can be shown that the function L defined by

$$L(t, \mathbf{q}, \dot{\mathbf{q}}) = \int_0^1 \sum_{k=1}^n q_k A_k(t, \xi\mathbf{q}, \xi\dot{\mathbf{q}}, \xi\ddot{\mathbf{q}})\, d\xi \tag{3.39}$$

satisfies equation (3.37). The Helmholtz conditions are discussed in more detail in [57].

Note that failure of the Helmholtz conditions does not preclude the possibility of a system having a variational formulation. Although these conditions preclude direct relationships such as (3.37), it may be that there is a multiplier matrix **B**, for example, such that

$$\sum_{i=1}^n b_{ij} A_i = E_j(L).$$

Here, **B** is a nonsingular $n \times n$ matrix with entries $b_{ij} = b_{ij}(t, \mathbf{q}, \dot{\mathbf{q}})$. For example, consider the simple case $n = 1$, $A(x, y, y', y'') = y'' - F(x, y, y')$. For this case, the Helmholtz conditions reduce to the condition $F_{y'} = 0$. But we know that *all* the second-order equations of the form (3.31) have a variational formulation. Suppose now that we introduce a multiplier $B = B(x, y, y')$ and apply the Helmholz conditions to $B(y'' - F)$. Then, the Helmholtz condition reduces to

$$\frac{d}{dx}B = \frac{\partial}{\partial y'}\left(B(y'' - F)\right).$$

Expanding the above relation gives

$$FB_{y'} + y'B_y + B_x + F_{y'}B = 0,$$

which is the same as differential equation (3.36).

The determination of a matrix **B** such that **BA** satisfies the Helmholz condition is called the "multiplier problem" in the calculus of variations. The difficulties and conditions on the b_{ij} escalate substantially for $n \geq 2$. The reader can find a summary of the problem, generalizations, and further results in the monograph by Anderson and Thompson [4].

4

Isoperimetric Problems

Variational problems are often accompanied by one or more constraints. The presence of a constraint further limits the space S in which we search for extremals. Constraints may be prescribed in any number of ways. For example, one might require the functions $\mathbf{q} \in S$ to satisfy an algebraic condition, a differential equation, or an inequality. Often there are different ways to impose the same constraint. In this chapter we discuss problems that have isoperimetric constraints. Problems that have algebraic equations or differential equations as constraints are discussed in Chapter 6.

4.1 The Finite-Dimensional Case and Lagrange Multipliers

It is useful to investigate a simple finite-dimensional example of a constrained optimization problem to gain some insight into the infinite-dimensional case. Moreover, the theory underlying the Lagrange multiplier technique for variational problems rests on that for finite-dimensional problems. In this section we review Lagrange multipliers for finite-dimensional optimization problems.

4.1.1 Single Constraint

Consider the problem of determining local extrema for a function $f : \mathbb{R}^2 \to \mathbb{R}$ subject to the condition that the values of f are sampled on a curve $\gamma \subset \mathbb{R}^2$. In other words, determine the points on γ at which f has a local extremum relative to values of f sampled at nearby points on γ. This problem is inherently one-dimensional in character, but the approach to locating the extrema really depends on the constraint used to define γ. We assume for simplicity that f is a smooth function and that γ is a smooth curve.

There are many ways to define a curve. Suppose, for example, that γ is defined parametrically by some function $\mathbf{r} : I \to \mathbb{R}^2$, where $I \subseteq \mathbb{R}$ is an interval, and for $t \in I$,

$$\mathbf{r}(t) = (x(t), y(t)).$$

Then we can build the constraint directly into the problem by constructing the function $F : I \to \mathbb{R}$ defined by $F(t) = f(x(t), y(t))$. Given that the parametrization is smooth, a necessary condition for a local extremum at t is

$$\frac{d}{dt}F(t) = \frac{\partial f}{\partial x}x'(t) + \frac{\partial f}{\partial y}y'(t) = 0.$$

Now, $x(t)$ and $y(t)$ are known and thus $\partial f/\partial x$ and $\partial f/\partial x$ are known functions of t. In principle we can thus solve the above equation for the values of t (if any) that make F an extremum. Note that a special case of the parametric representation is the "graphical" representations $\mathbf{r}(x) = (x, y(x))$ and $\mathbf{r}(y) = (x(y), y)$.

A curve may be defined implicitly by an equation of the form $g(x, y) = 0$. If g is a smooth function and $\nabla g \neq \mathbf{0}$, then in principle we could solve the equation for one of the variables and proceed as described above.[1] In practice, however, finding an explicit solution might not be possible or convenient. Moreover, even if g is smooth for all values of x and y, the resulting solution for x or y may not be. Consider, for example, the equation $g(x, y) = x^2 + y^2 - 1 = 0$ that describes the unit circle centred at $(0, 0)$. If we solve this equation for, say y, we get $y(x) = \sqrt{1 - x^2}$, and y is not smooth at $x = \pm 1$. Yet another concern with this approach is that it often leads to an artificial distinction of dependent variables. In many problems in geometry and physics the variables are all on the same footing and it is not desirable to make such a distinction for the purposes of analysis.

An elegant technique that avoids the problem of directly solving implicit equations involves the introduction of a constant called a Lagrange multiplier. The technique has a simple geometrical interpretation. Suppose that f and g are smooth functions. We wish to find a necessary condition for f to have a local extremum subject to the constraint

$$g(x, y) = 0. \tag{4.1}$$

We suppose further that

$$\nabla g(x, y) \neq \mathbf{0}. \tag{4.2}$$

The equation (4.1) defines a curve γ implicitly, and since $\nabla g \neq \mathbf{0}$ the curve is smooth; i.e., γ has a well-defined unit tangent vector at each point that varies smoothly along γ. This means that locally γ can be represented parametrically by a smooth vector function $\mathbf{r}(t) = (x(t), y(t))$, $t \in I$, such that $\mathbf{r}'(t) \neq \mathbf{0}$ for all $t \in I$. A necessary condition for f to have a local extremum on γ at $(x(t), y(t))$ is

---

[1] If one of the derivatives is nonzero, then we can use the implicit function theorem to assert the existence of a solution to the equation. Unfortunately, the theorem does not actually provide a means of obtaining the solution.

$$\frac{d}{dt}f(x(t), y(t)) = \frac{\partial f}{\partial x}x'(t) + \frac{\partial f}{\partial y}y'(t) = 0. \tag{4.3}$$

Since $g(x(t), y(t)) = 0$ for any $(x(t), y(t)) \in \gamma$ we also have

$$\frac{d}{dt}g(x(t), y(t)) = \frac{\partial g}{\partial x}x'(t) + \frac{\partial g}{\partial y}y'(t) = 0, \tag{4.4}$$

for all $t \in I$. Equation (4.2) implies that at any point on γ at least one of the derivatives $\partial g/\partial x$, $\partial g/\partial y$ is nonzero. Suppose for definiteness that $\partial g/\partial y \neq 0$. Then equation (4.4) implies that

$$y'(t) = -\frac{\frac{\partial g}{\partial x}}{\frac{\partial g}{\partial y}}x'(t) = 0, \tag{4.5}$$

and consequently equation (4.3) can be replaced by

$$\frac{x'(t)}{\frac{\partial g}{\partial y}}\left\{\frac{\partial f}{\partial x}\frac{\partial g}{\partial y} - \frac{\partial f}{\partial y}\frac{\partial g}{\partial x}\right\} = 0.$$

Now $\mathbf{r}'(t) = (x'(t), y'(t)) \neq \mathbf{0}$ so that $x'(t)$ and $y'(t)$ cannot both be zero; hence, equation (4.5) precludes the possibility that $x'(t) = 0$. Equation (4.3) thus reduces to the condition

$$\frac{\partial f}{\partial x}\frac{\partial g}{\partial y} - \frac{\partial f}{\partial y}\frac{\partial g}{\partial x} = 0,$$

which is equivalent to the condition

$$\nabla f \wedge \nabla g = \mathbf{0}, \tag{4.6}$$

where $\wedge$ denotes the exterior (cross) product. Recall that for any vectors $\mathbf{v}, \mathbf{w} \in \mathbb{R}^2$,

$$|\mathbf{v} \wedge \mathbf{w}| = |\mathbf{v}|\,|\mathbf{w}|\sin\phi,$$

where ϕ is the angle between $\mathbf{v}$ and $\mathbf{w}$. Equation (4.6) indicates that ∇f is parallel to ∇g at an extremum (i.e., $\phi = 0$). Since ∇f and ∇g are parallel, there is a constant λ such that $\nabla f = \lambda \nabla g$. The necessary condition (4.3) thus reduces to the condition

$$\nabla(f - \lambda g) = \mathbf{0}. \tag{4.7}$$

The constant λ is called a **Lagrange multiplier.**

It is evident graphically that ∇f is parallel to ∇g at an extremum. Figure 4.1 depicts level curves of f and the curve γ. The conditions on g ensure that γ does not have any discontinuities or "corners," and since f is smooth f has smooth level curves. Suppose that f has an extremum on γ at (x, y). If the level curve of f through the point (x, y) intersects γ transversally, then f is increasing/decreasing along γ at (x, y) and hence (x, y) will not yield an

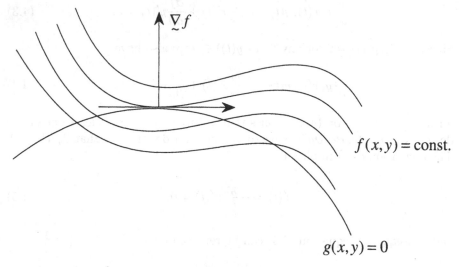

∇f

$f(x,y) = \text{const.}$

$g(x,y) = 0$

Fig. 4.1.

extremum for f on γ. The level curve of f through (x,y) must therefore be tangent to γ at (x,y) and consequently the unit normal to γ must be parallel to the unit normal to the level curve at (x,y). In other words, ∇f is parallel to ∇g at (x,y).

Under the above conditions, if f has an extremum subject to condition (4.1), then equation (4.7) must be satisfied. This vector equation provides two scalar equations for the three unknown quantities x, y, and λ. Equation (4.1) provides the third equation.

Example 4.1.1:   Find the local extrema for the function defined by $f(x,y) = x^2 - y^2$ subject to the condition $g(x,y) = x^2 + y^2 - 1 = 0$.

Equation (4.7) implies that

$$\nabla \left(x^2 - y^2 - \lambda(x^2 + y^2 - 1) \right) = \mathbf{0},$$

i.e.,

$$x(1 + \lambda) = 0,$$
$$y(-1 + \lambda) = 0.$$

The first equation indicates that either $x = 0$ or $\lambda = -1$. Suppose that $x = 0$. Then the second equation implies that either $y = 0$ or $\lambda = 1$, but $x = 0$ and the condition $x^2 + y^2 - 1 = 0$ implies that $y = \pm 1$. Thus, there are critical points at $(0, 1)$ and $(0, -1)$. Suppose instead that $x \neq 0$ and $\lambda = -1$. Then the second equation implies that $-2y = 0$, i.e., $y = 0$, so that the constraint implies that $x = \pm 1$. Hence there are critical points at $(1, 0)$ and $(-1, 0)$.

The Lagrange multiplier technique can be adapted to problems in higher dimensions. For example, to find the extrema for a function of the form $f(x, y, z)$ subject to a constraint of the form $g(x, y, z) = 0$, we can form the function $F = f - \lambda g$, where λ is an unknown constant, and look for solutions to the three equations given by $\nabla F = 0$, where ∇ is the operator $(\partial/\partial x, \partial/\partial y, \partial/\partial z)$. The constraint provides the fourth equation for the unknown quantities x, y, z, and λ. This approach is valid provided $\nabla g \neq 0$. In summary, we have the following result.

Theorem 4.1.1 (Lagrange Multiplier Rule) *Let $\Omega \subset \mathbb{R}^n$ be a region and let $f : \Omega \to \mathbb{R}$ and $g : \Omega \to \mathbb{R}$ be smooth functions. If f has a local extremum at $\mathbf{x} \in \Omega$ subject to the condition that $g(\mathbf{x}) = 0$ and if $\nabla g(\mathbf{x}) \neq 0$, then there is a number λ such that*

$$\nabla(f(\mathbf{x}) - \lambda g(\mathbf{x})) = 0.$$

4.1.2 Multiple Constraints

Let $\mathbf{x} = (x_1, x_2, \ldots, x_n)$ and let $f : \Omega \to \mathbb{R}$ be a smooth function defined on a region $\Omega \subseteq \mathbb{R}^n$. If $n > 2$ it is possible to impose more than one constraint. Suppose that $m < n$ and consider the problem of finding the local extrema of f in Ω subject to the m constraints

$$g_k(\mathbf{x}) = 0, \tag{4.8}$$

where $k = 1, 2, \ldots, m$ and the functions $g_k : \Omega \to \mathbb{R}$ are smooth. For the simple case where $n = 2$ and $m = 1$, we saw that f and g share the same tangent line at an extremum. In higher dimensions the analogue of this condition is that the tangent space (hyperplane) of f at a critical point $\mathbf{x}$ is contained in the tangent space defined by the g_k at $\mathbf{x}$. Geometrically, this means that the normal vector $\nabla f(\mathbf{x})$ lies in the normal space $N_g(\mathbf{x})$ spanned by the vectors $\nabla g_k(\mathbf{x})$. In terms of linear algebra, the vector $\nabla f(\mathbf{x})$ is linearly dependent on the set of vectors $\{\nabla g_k(\mathbf{x}), k = 1, 2, \ldots, m\}$. Thus, if f has a local extremum at $\mathbf{x}$, then there exist constants $\lambda_1, \lambda_2, \ldots, \lambda_m$ such that

$$\nabla f(\mathbf{x}) = \sum_{k=1}^{m} \lambda_k \nabla g_k(\mathbf{x});$$

i.e.,

$$\nabla \left(f(\mathbf{x}) - \sum_{k=1}^{m} \lambda_k g_k(\mathbf{x}) \right) = 0. \tag{4.9}$$

Equation (4.9) is the m constraint analogue of equation (4.7).

The approach is valid provided $\nabla f(\mathbf{x})$ is linearly dependent on the $\nabla g_k(\mathbf{x})$, and this condition leads to the generalization of the condition $\nabla g(x, y) \neq \mathbf{0}$. Let $\mathbf{M}(\mathbf{x})$ be the $n \times m$ matrix

$$\mathbf{M}(\mathbf{x}) = \begin{pmatrix} \nabla g_1(\mathbf{x}) \\ \vdots \\ \nabla g_m(\mathbf{x}) \end{pmatrix} = \begin{pmatrix} \frac{\partial g_1}{\partial x_1} & \frac{\partial g_1}{\partial x_2} & \cdots & \frac{\partial g_1}{\partial x_n} \\ \vdots & & & \\ \frac{\partial g_m}{\partial x_1} & \frac{\partial g_m}{\partial x_2} & \cdots & \frac{\partial g_m}{\partial x_n} \end{pmatrix},$$

and let $\mathbf{M}_f(\mathbf{x})$ be the augmented matrix

$$\mathbf{M}_f(\mathbf{x}) = \begin{pmatrix} \mathbf{M}(\mathbf{x}) \\ \nabla f \end{pmatrix}.$$

The linear dependence of ∇f is assured if

$$\text{Rank}\, \mathbf{M}_f(\mathbf{x}) \leq \text{Rank}\, \mathbf{M}(\mathbf{x}). \tag{4.10}$$

Condition (4.10) provides the analogue of the gradient condition (4.2). In summary, we have the following extension of Theorem 4.1.1.

Theorem 4.1.2 *Let $\Omega \subset \mathbb{R}^n$ be a region and let $f : \Omega \to \mathbb{R}$ and $g_k : \Omega \to \mathbb{R}$ be smooth functions for $k = 1, \ldots, m$. If f has a local extremum at $\mathbf{x} \in \Omega$ subject to the m constraints that $g_k(\mathbf{x}) = 0$, and if inequality (4.10) is satisfied at $\mathbf{x}$, then there exist constants $\lambda_1, \lambda_2, \ldots, \lambda_m$ such that*

$$\nabla \left(f(\mathbf{x}) - \sum_{k=1}^{m} \lambda_k g_k(\mathbf{x}) \right) = \mathbf{0}.$$

Example 4.1.2: Find the local extrema for the function defined by

$$f(\mathbf{x}) = x_3^2/2 - x_1 x_2$$

subject to the conditions

$$g_1(\mathbf{x}) = x_1^2 + x_2 - 1 = 0,$$
$$g_2(\mathbf{x}) = x_1 + x_3 - 1 = 0.$$

Here, $n = 3$ and $m = 2$. Equation (4.9) produces the equations

$$x_2 + 2\lambda_1 x_1 + \lambda_2 = 0,$$
$$x_1 + \lambda_1 = 0,$$
$$x_3 - \lambda_2 = 0,$$

that along with the constraints provide five equations for the five quantities $x_1, x_2, x_3, \lambda_1,$ and λ_2. This system of equations has the two solutions $\mathbf{w} =$

$(-1, 0, 2)$ with $\lambda_1 = 1, \lambda_2 = 2$, and $\mathbf{z} = (2/3, 5/9, 1/3)$ with $\lambda_1 = -2/3, \lambda_2 = 1/3$. For the first solution the matrix $\mathbf{M}$ is

$$\mathbf{M}(\mathbf{w}) = \begin{pmatrix} -2 & 1 & 0 \\ 1 & 0 & 1 \end{pmatrix},$$

which has rank 2. The augmented matrix is

$$\mathbf{M}_f(\mathbf{w}) = \begin{pmatrix} -2 & 1 & 0 \\ 1 & 0 & 1 \\ 0 & 1 & 2 \end{pmatrix},$$

and since the determinant of $\mathbf{M}_f(\mathbf{x})$ is zero, we must have that Rank $\mathbf{M}_f(\mathbf{w}) \leq$ Rank $\mathbf{M}(\mathbf{w})$. A similar calculation indicates that condition (4.10) is also satisfied for the second solution. Hence, if f has any local extrema under the given constraints then they must occur at either $\mathbf{w}$ or $\mathbf{z}$.

4.1.3 Abnormal Problems

The Lagrange multiplier technique breaks down if condition (4.2) (or condition (4.10)) is not satisfied. The technique, however, can be adapted to cope with these cases.

We consider here only the optimization problem of finding the local extrema for a function f of two independent variables subject to a single constraint $g = 0$. We assume (as always) that f and g are smooth functions. If (x, y) is a local extremum for this problem and $\nabla g(x, y) \neq \mathbf{0}$, then we have the existence of a number λ such that $\nabla \left(f(x, y) - \lambda g(x, y) \right) = \mathbf{0}$. We call a problem of this type **normal**. In contrast, if (x, y) is a local extremum for the problem and $\nabla g(x, y) = \mathbf{0}$, then the existence of a Lagrange multiplier is not assured. This type of problem is called **abnormal**.

If $g(x, y) = 0$ and $\nabla g(x, y) = \mathbf{0}$ then the implicit function theorem cannot be invoked to deduce that the equation $g = 0$ can be solved uniquely for x in terms of y or vice versa. Geometrically, this means that the set of solutions to $g = 0$ need not form a smooth curve in a neighbourhood of (x, y). This does not mean that the curve must have some singularity at (x, y) so that the tangent vector to the curve is not well-defined, only that it is a possibility. Various nasty things can happen to "curves" defined by an implicit relation when the gradient vanishes. For example, it may be that the curve has a "corner" or a cusp at this point. Another possibility is that the curve has a self-intersection, or that two distinct solution curves intersect at (x, y). An even more degenerate possibility is that (x, y) represents an isolated point in the set of solutions to the equation. In these cases it is clear that the geometrical arguments leading to the existence of a Lagrange multiplier are not applicable. The following barrage of examples illustrates these pathologies.

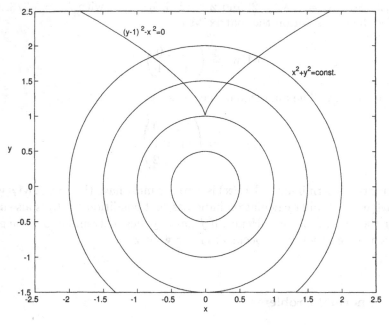

Fig. 4.2.

Example 4.1.3: Let $f(x, y) = x^2 + y^2$ and $g(x, y) = (y - 1)^3 - x^2$. We seek the local extrema of f subject to the constraint $g = 0$. Proceeding formally with Lagrange multipliers, the equation $\nabla(f - \lambda g) = \mathbf{0}$ yields the equations

$$x(1 + \lambda) = 0 \tag{4.11}$$

$$2y - 3\lambda(y - 1)^2 = 0. \tag{4.12}$$

Equation (4.11) implies that either $x = 0$ or $\lambda = -1$. If $x = 0$, then the condition $g = 0$ indicates that $y = 1$. If $\lambda = 1$, then equation (4.12) shows that y must be a solution of the quadratic equation

$$3y^2 - 4y + 3 = 0.$$

This equation, however, has no real solutions so that we have only the "solution" $(0, 1)$. But $\nabla g(0, 1) = 0$, so that the problem is abnormal. It is easy to verify that there is no λ such that equation (4.12) is satisfied at $(0, 1)$.

Geometrically, the equation $g = 0$ describes a semicubical parabola with a singularity at $(0, 1)$, where the tangent vector is not well defined (figure 4.2). If the semicubical parabola is plotted with the level curves of f, we see that $(0, 1)$ is in fact a minimum for f on the curve defined by $g = 0$.

Example 4.1.4: Let us look at the same problem as in Example 4.1.3 but change the functions to $f(x, y) = x^2 + y^2$ and $g(x, y) = x^2 - y^2$. The Lagrange

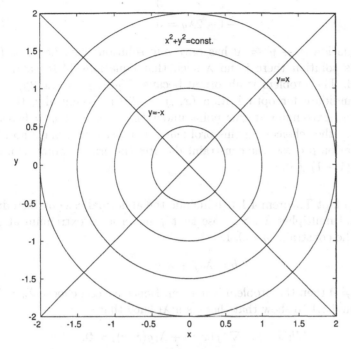

Fig. 4.3.

multiplier approach yields the equations

$$x(1 - \lambda) = 0 \tag{4.13}$$
$$y(1 + \lambda) = 0. \tag{4.14}$$

Equation (4.13) implies that either $x = 0$ or $\lambda = 1$. If $x = 0$ then the condition $g = 0$ implies that $y = 0$. If $\lambda = 1$ then equation (4.14) implies that $y = 0$ and the constraint then gives $x = 0$. In either case we have only the solution $(x, y) = (0, 0)$. Now, $\nabla g(x, y) = (2x, -2y)$, so that $\nabla g(0, 0) = \mathbf{0}$ and the problem is thus abnormal. Unlike Example 4.1.3, however, *any* choice of λ will satisfy equations (4.13) and (4.14) for $(x, y) = (0, 0)$, so that λ is indeterminate.

Note that in this case $\nabla f(x, y) = (2x, 2y) = \mathbf{0}$, so that $(0, 0)$ is a critical point for f. It is easy to see that f has a (global) minimum at $(0, 0)$ and hence the constrained problem will always have this critical point for *any* constraint that has $(0, 0)$ among its solutions. In this sense the constraint is passive. Near $(0, 0)$ the condition $g = 0$ defines two lines $x = \pm y$ (figure 4.3).

Example 4.1.5:   Let us look at the same problem again but change the functions to $f(x, y) = x - y$ and $g(x, y) = x^2 + y^2$. The Lagrange multiplier approach yields the equations

$$1 - 2\lambda x = 0 \tag{4.15}$$
$$-1 - 2\lambda y = 0. \tag{4.16}$$

But the equation $x^2 + y^2 = 0$ has only *one* solution, viz., $(x, y) = (0, 0)$, and for this solution there is no λ such that equations (4.15) and (4.16) are satisfied. The problem is abnormal because $\nabla g(x, y) = (2x, 2y) = \mathbf{0}$ at the only candidate for optimization $(x, y) = (0, 0)$. Technically, the function f has an extremum at this point under the constraint $g = 0$ because there are no other choices. In this problem f is passive and plays no rôle in the optimization process: the constraint dictates the critical point. Note that $\nabla f(0, 0) = (1, -1) \neq \mathbf{0}$.

We can adapt Theorem 4.1.1 to include the abnormal case by introducing an additional multiplier λ_0. Suppose that f has a local extremum at (x, y) subject to the constraint $g = 0$. Let

$$h = \lambda_0 f + \lambda_1 g.$$

If $\nabla g(x, y) \neq \mathbf{0}$ then the problem is normal. Hence we can choose $\lambda_0 = 1$ and use Theorem 4.1.1 to show that there is a λ_1 such that

$$\nabla h(x, y) = \nabla \left(f(x, y) + \lambda_1 g(x, y) \right) = \mathbf{0}.$$

Suppose now that the problem is abnormal so that $g(x, y) = 0$ and $\nabla g(x, y) = \mathbf{0}$. Then we can salvage the condition $\nabla h(x, y) = \mathbf{0}$ by requiring that

$$\lambda_0 \nabla f(x, y) = \mathbf{0}.$$

If $\nabla f(x, y) \neq \mathbf{0}$, then we must choose $\lambda_0 = 0$. The other constant λ_1 in this case is not determined. If $\nabla f(x, y) = \mathbf{0}$, then any choices of λ_0 and λ_1 will suffice.

Example 4.1.5 illustrates the case where $\lambda_0 = 0$. If we must choose $\lambda_0 = 0$, then the function f does not participate in the optimization. Example 4.1.4 illustrates the case where ∇f and ∇g are both zero and we are at liberty to choose any values for λ_0 and λ_1.

The above discussion shows that, for any scenario, we can always find two numbers λ_0, λ_1 such that at least one of them is nonzero and $\nabla h(x, y) = \mathbf{0}$. We summarize this formally in the next theorem. A similar extension can be made to Theorem 4.1.2.

Theorem 4.1.3 (Extended Multiplier Rule) *Let $\Omega \subset \mathbb{R}^n$ be a region and let $f : \Omega \to \mathbb{R}$ and $g : \Omega \to \mathbb{R}$ be smooth functions. If f has a local extremum at $\mathbf{x} \in \Omega$ subject to the condition that $g(\mathbf{x}) = 0$ then there are numbers λ_0 and λ_1 not both zero such that*

$$\nabla \left(\lambda_0 f(x, y) - \lambda_1 g(x, y) \right) = \mathbf{0}.$$

4.2 The Isoperimetric Problem

Let $J : C^2[x_0, x_1] \to \mathbb{R}$ be a functional of the form

$$J(y) = \int_{x_0}^{x_1} f(x, y, y') \, dx, \qquad (4.17)$$

where f is a smooth function of x, y, and y'. The **isoperimetric problem** consists of determining the extremals of J satisfying boundary conditions of the form

$$y(x_0) = y_0, \quad y(x_1) = y_1 \qquad (4.18)$$

and a constraint of the form

$$I(y) = \int_{x_0}^{x_1} g(x, y, y') \, dx = L, \qquad (4.19)$$

where g is a given function of x, y, and y', and L is a specified constant. Conditions of the form (4.19) are called **isoperimetric constraints**.[2] In this section we derive a necessary condition for a function to be a smooth extremal to the isoperimetric problem.

Suppose that J has an extremum at y, subject to the boundary and isoperimetric conditions. We can proceed as we would for the unconstrained problem and consider neighbouring functions of the form $\hat{y} = y + \epsilon\eta$, where $\eta \in C^2[x_0, x_1]$ and $\eta(x_0) = \eta(x_1) = 0$, but the constraint (4.19) complicates matters because it places an additional restriction on the term $\epsilon\eta$ and therefore results that are based on the arbitrary character of η such as Lemma 2.2.2 are not valid without further modifying the function space H. If we proceed in this manner we will have to determine the class of functions in H such that $\hat{y}$ satisfies the isoperimetric condition. We can avoid this problem by introducing another function and parameter. We thus consider neighbouring functions of the form

$$\hat{y} = y + \epsilon_1\eta_1 + \epsilon_2\eta_2, \qquad (4.20)$$

where the ϵ_ks are small parameters, $\eta_k(x) \in C^2[x_0, x_1]$, and $\eta_k(x_0) = \eta_k(x_1) = 0$ for $k = 1, 2$. Roughly speaking, the introduction of the additional term $\epsilon_2\eta_2$ can be viewed as a "correction term." The function η_1 can be regarded as arbitrary, but the term $\epsilon_2\eta_2$ must be selected so that $\hat{y}$ satisfies the condition (4.19).

---

[2] Literally, the word *isoperimetric* means same perimeter. The most famous isoperimetric problem is Dido's problem (Section 1.4), where the constraint took the form of a specified arclength. Indeed, many isoperimetric problems have arclength constraints. The usage of the term "isoperimetric constraint" in the literature has simply come to mean conditions of the form (4.19) of which arclength is a prominent example.

Even with the introduction of the extra term $\epsilon_2\eta_2$, it is not immediately obvious that we can always choose an arbitrary η_1 and then find an appropriate term to meet the isoperimetric problem. Consider, for example, the constraint

$$I(y) = \int_0^1 \sqrt{1 + y'^2}\, dx = \sqrt{2},$$

along with the boundary conditions $y(0) = 0$ and $y(1) = 1$. There is only one smooth function that will meet this constraint, viz., the function $y(x) = x$, and therefore there are no variations of the form (4.20) available (apart from $\hat{y} = y$). This situation arises because the choice $L = \sqrt{2}$ happens to be the minimal value the functional I can take. Note that $y(x) = x$ must also be an extremal for the functional I. Extremals such as the above one that cannot be varied owing to the constraint are called **rigid extremals**.

Although rigid extremals are a concern, it turns out that for the isoperimetric problem they have a tractable characterization. Consider the quantity

$$I(\hat{y}) = \int_{x_0}^{x_1} g(x, y + \epsilon_1\eta_1 + \epsilon_2\eta_2, y' + \epsilon_1\eta_1' + \epsilon_2\eta_2')\, dx.$$

For a fixed choice of η_k we can regard $I(\hat{y})$ as a function of the parameters ϵ_1, ϵ_2, say $I(\hat{y}) = \Xi(\epsilon_1, \epsilon_2)$. Since g is a smooth function we have that Ξ is also a smooth function. Moreover, if J has an extremum at y subject to the boundary and isoperimetric condition, we have that $\Xi(0,0) = L$. We can appeal to the implicit function theorem to assert that for $\|\epsilon\| = \max(|\epsilon_1|, |\epsilon_2|)$ sufficiently small there exists a curve $\epsilon_2 = \epsilon_2(\epsilon_1)$ (or $\epsilon_1 = \epsilon_1(\epsilon_2)$) such that $\Xi(\epsilon_1, \epsilon_2(\epsilon_1)) = L$, provided

$$\nabla\Xi \neq \mathbf{0} \tag{4.21}$$

at $(0,0)$. Thus, if y is a rigid extremal, then $\nabla\Xi = \mathbf{0}$ at $(0,0)$. We return to an interpretation of this condition later in this section. For the present, we shall suppose that condition (4.21) is satisfied so that we avoid rigid extremals.

Rather than use the Taylor series approach of Chapter 4, it is easier here to convert the problem to a finite-dimensional constrained optimization problem as discussed in Section 4.1. Suppose that y is a smooth extremal to the isoperimetric problem and that condition (4.21) is satisfied. Then there are neighbouring functions of the form (4.20) which meet the boundary conditions (4.18) and the isoperimetric condition (4.19), where η_1 is an arbitrary function.

The quantity $J(\hat{y})$ can be regarded as a function of the parameters ϵ_1, ϵ_2. Let $J(\hat{y}) = \Theta(\epsilon_1, \epsilon_2)$. Since J has an extremum at y subject to the constraint $I(y) = L$, the function Θ must have an extremum at $(0,0)$ subject to the constraint $\Xi(\epsilon_1, \epsilon_2) - L = 0$. The results of the previous section indicate that for any critical point (ϵ_1, ϵ_2) there is a constant λ such that

$$\nabla\left(\Theta(\epsilon_1, \epsilon_2) - \lambda(\Xi(\epsilon_1, \epsilon_2) - L)\right) = \mathbf{0}.$$

Here, ∇ denotes the operator $(\partial/\partial\epsilon_1, \partial/\partial\epsilon_2)$. In particular, since $(0,0)$ is a critical point,

$$\frac{\partial}{\partial\epsilon_1}\left(\Theta(\epsilon_1,\epsilon_2) - \lambda\Xi(\epsilon_1,\epsilon_2)\right)\Big|_{\epsilon=0} = 0, \qquad (4.22)$$

and

$$\frac{\partial}{\partial\epsilon_2}\left(\Theta(\epsilon_1,\epsilon_2) - \lambda\Xi(\epsilon_1,\epsilon_2)\right)\Big|_{\epsilon=0} = 0. \qquad (4.23)$$

Now,

$$\begin{aligned}
\frac{\partial}{\partial\epsilon_1}\Theta(\epsilon_1,\epsilon_2)\Big|_{\epsilon=0} &= \frac{\partial}{\partial\epsilon_1}\int_{x_0}^{x_1} f(x, y + \epsilon_1\eta_1 + \epsilon_2\eta_2, y' + \epsilon_1\eta_1' + \epsilon_2\eta_2')\, dx\Big|_{\epsilon=0} \\
&= \int_{x_0}^{x_1} \frac{\partial}{\partial\epsilon_1} f(x, y + \epsilon_1\eta_1 + \epsilon_2\eta_2, y' + \epsilon_1\eta_1' + \epsilon_2\eta_2')\, dx\Big|_{\epsilon=0} \\
&= \int_{x_0}^{x_1} \left(\eta_1\frac{\partial f}{\partial y} + \eta_1'\frac{\partial f}{\partial y'}\right) dx,
\end{aligned}$$

and integrating by parts we see that

$$\frac{\partial}{\partial\epsilon_1}\Theta(\epsilon_1,\epsilon_2)\Big|_{\epsilon=0} = \int_{x_0}^{x_1} \eta_1\left(\frac{\partial f}{\partial y} - \frac{d}{dx}\frac{\partial f}{\partial y'}\right) dx.$$

Similarly, we have that

$$\frac{\partial}{\partial\epsilon_1}\Xi(\epsilon_1,\epsilon_2)\Big|_{\epsilon=0} = \int_{x_0}^{x_1} \eta_1\left(\frac{\partial g}{\partial y} - \frac{d}{dx}\frac{\partial g}{\partial y'}\right) dx.$$

Equation (4.22) can thus be written

$$\int_{x_0}^{x_1} \eta_1\left\{\frac{d}{dx}\frac{\partial f}{\partial y'} - \frac{\partial f}{\partial y} - \lambda\left(\frac{d}{dx}\frac{\partial g}{\partial y'} - \frac{\partial g}{\partial y}\right)\right\} dx = 0.$$

The function η_1 is arbitrary and Lemma 2.2.2 implies that

$$\frac{d}{dx}\frac{\partial F}{\partial y'} - \frac{\partial F}{\partial y} = 0, \qquad (4.24)$$

where

$$F = f - \lambda g.$$

The extremal y must therefore satisfy the Euler-Lagrange equation (4.24). The concern now is that equation (4.23) might overdetermine the problem. In fact, the same arguments used to derive equation (4.24) lead to the expression

$$\int_{x_0}^{x_1} \eta_2\left(\frac{d}{dx}\frac{\partial F}{\partial y'} - \frac{\partial F}{\partial y}\right) dx = 0,$$

which is always satisfied for any η_2 provided equation (4.24) is satisfied.

The above analysis is valid provided condition (4.21) is satisfied. Suppose that $\nabla \Xi = \mathbf{0}$ at $\epsilon = 0$. The above calculations show that in this case

$$\int_{x_0}^{x_1} \eta_1 \left(\frac{\partial g}{\partial y} - \frac{d}{dx} \frac{\partial g}{\partial y'} \right) dx = 0$$

and

$$\int_{x_0}^{x_1} \eta_2 \left(\frac{\partial g}{\partial y} - \frac{d}{dx} \frac{\partial g}{\partial y'} \right) dx = 0.$$

The former equation must be valid for arbitrary η_1; hence, by Lemma 2.2.2 we have that

$$\frac{d}{dx} \frac{\partial g}{\partial y'} - \frac{\partial g}{\partial y} = 0. \tag{4.25}$$

The latter equation is automatically satisfied if equation (4.25) is satisfied. Hence the condition that $\nabla \Xi = \mathbf{0}$ at $\epsilon = 0$ reduces to condition (4.25), and this means that y is an extremal for the functional I. Rigid extremals for the isoperimetric problem are thus characterized as functions that are also extremals for the functional defining the isoperimetric condition.

In summary, we have the following result.

Theorem 4.2.1 *Suppose that J has an extremum at $y \in C^2[x_0, x_1]$ subject to the boundary conditions (4.18) and the isoperimetric constraint (4.19). Suppose further that y is not an extremal for the functional I. Then there exists a constant λ such that y satisfies equation (4.24).*

In light of the above theorem, the isoperimetric problem reduces to the unconstrained fixed endpoint problem with f replaced by F. The general solution to the Euler-Lagrange equation (4.24) will contain two constants of integration along with the constant λ. The boundary conditions (4.18) and the constraint (4.19) provide three equations for these constants. In this sense, the isoperimetric problem is more complicated than the unconstrained problem of Section 2.2. Another complication with the isoperimetric problem is the possibility of rigid extremals. To validate the method we must verify that the solution to the Euler-Lagrange equation (4.24) is not also a solution to equation (4.25), i.e., an extremal for I.

If y is an extremal for J subject to the isoperimetric condition $I = L$, and y is not an extremal for I, then the problem is called **normal**. The "normality" of this problem is inherited from condition (4.21), which indicates that the finite-dimensional problem of determining the local extrema for Θ subject to the condition $\Xi = 0$ is normal. In the same spirit, if y is an extremal for I, then $\nabla \Xi(0,0) = \mathbf{0}$ and the problem is called **abnormal**. Because we can relate the isoperimetric problem back to a finite-dimensional optimization problem, we can readily extend Theorem 4.2.1 to cope with abnormal problems by introducing an additional multiplier λ_0.

Theorem 4.2.2 *Suppose that J has an extremum at $y \in C^2[x_0, x_1]$ subject to the boundary conditions* (4.18) *and the isoperimetric constraint* (4.19). *Then there exist two numbers λ_0 and λ_1 not both zero such that*

$$\frac{d}{dx}\frac{\partial K}{\partial y'} - \frac{\partial K}{\partial y} = 0,$$

where $K = \lambda_0 f - \lambda_1 g$. If y is not an extremal for I then we may take $\lambda_0 = 1$. If y is an extremal for I then we take $\lambda_0 = 0$, unless y is also an extremal for J. In the latter case neither λ_0 nor λ_1 is determined.

Example 4.2.1:   Catenary

Consider the catenary problem discussed in Section 1.2 and Example 2.3.3, but now suppose the length of the cable is specified. This leads to an isoperimetric problem. For simplicity, let $x_0 = 0$, $x_1 = 1$, and let the poles be of the same height $h > 0$. We thus seek an extremal to the functional

$$J(y) = \int_0^1 y\sqrt{1 + y'^2}\, dx,$$

subject to

$$I(y) = \int_0^1 \sqrt{1 + y'^2}\, dx = L,$$

and the boundary conditions $y(0) = y(1) = h$. Here $L > 1$ denotes the length of the cable.

The extremals for I consist of line segments (Example 2.2.1). Since $L > 1$, no solution to the Euler-Lagrange equation (4.24) that satisfies the boundary and isoperimetric conditions can be an extremal for I. Thus, if J has a local minimum at y, then Theorem 4.2.1 implies that y is a solution to equation (4.24) with $F = (y - \lambda)\sqrt{1 + y'^2}$.

Now, the function F does not contain x explicitly; hence, we have the first integral

$$\begin{aligned} H &= y'F_{y'} - F \\ &= \frac{(y - \lambda)y'^2}{\sqrt{1 + y'^2}} - (y - \lambda)\sqrt{1 + y'^2} \\ &= const. \end{aligned}$$

(cf. Section 2.3). Let $u = y - \lambda$. Then $u' = y'$, and the above equation reduces to

$$\frac{u^2}{1 + u'^2} = c_1^2,$$

where c_1 is a constant. Since $L > 1$ we know that y is not a constant and hence $c_1 \neq 0$. This equation was solved in Example 2.3.3, and it was shown that

$$u(x) = c_1 \cosh\left(\frac{x - c_2}{c_1}\right),$$

where c_2 is a constant. The extremals to this problem are thus of the form

$$y(x) = \lambda + c_1 \cosh\left(\frac{x - c_2}{c_1}\right). \tag{4.26}$$

Let $\kappa_1 = c_1$ and $\kappa_2 = -c_2/c_1$. The boundary conditions imply that

$$h - \lambda = \kappa_1 \cosh(\kappa_2) \tag{4.27}$$

and

$$h - \lambda = \kappa_1 \cosh(\frac{1}{\kappa_1} + \kappa_2); \tag{4.28}$$

therefore,

$$\cosh(\kappa_2) = \cosh(\kappa_2 + \frac{1}{\kappa_1});$$

i.e.,

$$\kappa_2 = -\frac{1}{2\kappa_1}. \tag{4.29}$$

The isoperimetric condition implies that

$$L = \int_0^1 \sqrt{1 + y'^2}\, dx = \int_0^1 \sqrt{1 + \sinh^2\left(\frac{x}{\kappa_1} + \kappa_2\right)}\, dx$$

$$= \int_0^1 \cosh\left(\frac{x}{\kappa_1} + \kappa_2\right) dx$$

$$= \kappa_1 \sinh\left(\frac{x}{\kappa_1} + \kappa_2\right)\Big|_0^1.$$

The isoperimetric condition thus reduces to

$$L = 2\kappa_1 \sinh\left(\frac{1}{2\kappa_1}\right), \tag{4.30}$$

upon using equation (4.29). Let $\xi = 1/2\kappa_1$. Equation (4.30) is equivalent to

$$L\xi = \sinh(\xi). \tag{4.31}$$

Equation (4.31) is evidently satisfied for $\xi = 0$, but this solution corresponds to an infinite value for κ_1 and thus produces the function $y = \lambda + \cosh(0) = const.$, which cannot be a solution to the isoperimetric problem. Since $L > 1$, however, there are precisely two nonzero solutions $\hat{\xi}$ and $-\hat{\xi}$ to equation (4.31) (see figure 4.4). We always have two solutions to the equations generated by the boundary conditions and the isoperimetric constraint. For the nonzero solution $\hat{\xi}$, we have that $\kappa_1 = 1/2\hat{\xi}$, $\kappa_2 = -\hat{\xi}$, and therefore equation (4.27) yields

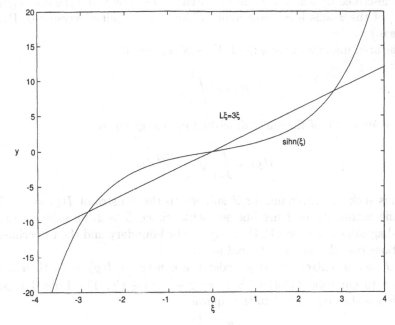

Fig. 4.4.

$$\lambda = h - \frac{1}{2\hat{\xi}}\cosh(\hat{\xi}).$$

The extremal is thus of the form

$$y(x) = h + \frac{1}{2\hat{\xi}}\left\{\cosh(\hat{\xi}(2x - 1)) - \cosh(\hat{\xi})\right\}. \tag{4.32}$$

Note that $\cosh(\hat{\xi}(2x - 1)) - \cosh(\hat{\xi}) \le 0$ for any choice of $\hat{\xi}$ since $x \in [0, 1]$. For physically sensible solutions we expect the cable to "hang down" below the points of suspension, and this means that $\hat{\xi} > 0$. There is always a unique positive solution to equation (4.31) so that this requirement can always be met.

It is intuitively obvious that for L sufficiently large there will be too much cable between the poles with the result that some of it must rest on the ground. In this case the model breaks down. Assuming the ground is level between the poles we need to further add the condition $y(x) > 0$ for all $x \in (0, 1)$ to avoid this problem. This inequality places a restriction on L in terms of h. We leave it as an exercise to show that for L sufficiently large there is an $x \in (0, 1)$ such that $y(x) < 0$.

Example 4.2.2: Dido's Problem
Determine the function y such that $y(-1) = y(1) = 0$, the perimeter of the

curve described by y is $L > 2$, and the area enclosed by y and the line segment $[-1, 1]$ of the x-axis is an extremum. (This is a simplified version of Dido's problem.)

The area under a curve $y : [-1, 1] \to \mathbb{R}$ is given by

$$J(y) = \int_{-1}^{1} y \, dx,$$

and the arclength of the curve described by y is given by

$$I(y) = \int_{-1}^{1} \sqrt{1 + y'^2} \, dx.$$

We thus seek an extremum for J subject to the constraint $I(y) = L$. Note that the extremals for I are line segments. Since $L > 2$, no solution to the Euler-Lagrange equation (4.24) satisfying the boundary and the isoperimetric conditions can also be an extremal for I.

If J has an extremum at y under the constraint $I(y) = L$, then y is a solution to equation (4.24) with $F = y - \lambda\sqrt{1 + y'^2}$. The Euler-Lagrange equation is thus equivalent to the equation

$$\frac{y''}{(1 + y'^2)^{3/2}} + \frac{1}{\lambda} = 0. \tag{4.33}$$

Recall that the curvature κ of a plane curve described in "graphical coordinates" by $\mathbf{r}(x) = (x, y(x))$ is given by

$$\kappa = \frac{|y''|}{(1 + y'^2)^{3/2}}. \tag{4.34}$$

Equation (4.33) indicates that the curve described by y must be of constant curvature $\kappa = 1/|\lambda|$; i.e., the curve must be an arc of a circle of radius $|\lambda|$.

Another way of deducing the shape of the extremal curve is to note that F does not contain x explicitly and hence the quantity $H = y'F_{y'} - F$ must be constant along any extremal. Therefore,

$$H = \frac{-\lambda y'^2}{\sqrt{1 + y'^2}} - (y - \lambda\sqrt{1 + y'^2}) = c_1,$$

where c_1 is a constant. The above equation simplifies to

$$(y + c_1)\sqrt{1 + y'^2} = \lambda;$$

i.e.,

$$y' = \sqrt{\frac{\lambda^2}{(y + c_1)^2} - 1}.$$

We thus have that

$$\int \frac{y + c_1}{\sqrt{\lambda^2 - (y + c_1)^2}} \, dy = x + c_2, \tag{4.35}$$

where c_2 is a constant. Let

$$y + c_1 = \lambda \sin \phi. \tag{4.36}$$

Then $dy = \lambda \cos \phi \, d\phi$, and hence

$$x + c_2 = \lambda \int \sin \phi \, d\phi = \lambda \cos \phi. \tag{4.37}$$

Equations (4.36) and (4.37) are the parametric equations for a circle of radius $|\lambda|$ centred at $(-c_2, -c_1)$.

The extremals are thus of the form

$$(x + c_2)^2 + (y + c_1)^2 = \lambda^2.$$

The boundary conditions $y(-1) = y(1) = 0$ imply that

$$(-1 + c_2)^2 + c_1^2 = \lambda^2 \tag{4.38}$$

and

$$(1 + c_2)^2 + c_1^2 = \lambda^2; \tag{4.39}$$

hence, $c_2 = 0$. The isoperimetric condition reduces to the equation

$$L = 2|\lambda||\phi|, \tag{4.40}$$

where ϕ denotes the angle between the y-axis and the line containing the points $(0, -c_1)$, $(1, 0)$ (figure 4.5). In terms of the constant c_1, the isoperimetric condition and equation (4.38) imply that

$$L = 2\sqrt{c_1^2 + 1} \arctan(\frac{1}{c_1}). \tag{4.41}$$

The condition that y is a function (i.e., single-valued) on $[-1, 1]$ places the somewhat artificial restriction that $c_1 \geq 0$, so that the centre of the circle is not above the x-axis. This in turn places a restriction on L for solutions of this type. It is easy to see geometrically that we must have $2 < L \leq \pi$ under these circumstances. With these conditions it can be argued geometrically (and analytically) that equation (4.41) has a unique solution for c_1 in terms of L, and that equation (4.38) has a unique positive solution for λ. We revisit this problem in Example 4.3.3, where we lift the restriction that y be single-valued.

The Lagrange multiplier λ plays a seemingly formal but useful rôle in the solution of isoperimetric problems. Example 4.2.2 shows that λ can correspond

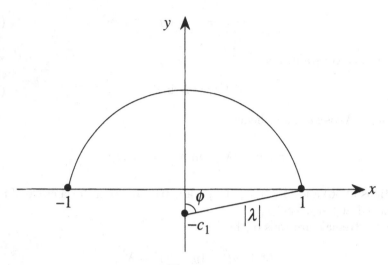

Fig. 4.5.

to a physically/geometrically significant parameter in the problem and this prompts us to look a bit deeper into the rôle of λ. The functional J can be written in the form

$$J(y) = \int_{x_0}^{x_1} \left\{ f(x,y,y') + \lambda \left(\frac{L}{x_1 - x_0} - g(x,y,y') \right) \right\} dx.$$

Suppose that J has an extremum at y. The general solution to the corresponding Euler-Lagrange equation will depend on x_0, x_1, y_0, y_1, and L. The Lagrange multiplier λ will also depend on these parameters. Suppose now that the boundary conditions are fixed. Then we may regard J as a function of the parameter L. Now

$$\frac{\partial J}{\partial L} = \int_{x_0}^{x_1} \frac{\partial}{\partial L} \left\{ (f(x,y,y') - \lambda g(x,y,y')) + \frac{\lambda L}{x_1 - x_0} \right\} dx$$

$$= \int_{x_0}^{x_1} \left\{ \left(\frac{\partial F}{\partial y} \frac{\partial y}{\partial L} + \frac{\partial F}{\partial y'} \frac{\partial y'}{\partial L} \right) + \frac{\partial \lambda}{\partial L} \left(\frac{L}{x_1 - x_0} - g(x,y,y') \right) + \frac{\lambda}{x_1 - x_0} \right\} dx$$

$$= \int_{x_0}^{x_1} \left(\frac{\partial F}{\partial y} - \frac{d}{dx} \frac{\partial F}{\partial y'} \right) \frac{\partial y}{\partial L} dx + \frac{\partial \lambda}{\partial L} \left(L - \int_{x_0}^{x_1} g(x,y,y') dx \right) + \lambda,$$

where the first integral on the last line was derived by integration by parts. Since y is an extremal for the problem, the first integral on the last line must vanish. The second term on the last line vanishes because y must satisfy the isoperimetric constraint. We therefore have that

$$\frac{\partial J}{\partial L} = \lambda.$$

The Lagrange multiplier therefore corresponds to the rate of change of the extremum $J(y)$ with respect to the isoperimetric parameter L.

We note a certain duality that exists for the isoperimetric problem. Suppose that $\lambda \neq 0$; then any extremal y to the problem with $F = f - \lambda g$ must also be an extremal to a problem with $G = g - \hat{\lambda} f$, where $\hat{\lambda} = 1/\lambda$. More specifically, suppose that y minimizes J subject to the isoperimetric constraint $I(y) = L$. Let K denote the minimum value $J(y)$. Then

$$K = J(y) - \lambda(I(y) - L),$$

and thus

$$L = I(y) - \hat{\lambda}(J(y) - K).$$

We have that $J - \lambda I = -\lambda(I - \hat{\lambda}J)$, and this indicates that the minimum for the functional $\int_{x_0}^{x_1} F \, dx$ corresponds to the maximum for the functional $\int_{x_0}^{x_1} G \, dx$. A similar statement can be made if y produces a maximum for J subject to $I(y) = L$. We thus have the following result.

Theorem 4.2.3 *Suppose that y produces a minimum (maximum) value for J subject to the constraint $I(y) = L$ and that $\lambda \neq 0$. Let $K = J(y)$. Then y produces a maximum (minimum) for I subject to the constraint $J(y) = K$, and $I(y) = L$.*

In view of the above result, suppose we revisit, for example, the catenary problem of Example 4.2.1. We saw that the catenary is the curve along which the potential energy is an extremum subject to the condition that the cable is of length L. In fact, it can be shown that the potential energy is minimum along a catenary for the appropriate choice of $\hat{\xi}$. Theorem 4.2.3 shows that, for a fixed value of potential energy, the catenary is the curve along which the arclength is maximized.

The duality relationship also helps to elucidate the condition that y not be an extremal for I in Theorem 4.2.1. If y is an extremal for I, then in the dual problem $\hat{\lambda} = 0$. This means that $I(y) = L$, independent of the constraint $J(y) = K$, so that K can be prescribed without changing the extremum for I. Alternatively, if $\lambda = 0$, then $J(y) = K$, independent of the constraint $I(y) = L$, so that the problem does not depend on the constraint. At any rate, if λ is not finite or if $\lambda = 0$ the problem is degenerate.

Exercises 4.2:

1. Let J and I be the functionals defined by

$$J(y) = \int_0^1 y'^2 \, dx, \quad I(y) = \int_0^1 y \, dx.$$

Find the extremals for J subject to the conditions $y(0) = 0$, $y(1) = 2$, and $I(y) = L$.

2. **Dido's Problem in Polar Coördinates**: Let J and I be functionals defined by

$$J(r) = \frac{1}{2} \int_0^\pi r^2 \, d\theta, \quad I(r) = \int_0^\pi \sqrt{r^2 + r'^2} \, d\theta,$$

where $r' = dr/d\theta$. Find an extremal for J subject to the conditions $r(0) = 0$, $r(\pi) = 0$, and $I(r) = L > 0$.

3. Let J and I be the functionals defined by

$$J(y) = \int_0^1 (yy')^2 \, dx, \quad I(y) = int_0^1 y^2 \, dx.$$

Suppose that y is an extremal for J subject to the conditions $y(0) = 1$, $y(1) = 2$, and $I(y) = L$.

(a) Find a first integral for the Euler-Lagrange equations for this problem and show that for $L = 3$,

$$y(x) = \sqrt{4 - 3(x - 1)^2}.$$

(b) For $L = 7/3$ show that there exists a linear function that is an extremal for the problem.

(c) For $L = 5/2$ show that this problem admits the solution $\lambda = 0$. Find the extremal corresponding to this value.

4. Let $A(y)$ be a smooth function and let

$$J(y) = \int_0^1 A(y)y' \, dx$$

and

$$I(y) = \int_0^1 \sqrt{1 + y'^2} \, dx.$$

Formulate the Euler-Lagrange equations for the isoperimetric problem with $y(0) = 0$, $y(1) = 1$, and $I(y) = L > \sqrt{2}$. Show that $\lambda = 0$, and that there are an infinite number of solutions to the problem. Explain without using the Euler-Lagrange equations (or any conservation laws) why there must be an infinite number of solutions to this problem.

5. Let y be the extremal to the catenary problem of Example 4.2.1. Show that for L sufficiently large there is an $x \in (0, 1)$ such that $y(x) < 0$.

4.3 Some Generalizations on the Isoperimetric Problem

In this section we present some modest generalizations on the isoperimetric problem discussed in Section 4.2. Most of the details are left to the reader.

4.3.1 Problems Containing Higher-Order Derivatives

Suppose that J and I are functionals of the form

$$J(y) = \int_{x_0}^{x_1} f(x, y, y', y'') \, dx,$$

$$I(y) = \int_{x_0}^{x_1} g(x, y, y', y'') \, dx,$$

where f and g are smooth functions. The same analysis used in the previous section can be used to show that any smooth extremal to J subject to the isoperimetric constraint $I(y) = L$ must satisfy the Euler-Lagrange equation

$$\frac{d^2}{dx^2} \frac{\partial F}{\partial y''} - \frac{d}{dx} \frac{\partial F}{\partial y'} + \frac{\partial F}{\partial y} = 0, \qquad (4.42)$$

where, for some constant λ,

$$F = f - \lambda g.$$

The existence of the constant λ is assured provided y is not also an extremal for the functional I. Indeed, it is straightforward to prove an analogous result for functionals containing derivatives of order higher than two. Abnormal problems can be treated in a manner similar to that used for the basic problem in Section 4.2.

Example 4.3.1:   Let

$$J(y) = \int_0^1 y''^2 \, dx,$$

and

$$I(y) = \int_0^1 y \, dx.$$

Find the extremals to J subject to the condition $I(y) = 1$, and the boundary conditions $y(0) = y(1) = 0$, $y'(0) = y'(1) = 0$.

For this problem $f = y''^2$ and $g = y$. The extremals for I must satisfy the Euler-Lagrange equation

$$\frac{d}{dx} \frac{\partial g}{\partial y'} - \frac{\partial g}{\partial y} = 0,$$

but $\partial g / \partial y' = 0$ and $\partial g / \partial y = 1$, so that I has no extremals. Any smooth extremal to the problem must therefore satisfy equation (4.42), which reduces to

$$2y^{(iv)}(x) - \lambda = 0.$$

The above differential equation has a general solution of the form

$$y(x) = \frac{\lambda x^4}{4!} + c_3 x^3 + c_2 x^2 + c_1 x + c_0,$$

where the c_ks are constants of integration. The boundary conditions $y(0) = 0$ and $y'(0) = 0$ imply that $c_0 = 0$ and $c_1 = 0$, respectively. At $x = 1$, the boundary conditions yield the equations

$$0 = \frac{\lambda}{4!} + c_3 + c_2$$

$$0 = \frac{\lambda}{3!} + 3c_3 + 2c_2.$$

The isoperimetric constraint implies that

$$\frac{\lambda}{5!} + \frac{c_3}{4} + \frac{c_2}{3} = 1.$$

The solution to this linear system of equations is $\lambda = 6!$, $c_2 = 30$, and $c_3 = -60$. The extremal is thus

$$y(x) = 30x^4 - 60x^3 + 30x^2.$$

4.3.2 Multiple Isoperimetric Constraints

Let J be a functional of the form

$$J(y) = \int_{x_0}^{x_1} f(x, y, y') \, dx,$$

and suppose $I_1, I_2, \ldots, I_m$ are functionals of the form

$$I_k(y) = \int_{x_0}^{x_1} g_k(x, y, y') \, dx,$$

for $k = 1, 2, \ldots, m$. Here, we assume that the functions f and g_k are smooth, and that some boundary conditions $y(x_0) = y_0$, $y(x_1) = y_1$ are prescribed. A generalization of the isoperimetric problem consists of determining the extremals to J subject to the m isoperimetric constraints $I_k(y) = L_k$, where the L_k are specified numbers. The Lagrange multiplier technique can be adapted to this type of problem, but the analogue of the condition that y is not an extremal for the isoperimetric functional is less tractable. We discuss the case $m = 2$.

Suppose that y is an extremal for J subject to the constraints $I_1(y) = L_1$ and $I_2(y) = L_2$. In order to meet both constraints and still have an arbitrary term in our variation of y we use a neighbouring function of the form

$$\hat{y} = y + \epsilon_1 \eta_1 + \epsilon_2 \eta_2 + \epsilon_3 \eta_3$$
$$= y + \langle \epsilon, \eta \rangle,$$

where $\epsilon = (\epsilon_1, \epsilon_2, \epsilon_3)$ and $\eta = (\eta_1, \eta_2, \eta_3)$. In addition, we require that $\eta_k \in C^2[x_0, x_1]$ and $\eta(x_0) = \eta(x_1) = 0$. There is still the problem of rigid extremals. We do not enter into a general discussion of this problem. Instead, we use the conditions developed at the end of Section 4.1 for finite-dimensional problems with multiple constraints.

As in the single constraint case we can regard $J(\hat{y})$, $I_1(\hat{y})$, and $I_2(\hat{y})$ as functions of ϵ. Let

$$\Theta(\epsilon) = \int_{x_0}^{x_1} f(x, y + \langle \epsilon, \eta \rangle, y' + \langle \epsilon, \eta' \rangle) \, dx,$$

and for $k = 1, 2$ let

$$\Xi_k(\epsilon) = \int_{x_0}^{x_1} g_k(x, y + \langle \epsilon, \eta \rangle, y' + \langle \epsilon, \eta' \rangle) \, dx.$$

If y is an extremal for the problem, then $\mathbf{0}$ is an extremal for the function Θ subject to the constraints $\Xi_k = L_k$. We know from Section 4.1 that there exist constants λ_1 and λ_2 such that the critical point must satisfy the vector equation

$$\nabla \left(\Theta - \lambda_1 \Xi_1 - \lambda_2 \Xi_2 \right) \Big|_{\epsilon=0} = \mathbf{0}, \tag{4.43}$$

provided condition (4.10) is satisfied. For $j = 1, 2, 3$, we have

$$\frac{\partial \Theta}{\partial \epsilon_j} \Big|_{\epsilon=0} = 0 = \int_{x_0}^{x_1} \eta_j \left(\frac{\partial f}{\partial y} - \frac{d}{dx} \frac{\partial f}{\partial y'} \right) dx$$

and

$$\frac{\partial \Xi_k}{\partial \epsilon_j} \Big|_{\epsilon=0} = 0 = \int_{x_0}^{x_1} \eta_j \left(\frac{\partial g_k}{\partial y} - \frac{d}{dx} \frac{\partial g_k}{\partial y'} \right) dx,$$

so that equation (4.43) produces the three equations

$$\int_{x_0}^{x_1} \eta_j \left(\frac{\partial F}{\partial y} - \frac{d}{dx} \frac{\partial F}{\partial y'} \right) dx = 0, \tag{4.44}$$

where

$$F = f - \lambda_1 g_1 - \lambda_2 g_2.$$

Now we can regard the term $\epsilon_1 \eta_1$ as an arbitrary function with the terms $\epsilon_2 \eta_2$ and $\epsilon_3 \eta_3$ used as "correction" terms so that the constraints are met. We can thus apply Lemma 2.2.2 to equation (4.44) with $j = 1$, and this gives the Euler-Lagrange equation

$$\frac{d}{dx} \frac{\partial F}{\partial y'} - \frac{\partial F}{\partial y} = 0. \tag{4.45}$$

As with the single constraint case, the other equations for $j = 2, 3$ are satisfied automatically if equation (4.45) is satisfied regardless of what functions the η_k might be. The general solution to equation (4.45) will contain two constants of integration along with the constants λ_1 and λ_2. The boundary conditions and the isoperimetric constraints can be used to determine these constants.

The condition that ensures the existence of constants λ_1, λ_2 such that equation (4.45) produces an extremal to the problem is less easy to interpret than the single constraint case. We know from Section 4.1 that this condition translates to inequality (4.10). For this problem we have that

$$\mathbf{M}(0) = \begin{pmatrix} \nabla \Xi_1(0) \\ \nabla \Xi_2(0) \end{pmatrix} = \begin{pmatrix} \alpha_{11} \ \alpha_{12} \ \alpha_{13} \\ \alpha_{21} \ \alpha_{22} \ \alpha_{23} \end{pmatrix},$$

where

$$\alpha_{ij} = \int_{x_0}^{x_1} \eta_j \left\{ \frac{\partial g_i}{\partial y} - \frac{d}{dx} \frac{\partial g_i}{\partial y'} \right\} dx,$$

and the augmented matrix $\mathbf{M}_f(0)$ is

$$\mathbf{M}_f(0) = \begin{pmatrix} \mathbf{M}(0) \\ \nabla \Theta(0) \end{pmatrix} = \begin{pmatrix} \alpha_{11} \ \alpha_{12} \ \alpha_{13} \\ \alpha_{21} \ \alpha_{22} \ \alpha_{23} \\ \beta_{31} \ \beta_{32} \ \beta_{33} \end{pmatrix},$$

where

$$\beta_{3j} = \int_{x_0}^{x_1} \eta_j \left\{ \frac{\partial f}{\partial y} - \frac{d}{dx} \frac{\partial f}{\partial y'} \right\} dx.$$

The Lagrange multiplier technique will be valid provided there exist smooth functions η_k such that:

(a) $\eta_k(x_0) = \eta_k(x_1) = 0$;
(b) $y + \langle \epsilon, \eta \rangle$ satisfies the isoperimetric constraints for $\|\epsilon\|$ small; and
(c) $\text{Rank}\,\mathbf{M}_f(0) \leq \text{Rank}\,\mathbf{M}(0)$.

Example 4.3.2: Let

$$J(y) = \int_0^1 y'^2 \, dx,$$

and

$$I_1(y) = \int_0^1 y \, dx,$$

$$I_2(y) = \int_0^1 xy \, dx.$$

Find the extremals for J subject to the constraints $I_1 = 2$, $I_2 = 1/2$ and the boundary conditions $y(0) = y(1) = 0$.

Let

$$F = y'^2 - \lambda_1 y - \lambda_2 xy.$$

The Euler-Lagrange equation for this choice of F is

$$2y'' + \lambda_1 + \lambda_2 x = 0,$$

which has the general solution

$$y = -\frac{\lambda_2 x^3}{6} - \frac{\lambda_1 x^2}{4} + c_1 x + c_0,$$

where c_0 and c_1 are constants of integration. The boundary condition $y(0) = 0$ implies that $c_0 = 0$, and the boundary condition $y(1) = 0$ yields the equation

$$c_1 = \frac{\lambda_1}{4} + \frac{\lambda_2}{6}.$$

The isoperimetric constraints provide the equations

$$2 = -\frac{\lambda_1}{12} - \frac{\lambda_2}{24} + \frac{c_1}{12},$$

$$\frac{1}{2} = -\frac{\lambda_1}{16} - \frac{\lambda_2}{30} + \frac{c_1}{3}.$$

The system of linear equations for λ_1, λ_2, and c_1 has the solution $\lambda_1 = 408$, $\lambda_2 = -360$, and $c_1 = 42$; thus, the function y defined by

$$y = 60x^3 - 102x^2 + 42x,$$

is a solution to the Euler-Lagrange equation that satisfies the isoperimetric and boundary conditions. Note that, for arbitrary η,

$$\alpha_{1j} = \int_0^1 \eta_j\, dx, \qquad \alpha_{2j} = \int_0^1 x\eta_j\, dx,$$

and

$$\beta_{3j} = \int_0^1 2y'\eta_j\, dx$$

$$= -\lambda_1 \int_0^1 \eta_j\, dx - \lambda_2 \int_0^1 x\eta_j\, dx$$

$$= -\lambda_1\alpha_{1j} + -\lambda_2\alpha_{2j},$$

so that $\operatorname{Rank} \mathbf{M}_f(0) \leq \operatorname{Rank} \mathbf{M}(0)$.

4.3.3 Several Dependent Variables

The Lagrange multiplier technique extends readily to isoperimetric problems involving several dependent variables. Let J and I be functionals of the form

$$J(\mathbf{q}) = \int_{t_0}^{t_1} L(t, \mathbf{q}, \dot{\mathbf{q}})\, dt,$$

$$I(\mathbf{q}) = \int_{t_0}^{t_1} g(t, \mathbf{q}, \dot{\mathbf{q}})\, dt,$$

where $\mathbf{q} = (q_1, q_2, \ldots, q_n)$, $\dot{}$ denotes d/dt, and L and g are smooth functions. If $\mathbf{q}$ is a smooth extremal for J subject to the boundary conditions $\mathbf{q}(t_0) = \mathbf{q}_0$, $\mathbf{q}(t_1) = \mathbf{q}_1$ and the isoperimetric condition $I(\mathbf{q}) = \ell$, and $\mathbf{q}$ is not an extremal for I, then there exists a constant λ such that $\mathbf{q}$ satisfies the n Euler-Lagrange equations

$$\frac{d}{dt}\frac{\partial F}{\partial \dot{q}_j} - \frac{\partial F}{\partial q_j} = 0, \tag{4.46}$$

where $j = 1, 2, \ldots, n$, and

$$F = L - \lambda g.$$

The technique can also be adapted *mutatis mutandis* to problems with several isoperimetric constraints.

Example 4.3.3:   Let us revisit the problem of determining a curve γ of length $\ell > 2$ containing the points $\mathbf{P}_{-1} = (-1, 0)$ and $\mathbf{P}_1 = (1, 0)$ such that the closed curve formed by γ and the line segment from $\mathbf{P}_{-1}$ to $\mathbf{P}_1$ encloses maximum area. We show that any smooth extremal for this problem must correspond to a circular arc, but we lift the restriction that γ must be described by a scalar function y.

Suppose that $\mathbf{q}(t) = (x(t), y(t))$, $t \in [t_0, t_1]$ is an extremal for the problem. Green's theorem implies that the area under the curve is

$$J(\mathbf{q}) = \frac{1}{2} \int_{t_0}^{t_1} (x\dot{y} - y\dot{x}) \, dt,$$

and the isoperimetric condition is

$$I(\mathbf{q}) = \int_{t_0}^{t_1} \sqrt{\dot{x}^2 + \dot{y}^2} \, dt = \ell.$$

The extremal must thus satisfy equations (4.46) with

$$F = \frac{1}{2} (x\dot{y} - y\dot{x}) - \lambda\sqrt{\dot{x}^2 + \dot{y}^2};$$

i.e.,

$$\frac{d}{dt}\left(\frac{-\lambda\dot{x}}{\sqrt{\dot{x}^2 + \dot{y}^2}} - \frac{1}{2}y \right) - \frac{1}{2}\dot{y} = 0,$$

$$\frac{d}{dt}\left(\frac{-\lambda\dot{y}}{\sqrt{\dot{x}^2 + \dot{y}^2}} + \frac{1}{2}x \right) + \frac{1}{2}\dot{x} = 0.$$

We therefore have that

$$\frac{-\lambda\dot{x}}{\sqrt{\dot{x}^2 + \dot{y}^2}} + y = c_0,$$

$$\frac{-\lambda\dot{y}}{\sqrt{\dot{x}^2 + \dot{y}^2}} + x = c_1,$$

where c_0 and c_1 are constants of integration. The above equations imply that

$$(x - c_1)^2 + (y - c_0)^2 = \frac{\lambda^2 \dot{x}^2}{\dot{x}^2 + \dot{y}^2} + \frac{\lambda^2 \dot{y}^2}{\dot{x}^2 + \dot{y}^2}$$
$$= \lambda^2;$$

hence, the extremal must be a circular arc of radius λ with centre at (c_1, c_0). It follows readily from the boundary conditions that $c_1 = 0$. The other constants require more effort, but can be obtained essentially as described in Example 4.2.2. Note that now the constant c_2 may be negative or positive depending on ℓ.

Exercises 4.3:

1. Let J and I be functionals defined by

$$J(x, y) = \int_{t_0}^{t_1} \left(1 - \frac{x}{\sqrt{x^2 + y^2}}\right) x' \, dt,$$

$$I(x, y) = \int_{t_0}^{t_1} y^2 x' \, dt,$$

where $x = x(t), y = y(t)$, and $'$ denotes d/dt. Suppose that (x, y) is an extremal for J subject to the constraint $I(x, y) = K$, where K is a positive constant. Prove that neither $x(t)$ nor $y(t)$ can be identically zero on the interval $[t_0, t_1]$ and that there is a constant Λ such that

$$x = \Lambda \left(x^2 + y^2\right)^{3/2}.$$

2. Let J, I_1 and I_2 be functionals defined by

$$J(y) = \int_0^1 y'^2 \, dx,$$

$$I_1(y) = \int_0^1 x^2 y'^2 \, dx,$$

$$I_2(y) = \int_0^1 y \, dx.$$

Find the extremals for J subject to the conditions $I_1(y) = \ell_1$, $I_2(y) = \ell_2$ and the boundary conditions $y(0) = 0$, $y(1) = 1$.

Applications to Eigenvalue Problems*

Eigenvalue problems infest applied mathematics. These problems consist of finding nontrivial solutions to a linear differential equation subject to boundary conditions that admit the trivial solution. The differential equation contains an eigenvalue parameter, and nontrivial solutions exist only for special values of this parameter, the eigenvalues. Generally, finding the eigenvalues and the corresponding nontrivial solutions poses a formidable task.

Certain eigenvalue problems can be recast as isoperimetric problems. Indeed, many of the eigenvalue problems have their origin in the calculus of variations. Although the Euler-Lagrange equation is essentially the original differential equation and thus of limited value for deriving solutions, the variational formulation is helpful for extracting results about the distribution of eigenvalues. In this chapter we discuss a few simple applications of the variational approach to Sturm-Liouville problems. The standard reference on this material is Courant and Hilbert [25]. Moiseiwitsch [54] also discusses at length eigenvalue problems in the framework of the calculus of variations. Our brief account is a blend of material from Courant and Hilbert *op. cit.* and Wan [71].

5.1 The Sturm-Liouville Problem

The (regular) Sturm-Liouville problem entails finding nontrivial solutions to differential equations of the form

$$(-p(x)y'(x))' + q(x)y(x) - \lambda r(x)y(x) = 0, \tag{5.1}$$

for the unknown function $y : [x_0, x_1] \to \mathbb{R}$ subject to boundary conditions of the form

$$\alpha_0 y(x_0) + \beta_0 y'(x_0) = 0,$$

$$\tag{5.2}$$

$$\alpha_1 y(x_1) + \beta_1 y'(x_1) = 0.$$

Here, q and r are functions continuous on the interval $[x_0, x_1]$, and $p \in C^1[x_0, x_1]$. In addition, $p(x) > 0$ and $r(x) > 0$ for all $x \in [x_0, x_1]$. The α_k and β_k in the boundary conditions are constants such that $\alpha_k^2 + \beta_k^2 \neq 0$, and λ is a parameter.

Generically, the only solution to equation (5.1) that satisfies the boundary conditions (5.2) is the trivial solution, $y(x) = 0$ for all $x \in [x_0, x_1]$. There are, however, certain values of λ that lead to nontrivial solutions. These special values are called **eigenvalues** and the corresponding nontrivial solutions are called **eigenfunctions**. The set of all eigenvalues for the problem is called the **spectrum**.

An extensive theory has been developed for the Sturm-Liouville problem. Here, we limit ourselves to citing a few basic results and direct the reader to standard works such as Birkhoff and Rota [9], Coddington and Levinson [24], and Titchmarsh [70] for further details.

The "natural" function space in which to study the Sturm-Liouville problem is the (real) Hilbert space $L^2[x_0, x_1]$, which consists of functions $f : [x_0, x_1] \to \mathbb{R}$ such that

$$\int_{x_0}^{x_1} f^2(x)\, dx < \infty.$$

The inner product on this Hilbert space is defined by

$$\langle f, g \rangle = \int_{x_0}^{x_1} r(x) f(x) g(x)\, dx,$$

for all $f, g \in L^2[x_0, x_1]$.[1] The norm induced by this inner product is defined by

$$_r\|f\|_2 = \sqrt{\langle f, f \rangle} = \sqrt{\int_{x_0}^{x_1} r(x) f^2(x)\, dx},$$

for all $f \in L^2[x_0, x_1]$. Note that the norm $_r\| \cdot \|_2$ is equivalent to the usual norm $\| \cdot \|_2$ defined by

$$\|f\|_2 = \sqrt{\int_{x_0}^{x_1} f^2(x)\, dx},$$

because r is continuous on $[x_0, x_1]$ and positive; hence, r is bounded above and below by positive numbers.[2]

Some notable results from the theory are:

---

[1] Strictly speaking, the integrals defining the Hilbert space are Lebesgue integrals and the elements of the space are equivalence classes of functions. We deal here with solutions to the Sturm-Liouville problem and these functions are continuous on $[x_0, x_1]$. For such functions the Lebesgue and Riemann integrals are equivalent. Note that $L^2[x_0, x_1]$ also includes much "rougher" functions that are not Riemann integrable.

[2] See Appendix B.1.

(a) There exist an infinite number of eigenvalues. All the eigenvalues are real and isolated. The spectrum can be represented as a monotonic increasing sequence $\{\lambda_n\}$ with $\lim_{n\to\infty} \lambda_n = \infty$. The least element in the spectrum is called the **first eigenvalue**.

(b) The eigenvalues are **simple**. This means that there exists precisely one eigenfunction (apart from multiplicative factors) corresponding to each eigenvalue.

(c) If λ_m and λ_n are distinct eigenvalues with corresponding eigenfunctions ϕ_m and ϕ_n, respectively, the orthogonality relation

$$\langle \phi_m, \phi_n \rangle = 0$$

is satisfied. (Note that $\langle \phi_m, \phi_m \rangle > 0$, since ϕ_m is a nontrivial solution.)

(d) The set of all eigenfunctions $\{\phi_n\}$ forms a basis for the space $L^2[x_0, x_1]$. In other words, for any function $f \in L^2[x_0, x_1]$ there exist constants $\{a_n\}$ such that the series

$$\mathcal{F}(f) = \sum_{n=1}^{\infty} a_n \phi_n$$

converges in the $_r\|\cdot\|_2$ norm to f; i.e.,

$$\lim_{k\to\infty} {}_r\|f - \sum_{n=1}^{\infty} a_n \phi_n\|_2 = 0.$$

The series representing f is called an **eigenfunction expansion** or **generalized Fourier series** of f.

The Sturm-Liouville problem can be recast as a variational problem. We do this for the case $\beta_0 = \beta_1 = 0$. The formulation for the general boundary conditions (5.2) can be found in Wan, *op. cit.*, p. 285. Let J be the functional defined by

$$J(y) = \int_{x_0}^{x_1} \left(py'^2 + qy^2 \right) dx, \tag{5.3}$$

and consider the problem of finding the extremals for J subject to boundary conditions of the form

$$y(x_0) = y(x_1) = 0, \tag{5.4}$$

and the isoperimetric constraint

$$I(y) = \int_{x_0}^{x_1} r(x)y^2(x)\, dx = 1. \tag{5.5}$$

The Euler-Lagrange equation for the functional I is

$$-2r(x)y(x) = 0,$$

which is satisfied only for the trivial solution $y = 0$, because r is positive. No extremals for I can therefore satisfy the isoperimetric condition (5.5). If y is

an extremal for the isoperimetric problem, then Theorem 4.2.1 implies that
there is a constant λ such that y satisfies the Euler-Lagrange equation

$$\frac{d}{dx}\frac{\partial F}{\partial y'} - \frac{\partial F}{\partial y} = 0, \tag{5.6}$$

for

$$F = py'^2 + qy^2 - \lambda ry^2.$$

But the Euler-Lagrange equation for this choice of F is equivalent to the
differential equation (5.1). The isoperimetric problem thus corresponds to the
Sturm-Liouville problem augmented by the normalizing condition (5.5), which
simply scales the eigenfunctions. Here, the Lagrange multiplier plays the rôle
of the eigenvalue parameter.

Example 5.1.1:   Let $p(x) = 1$, $q(x) = 0$, $r(x) = 1$, and $[x_0, x_1] = [0, \pi]$.
Then the Euler-Lagrange equation reduces to

$$y''(x) + \lambda y(x) = 0, \tag{5.7}$$

and the boundary conditions are

$$y(0) = y(\pi) = 0. \tag{5.8}$$

If $\lambda < 0$, then the general solution to equation (5.7) is

$$y(x) = Ae^{\sqrt{-\lambda}x} + Be^{-\sqrt{-\lambda}x},$$

where A and B are constants. The boundary conditions imply that $A = B = 0$,
and therefore there are only trivial solutions if $\lambda < 0$. If $\lambda = 0$, then equation
(5.7) has the general solution

$$y(x) = Ax + B.$$

Again the boundary conditions imply that $A = B = 0$, and therefore preclude
the possibility of nontrivial solutions. Hence, any eigenvalues for this problem
must be positive.
   If $\lambda > 0$, then the general solution to equation (5.7) is

$$y(x) = A\cos(\sqrt{\lambda}x) + B\sin(\sqrt{\lambda}x).$$

The condition $y(0) = 0$ implies that $A = 0$; the condition $y(\pi) = 0$ implies
that

$$B\sin(\sqrt{\lambda}\pi) = 0. \tag{5.9}$$

Equation (5.9) is satisfied for $B \neq 0$ provided $\sqrt{\lambda}$ is a positive integer, and
this leads to the nontrivial solution $y(x) = B\sin(\sqrt{\lambda}x)$. The eigenvalues for
this problem are therefore $\lambda_n = n^2$, and the first eigenvalue is $\lambda_1 = 1$. The
eigenfunctions corresponding to λ_n are of the form

$$\phi_n(x) = B\sin(nx), \tag{5.10}$$

where B is an arbitrary constant.

In terms of the isoperimetric problem, there are an infinite number of Lagrange multipliers that can be used. Each Lagrange multiplier corresponds to an eigenvalue, and the linearity of the Euler-Lagrange equation implies that any function of the form

$$f(x) = \sum_{n=1}^{\infty} a_n\sin(nx), \tag{5.11}$$

such that the Fourier series is convergent and twice term by term differentiable, is an extremal for the problem, provided f satisfies the isoperimetric condition (5.5). Now,

$$\int_0^{\pi} f^2(x)\,dx = \int_0^{\pi} \left(\sum_{n=1}^{\infty} a_n\sin(nx)\right)^2 dx$$
$$= \sum_{n=1}^{\infty} a_n^2 \int_0^{\pi} \sin^2(nx)\,dx$$
$$= \frac{\pi}{2}\sum_{n=1}^{\infty} a_n^2,$$

where we have used the orthogonality relation

$$\langle \sin(mx), \sin(nx)\rangle = \begin{cases} 0, & \text{if } m \neq n, \\ \frac{\pi}{2}, & \text{if } m = n. \end{cases}$$

Hence, any eigenfunction expansion of the form (5.11) having the requisite convergence properties and satisfying the condition

$$\sum_{n=1}^{\infty} a_n^2 = \frac{2}{\pi} \tag{5.12}$$

is an extremal for the problem. Any finite combination of the eigenfunctions such as

$$f(x) = a_1\sin(x) + a_2\sin(2x) + \cdots + a_m\sin(mx),$$

where

$$a_1^2 + \cdots + a_m^2 = \frac{2}{\pi},$$

for example, is an extremal.

If we are searching among the eigenfunction expansions for extremals that make J a minimum, then the situation changes considerably. Suppose that f is an eigenfunction extremal for the problem. Then

$$y'(x) = \sum_{n=1}^{\infty} na_n \cos(nx),$$

so that

$$J(y) = \int_0^{\pi} y'^2(x)\,dx = \int_0^{\pi} \left(\sum_{n=1}^{\infty} na_n \cos(nx) \right)^2 dx$$

$$= \sum_{n=1}^{\infty} n^2 a_n^2 \int_0^{\pi} \cos^2(nx)\,dx$$

$$= \frac{\pi}{2} \sum_{n=1}^{\infty} n^2 a_n^2.$$

Here, we have used the orthogonality relation

$$\langle \cos(mx), \cos(nx) \rangle = \begin{cases} 0, & \text{if } m \neq n, \\ \frac{\pi}{2}, & \text{if } m = n. \end{cases}$$

The eigenfunction extremal for the first eigenvalue is

$$y_1(x) = \sqrt{\frac{2}{\pi}} \sin(x),$$

and for this extremal

$$J(y_1) = 1.$$

In fact, y_1 produces the minimum value for J. To see this, let f be another extremal for the problem. Then the completeness property of the Fourier series implies that f can be expressed as an eigenfunction expansion of the form (5.11), where the coefficients a_n satisfy relation (5.12). If f is distinct from y_1 then there is an integer $m \geq 2$ such that $a_m \neq 0$. Now,

$$J(f) = \frac{\pi}{2} \sum_{n=1}^{\infty} n^2 a_n^2$$

$$\geq \frac{\pi}{2} \left((m^2 - 1)a_m^2 + \sum_{n=1}^{\infty} a_n^2 \right)$$

$$> \frac{\pi}{2} \sum_{n=1}^{\infty} a_n^2 = 1,$$

and hence $J(f) > J(y_1)$.

Exercises 5.1:

1. The **Cauchy-Euler equation** is

$$(xy'(x))' + \frac{\lambda}{x}y(x) = 0.$$

Show that

$$y(x) = c_1 \cosh\left(\sqrt{-\lambda}\ln x\right) + c_2 \sinh\left(\sqrt{-\lambda}\ln x\right),$$

where c_1 and c_2 are constants, is a general solution to this equation. Given the boundary conditions $y(0) = y(e^\pi) = 0$ find the eigenvalues.

2. Reformulate the differential equation

$$y^{(iv)}(x) + (\lambda + \rho(x))y(x) = 0$$

along with the boundary values $y(0) = y'(0) = 0$, $y(1) = y'(1) = 0$ as an isoperimetric problem.

5.2 The First Eigenvalue

The first eigenvalue in Example 5.1.1 has the notable property that the corresponding eigenfunction produced the minimum value for J. If fact, this relationship persists for the general Sturm-Liouville problem.

Theorem 5.2.1 *Let λ_1 be the first eigenvalue for the Sturm-Liouville problem (5.1) with boundary conditions (5.4), and let y_1 be the corresponding eigenfunction normalized to satisfy the isoperimetric constraint (5.5). Then, among functions in $C^2[x_0, x_1]$ that satisfy the boundary conditions (5.4) and the isoperimetric condition (5.5), the functional J defined by equation (5.3) is minimum at $y = y_1$. Moreover,*

$$J(y_1) = \lambda_1.$$

Proof: Suppose that J has a minimum at y. Then y is an extremal and thus satisfies equation (5.1) and conditions (5.4) and (5.5). Multiplying equation (5.1) by y and integrating from x_0 to x_1 gives

$$-pyy'\Big|_{x_0}^{x_1} + \int_{x_0}^{x_1} \left(py'^2 + qy^2\right)\,dx = \lambda \int_{x_0}^{x_1} ry^2\,dx.$$

The first term on the left-hand side of the above expression is zero since $y(x_0) = y(x_1) = 0$; the integral on the left-hand side of the equation is one by the isoperimetric condition. Hence we have

$$J(y) = \lambda.$$

Any extremal to the problem must be a nontrivial solution to equation (5.1) because of the isoperimetric condition; consequently, λ must be an eigenvalue. By property (a) there must be a least element in the spectrum, the first eigenvalue λ_1, and a corresponding eigenfunction y_1 normalized to meet the isoperimetric condition. Hence the minimum value for J is λ_1 and $J(y_1) = \lambda_1$.

$\square$

Eigenvalues for the Sturm-Liouville problem signal a bifurcation: in a deleted neighbourhood of an eigenvalue there is only the trivial solution available; at an eigenvalue there are nontrivial solutions (multiples of the eigenfunction) available in addition to the trivial solution. In applications such as those involving the stability of elastic bodies, eigenvalues indicate potential abrupt changes. Often the most vital piece of information in a model is the location of the first eigenvalue. For example, an engineer may wish to design a column so that the first eigenvalue in the problem modelling the deflection of the column is sufficiently high that it will not be attained under normal loadings.[3]

Theorem 5.2.1 suggests a characterization of the first eigenvalue in terms of the functionals J and I. Let R be the functional defined by

$$R(y) = \frac{J(y)}{I(y)}. \tag{5.13}$$

The functional R is called the **Rayleigh quotient** for the Sturm-Liouville problem. If $I(y) = 1$, then for any nontrivial solution y we have

$$\lambda = R(y). \tag{5.14}$$

We can, however, drop this normalization restriction on I because both J and I are homogeneous quadratic functions in y and y' so that any normalization factors cancel out in the quotient. Relation (5.14) is thus valid for any nontrivial solution, and we can make use of this observation to characterize the first eigenvalue as the minimum of the Rayleigh quotient.

Theorem 5.2.2 *Let S' denote the set of all functions in $C^2[x_0, x_1]$ that satisfy the boundary conditions (5.4) except the trivial solution $y \equiv 0$. The minimum of the Rayleigh quotient R for the Sturm-Liouville problem (5.1), (5.4) over all functions in S' is the first eigenvalue; i.e.,*

$$\min_{y \in S'} R(y) = \lambda_1. \tag{5.15}$$

---

[3] The governing differential equation for this model is in fact of fourth order, but similar comments apply. The variational formulation of this model is discussed in Courant and Hilbert, *op. cit.*, p. 272.

Proof: Suppose that R has minimum value Λ at $y \in S'$, and let

$$\hat{y} = y + \epsilon\eta,$$

where ϵ is small and η is a smooth function such that $\eta(x_0) = \eta(x_1) = 0$, to ensure that $\hat{y} \in S'$. Now,

$$I(y + \epsilon\eta) = I(y) + 2\epsilon \int_{x_0}^{x_1} \eta r y \, dx + O(\epsilon^2),$$

so that

$$\frac{1}{I(\hat{y})} = \frac{1}{I(y)} + O(\epsilon),$$

where $I(y) \neq 0$, and

$$J(\hat{y}) = J(y) + 2\epsilon \int_{x_0}^{x_1} \eta \left((-py')' + qy\right) dx + O(\epsilon^2)$$

$$= \Lambda I(y) + 2\epsilon \int_{x_0}^{x_1} \eta \left((-py')' + qy\right) dx + O(\epsilon^2),$$

so that

$$J(\hat{y}) - \Lambda I(y) = 2\epsilon \int_{x_0}^{x_1} \eta \left((-py')' + qy - \Lambda ry\right) dx$$
$$+ O(\epsilon^2).$$

We thus have

$$R(\hat{y}) - R(y) = \frac{J(\hat{y})}{I(\hat{y})} - \frac{J(y)}{I(y)} = \frac{J(\hat{y}) - \Lambda I(\hat{y})}{I(\hat{y})}$$

$$= 2\epsilon \frac{\int_{x_0}^{x_1} \eta \left((-py')' + qy - \Lambda ry\right) dx}{I(y)} + O(\epsilon^2),$$

and since R is minimum at y, the terms of order ϵ must vanish in the above expression for arbitrary η. We can apply Lemma 2.2.2 to the numerator of the order ϵ term and deduce that y must satisfy equation (5.1). Since $y \in S'$, the constant Λ must be an eigenvalue. Any extremal for R must therefore be a nontrivial solution to the Sturm-Liouville problem.

If λ_m is an eigenvalue for the problem and y_m is a corresponding eigenfunction, then the calculation in the proof of Theorem 5.2.1 can be used to show that

$$R(y_m) = \frac{J(y_m)}{I(y_m)} = \lambda_m.$$

Since R is minimum at y and Λ is the corresponding eigenvalue we have that

$$\lambda_m = R(y_m) \geq \Lambda$$

for all eigenvalues. Therefore we have $\Lambda = \lambda_1$. $\square$

Generally, the eigenvalues (and hence eigenfunctions) for a Sturm-Liouville problem cannot be determined explicitly. Bounds for the first eigenvalue, however, can be obtained using the Rayleigh quotient. Upper bounds for λ_1 can be readily obtained since λ_1 is a minimum value: for any function $\phi \in S'$ we have

$$R(\phi) \geq \lambda_1, \tag{5.16}$$

so that an upper bound can be derived by using any function in S'. Lower bounds require a bit more work.

To get a lower bound, the strategy is to construct a comparison problem that can be solved explicitly, the first eigenvalue $\bar{\lambda}_1$ of which is guaranteed to be no greater than λ_1. To construct a comparison problem, we make the following simple observations.

(a) Let $\bar{p} \in C^1[x_0, x_1]$ be any function such that $p(x) \geq \bar{p}(x) > 0$ for all $x \in [x_0, x_1]$, and let $\bar{q} \in C^0[x_0, x_1]$ be any function such that $q(x) \geq \bar{q}(x)$ for all $x \in [x_0, x_1]$. Then, for

$$\bar{J}(y) = \int_{x_0}^{x_1} \left(\bar{p} y'^2 + \bar{q} y^2 \right) dx,$$

we have

$$\bar{J}(y) \leq J(y).$$

(b) Let $\bar{r} \in C^0[x_0, x_1]$ be any function such that $\bar{r}(x) \geq r(x) > 0$ for all $x \in [x_0, x_1]$. Then, for

$$\bar{I}(y) = \int_{x_0}^{x_1} \bar{r} y^2 \, dx$$

we have

$$\bar{I}(y) \geq I(y).$$

If we choose $\bar{p}$, $\bar{q}$, and $\bar{r}$ as above, then

$$\bar{R}(y) = \frac{\bar{J}(y)}{\bar{I}(y)} \leq \frac{J(y)}{I(y)} = R(y),$$

and hence

$$\bar{\lambda}_1 \leq \lambda_1. \tag{5.17}$$

Inequality (5.17) is useful only if we can determine $\bar{\lambda}_1$ explicitly. We have considerable freedom, however, in our choices for $\bar{p}$, $\bar{q}$, and $\bar{r}$, and the simplest choice is when these functions are constants; i.e.,

$$\bar{p}(x) = \min_{x \in [x_0, x_1]} p(x) \equiv p_m,$$

$$\bar{q}(x) = \min_{x \in [x_0, x_1]} q(x) \equiv q_m,$$

$$\bar{r}(x) = \max_{x \in [x_0, x_1]} r(x) \equiv r_M.$$

For this choice, the differential equation is

$$(-p_m y')' + q_m y - \bar{\lambda} r_M y = 0;$$

i.e.,

$$y'' + \frac{1}{p_m} \left(\bar{\lambda} r_M - q_m \right) y = 0. \tag{5.18}$$

The solution of equation (5.18) subject to boundary conditions (5.4) follows essentially along the same lines as that given in Example 5.1.1. The eigenvalues for this problem are

$$\bar{\lambda}_n = \frac{1}{r_M} \left(\frac{p_m n^2 \pi^2}{(x_0 - x_1)^2} + q_m \right).$$

We thus get the lower bound

$$\bar{\lambda}_1 = \frac{1}{r_M} \left(\frac{p_m \pi^2}{(x_0 - x_1)^2} + q_m \right) \le \lambda_1. \tag{5.19}$$

Example 5.2.1: Mathieu's Equation
Let $p(x) = r(x) = 1$, and $q(x) = 2\theta \cos(2x)$, where $\theta \in \mathbb{R}$ is a constant. Let $x_0 = 0$ and $x_1 = \pi$. For this choice of functions equation (5.1) is equivalent to

$$y'' + (\lambda - 2\theta \cos(2x)) y = 0, \tag{5.20}$$

and the boundary conditions are

$$y(0) = y(\pi) = 0. \tag{5.21}$$

The expression (5.20) is called **Mathieu's equation**, and its solutions have been investigated in depth (cf. McLachlan [52] and Whittaker and Watson [74]). If $\theta = 0$, then the problem reduces to that studied in Example 5.1.1. If $\theta \ne 0$, then the nontrivial solutions to this problem cannot be expressed in closed form in terms of elementary functions. Indeed, this problem defines a new class of functions $\{se_n\}$ called **Mathieu functions**,[4] that correspond to the eigenfunctions of the problem. The determination of the eigenvalues for this problem is a more complicated affair compared to the simple problem of Example 5.1.1. Briefly, it can be shown that the first eigenvalue λ_1 and the corresponding eigenfunction se_1 are given asymptotically by

$$\lambda_1 = 1 - \theta - \frac{1}{8}\theta^2 + \frac{1}{64}\theta^3 - \frac{1}{1536}\theta^4 - \frac{11}{36864}\theta^5 + O(\theta^6), \tag{5.22}$$

and

---

[4] The notation se_n is an abbreviation for "sine-elliptic." There are also "cosine-elliptic" Mathieu functions ce_n.

$$se_1(x) = \sin(x) - \frac{1}{8}\theta\sin(3x) + \frac{1}{64}\theta^2\left(\sin(3x) + \frac{1}{3}\sin(5x)\right)$$
$$- \frac{1}{512}\theta^3\left(\frac{1}{3}\sin(3x) + \frac{4}{9}\sin(5x) + \frac{1}{18}\sin(7x)\right)$$
$$+ O(\theta^4),$$

for $|\theta|$ small (cf. McLachlan, *op. cit.* p. 10–14).

In contrast, a rough lower bound for λ_1 can be readily gleaned from inequality (5.19). Suppose that $\theta \geq 0$, and let $p_m = r_M = 1$, $q_m = -2\theta \leq 2\theta\cos(2x)$. Inequality (5.19) then implies

$$1 - 2\theta \leq \lambda_1. \tag{5.23}$$

Given the asymptotic expression (5.22), if $\theta \geq 0$ is small then the lower bound (5.23) can be verified directly. But inequality (5.23) is also valid for θ large, and this is not so obvious.

Note that if $\theta < 0$, we cannot use $q_m = 2\theta$ in our comparison problem since $-2\theta \geq 2\theta\cos(2x)$ for $x \in [x_0, x_1]$. For this case we can use $q_m = 2\theta$ and thus get the lower bound

$$1 + 2\theta \leq \lambda_1.$$

Exercises 5.2:

1. Mathieu's equation (5.20) can have a first eigenvalue λ_1 that is negative depending on the constant θ. Write out the Rayleigh quotient for Mathieu's equation. Now, $\phi = \sin(x)$ is in the space S'. Use this function and inequality (5.16) to get an upper bound for λ_1, and show that $\lambda_1 < 0$ whenever $\theta > 1$. Compare this with expression (5.22). (For the choice $\theta = 5$ the value of λ_1 is given in table 5.1 at the end of Section 5.3.)

2. **Halm's equation** is

$$(1 + x^2)^2 y''(x) + \lambda y(x) = 0.$$

Under the boundary conditions $y(0) = y(\pi) = 0$, find a lower bound for λ_1.

3. The **Titchmarsh equation** is

$$y''(x) + (\lambda - x^{2n})y(x) = 0,$$

where n is a nonnegative integer. Under the boundary conditions $y(0) = y(1) = 0$ show that the first eigenvalue λ_1 satisfies $\pi^2 < \lambda_1 < 11$. (The function $\phi = x(x - 1)$ can be used to get the upper bound.)

5.3 Higher Eigenvalues

The Rayleigh quotient can be used to frame a variational characterization of higher eigenvalues. The eigenfunctions for the Sturm-Liouville problem are mutually orthogonal, and this relationship can be exploited to give such a characterization. For example, it can be shown that the eigenvalue λ_2 corresponds to the minimum of R among functions in $y \in S'$ that also satisfy the orthogonality condition

$$\langle y, y_1 \rangle = 0,$$

where y_1 is an eigenfunction corresponding to λ_1. More generally, we have the following result the proof of which we omit.

Theorem 5.3.1 *Let y_k denote the eigenfunction associated with the eigenvalue λ_k, and let S'_n be the set of functions $y \in S'$ such that*

$$\langle y, y_k \rangle = 0 \tag{5.24}$$

for $k = 1, 2, \ldots, n - 1$. Then

$$\lambda_n = \min_{y \in S'_n} R(y). \tag{5.25}$$

The above theorem is of limited practical value because, in general, the eigenvalues $\lambda_1, \ldots, \lambda_{n-1}$ and corresponding eigenfunction $y_1, \ldots, y_{n-1}$ are not known explicitly. Constraints such as (5.24) require precise knowledge of the eigenfunctions as opposed to approximations. Fortunately, we can characterize higher eigenvalues with a "max-min" type principle involving the Rayleigh quotient, and circumvent the problem of finding eigenfunctions. The next results we state without proof. Some details can be found in Wan *op. cit.*, p. 284, and in Courant and Hilbert *op. cit.*, p. 406.

Lemma 5.3.2 *Let $z_1, \ldots, z_{n-1}$ be any functions in S' and let $\bar{\lambda}_n$ be the minimum of R subject to the $n - 1$ constraints*

$$\langle y, z_k \rangle = 0,$$

where $k = 1, \ldots n - 1$. Then

$$\bar{\lambda}_n \leq \lambda_n.$$

Lemma 5.3.2 is a key result used to establish the following "max-min" principle for higher eigenvalues.

Theorem 5.3.3 *Let Ω_{n-1} be the set of all functions $\mathbf{z} = (z_1, \ldots, z_{n-1})$ such that $z_k \in S'$ for $k = 1, \ldots, n - 1$. Then*

$$\lambda_n = \max_{\mathbf{z} \in \Omega_{n-1}} \{\bar{\lambda}_n(\mathbf{z})\},$$

where

$$\bar{\lambda}_n(\mathbf{z}) = \min_{y \in S'}\{R(y) : \langle y, z_k \rangle = 0, \quad k = 1, \ldots, n-1\}.$$

The "max-min" property of eigenvalues can be exploited to get a simple asymptotic estimate of the eigenvalues λ_n as $n \to \infty$. Note that the problem

$$(-py')' + qy - \lambda ry = 0,$$

$$y(0) = y(\pi) = 0,$$

can be converted into the problem

$$\phi''(t) - f(t)\phi(t) + \lambda\phi(t) = 0, \tag{5.26}$$

$$\phi(0) = \phi(\ell) = 0, \tag{5.27}$$

by the transformation

$$\phi = \sqrt[4]{rp}\,y, \quad t = \int_0^x \sqrt{\frac{r(\xi)}{p(\xi)}}\,d\xi, \quad \ell = \int_0^\pi \sqrt{\frac{r(x)}{p(x)}}\,dx.$$

Here, the function f is given by

$$f = \frac{g''}{g} + \frac{q}{r},$$

where $g = \sqrt[4]{rp}$. We can thus restrict our attention to the problem (5.26), (5.27).[5] The Rayleigh quotient for this problem is

$$R(\phi) = \frac{J(\phi)}{I(\phi)},$$

where

$$J(\phi) = \int_0^\ell \left(\phi'^2 + f(t)\phi^2\right)\,dt$$

and

$$I(\phi) = \int_0^\ell \phi^2\,dt.$$

Let

$$k = \max_{t \in [0,\ell]}|f(t)|, \tag{5.28}$$

and

---

[5] This formulation is called the **Liouville normal form** of the problem. Details on this transformation and extensions to more general intervals can be found in Birkhoff and Rota [9], p. 320.

$$J^+(\phi) = \int_0^\ell \left(\phi'^2 + k\phi^2\right) dt,$$

$$R^+(\phi) = \frac{J^+(\phi)}{I(\phi)},$$

$$J^-(\phi) = \int_0^\ell \left(\phi'^2 - k\phi^2\right) dt,$$

$$R^-(\phi) = \frac{J^-(\phi)}{I(\phi)}.$$

Then,

$$R^+(\phi) = \frac{\int_0^\ell \phi'^2 \, dt}{I(\phi)} + k,$$

$$R^-(\phi) = \frac{\int_0^\ell \phi'^2 \, dt}{I(\phi)} - k,$$

and, since $J^+(\phi) \geq J(\phi) \geq J^-(\phi)$,

$$R^+(\phi) \geq R(\phi) \geq R^-(\phi);$$

i.e.,

$$|\bar{R}(\phi) - R(\phi)| \leq k, \tag{5.29}$$

where

$$\bar{R}(\phi) = \frac{\int_0^\ell \phi'^2 \, dt}{I(\phi)}. \tag{5.30}$$

The Rayleigh quotient defined by equation (5.30) is associated with the Sturm-Liouville problem

$$\phi'' + \bar{\lambda}\phi = 0, \tag{5.31}$$

$$\phi(0) = \phi(\ell) = 0, \tag{5.32}$$

and the eigenvalues for this problem are given by

$$\bar{\lambda}_n = \frac{n^2\pi^2}{\ell^2}. \tag{5.33}$$

Inequality (5.29) indicates that $R(\phi)$ can differ from $\bar{R}(\phi)$ by no more than $\pm k$. By the "max-min" principle for higher eigenvalues we see that λ_n and $\bar{\lambda}_n$ can differ by no more than $\pm k$ and thus deduce the asymptotic relation

$$\lambda_n = \frac{n^2\pi^2}{\ell^2} + O(1), \tag{5.34}$$

as $n \to \infty$. The function f influences only the $O(1)$ term (a term that is bounded as $n \to \infty$); λ_n is approximately $n^2\pi^2/\ell^2$ for large values of n. If we return back to the original problem, the relation (5.34) can be recast as

n	n^2	λ_n
1	1	-5.790
2	4	2.099
3	9	9.236
4	16	16.648
5	25	25.511
6	36	36.359

Table 5.1. Eigenvalues for Mathieu's equation, $\theta = 5$

$$\lambda_n = n^2\pi^2 \left\{ \int_0^\pi \sqrt{\frac{r(x)}{p(x)}}\, dx \right\}^{-2} + O(1); \tag{5.35}$$

i.e.,

$$\lim_{n\to\infty} \frac{n^2}{\lambda_n} = \frac{1}{\pi^2} \left\{ \int_0^\pi \sqrt{\frac{r(x)}{p(x)}}\, dx \right\}^{2}. \tag{5.36}$$

Note that q does not influence the leading order behaviour for the asymptotic distribution of eigenvalues.

Equation (5.35), for example, predicts that the higher eigenvalues for Mathieu's equation (Example 5.2.1) are

$$\lambda_n = n^2 + O(1), \tag{5.37}$$

as $n \to \infty$. In fact, the approximation is not "too bad" for θ small even with n small (cf. Table 5.1).

In closing, we note that the results of this chapter can be extended for the general Sturm-Liouville boundary conditions (5.2). Some extensions can also be made to cope with singular Sturm-Liouville problems. The reader is directed to Courant and Hilbert, *op. cit.* Chapter 5, for a fuller discussion and a wealth of examples from mathematical physics.

Exercises 5.3:

1. For Mathieu's equation (5.20) show that $|\lambda_n - n^2| \le 2\theta$ for all n.
2. Determine a constant Λ such that for Halm's equation (Exercise 5.2-2)

$$\lambda_n = \Lambda n^2 + O(1).$$

Derive a number M such that $|\lambda_n - \Lambda n^2| \le M$ for all n.

6

Holonomic and Nonholonomic Constraints

6.1 Holonomic Constraints

A **holonomic**[1] constraint is a condition of the form

$$g(t, \mathbf{q}) = 0, \tag{6.1}$$

where $\mathbf{q} = (q_1, q_2, \ldots, q_n)$, $n \geq 2$, and g is a given function. In contrast, a **nonholonomic** constraint[2] is a condition of the form

$$g(t, \mathbf{q}, \dot{\mathbf{q}}) = 0.$$

The analysis underlying variational problems with holonomic constraints is noticeably simpler than that for problems with nonholonomic constraints. In this section we focus on holonomic constraints and postpone our discussion of nonholonomic constraints until the next section. For simplicity we consider the simplest case when $n = 2$.

Let J be a functional of the form

$$J(\mathbf{q}) = \int_{t_0}^{t_1} L(t, \mathbf{q}, \dot{\mathbf{q}}) \, dt,$$

and suppose that J has an extremum at $\mathbf{q}$ subject to the boundary conditions $\mathbf{q}(t_0) = \mathbf{q}_0$, $\mathbf{q}(t_1) = \mathbf{q}_1$, and the condition (6.1). For consistency we require that $g(t_0, \mathbf{q}_0) = 0$ and $g(t_1, \mathbf{q}_1) = 0$. We assume that L and g are smooth functions. We also make the assumption that

---

[1] The curious name for this type of constraint stems from the Greek word *holos* meaning whole or entire. In this context holonomic means "integrable." This type of constraint is also called "finite."

[2] Some authors use this term specifically to identify constraints that are not integrable, i.e., cannot be reduced to a holonomic condition. For our purposes we simply call any constraint involving derivatives nonholonomic.

$$\nabla g = \left(\frac{\partial g}{\partial q_1}, \frac{\partial g}{\partial q_2}\right) \neq \mathbf{0} \tag{6.2}$$

for the extremal $\mathbf{q}$ in the interval $[t_0, t_1]$. Given that $\nabla g \neq \mathbf{0}$, we could (at least in principle) solve equation (6.1) for one of the q_k.[3] We could thus apply the constraint immediately and reduce the problem to an unconstrained problem involving a single dependent variable. This approach, however, is fraught with the same problems as its finite-dimensional analogue discussed in Section 4.1. Fortunately, the Lagrange multiplier technique can be adapted to cope with these types of problems.

We seek a necessary condition on $\mathbf{q}$ for $J(\mathbf{q})$ to be an extremum. As before, we perturb $\mathbf{q}$ to get some nearby curve $\hat{\mathbf{q}} = \mathbf{q} + \epsilon\eta$ and use the condition $J(\hat{\mathbf{q}}) - J(\mathbf{q}) = O(\epsilon^2)$ to get a necessary condition. We assume that the q_k are in $C^2[t_0, t_1]$ for $k = 1, 2$. A function $\hat{\mathbf{q}} = (\hat{q}_1, \hat{q}_2)$ is called an **allowable variation** for the problem if $\hat{q}_k \in C^2[t_0, t_1]$, $\hat{\mathbf{q}}(t_0) = \mathbf{q}_0$, $\hat{\mathbf{q}}(t_1) = \mathbf{q}_1$, and $g(t, \hat{\mathbf{q}}) = 0$.

Our first concern is whether there are any allowable variations (apart from the trivial one $\hat{\mathbf{q}} = \mathbf{q}$). As with the isoperimetric problem, there may be rigid extremals. Let $\eta = (\eta_1, \eta_2)$. The conditions on an allowable variation require that the η_k be in the set $C^2[t_0, t_1]$ and that $\eta(t_0) = \eta(t_1) = \mathbf{0}$. In addition, we must also have that $g(t, \mathbf{q} + \epsilon\eta) = 0$. Now, $\mathbf{q}$ is a fixed function and $\nabla g \neq \mathbf{0}$ at $(t, \mathbf{q})$ for $t \in [t_0, t_1]$. Suppose for definiteness and simplicity that

$$\frac{\partial g}{\partial q_2} \neq 0 \tag{6.3}$$

for all $t \in [t_0, t_1]$. Then the implicit function theorem implies that the equation $g(t, \mathbf{q} + \epsilon\eta) = 0$ can be solved for η_2 in terms of η_1 and ϵ, provided $|\epsilon|$ is sufficiently small. The smoothness of the derivatives of g also ensures that η_2 is in the set $C^2[t_0, t_1]$. We can thus regard η_1 as an arbitrary function in $C^2[t_0, t_1]$ such that $\eta_1(t_0) = \eta_1(t_1) = 0$, and η_2 as the solution to the equation $g(t, \mathbf{q} + \epsilon\eta) = 0$. The implicit function theorem guarantees a unique solution to this equation. In particular, we know that at $t = t_0$ we have $\eta_1(t_0) = 0$ and that $(t_0, \mathbf{q}_0)$ is the solution to $g(t_0, \mathbf{q}_0 + \epsilon\eta) = 0$; hence, $\eta_2(t_0) = 0$. A similar argument can be framed to show that $\eta_2(t_1) = 0$. Thus we always have nontrivial allowable variations provided condition (6.3) is satisfied. This condition can be relaxed to condition (6.2), but we do not pursue this generalization.

Suppose that $\hat{\mathbf{q}}$ is an allowable variation. Since J is stationary at $\mathbf{q}$, the condition $J(\hat{\mathbf{q}}) - J(\mathbf{q}) = O(\epsilon^2)$ leads to the equation

$$\int_{t_0}^{t_1} \left\{ \left(\frac{\partial L}{\partial q_1} - \frac{d}{dx}\frac{\partial L}{\partial \dot{q}_1}\right)\eta_1 + \left(\frac{\partial L}{\partial q_2} - \frac{d}{dx}\frac{\partial L}{\partial \dot{q}_2}\right)\eta_2 \right\} dt = 0. \tag{6.4}$$

We cannot proceed as in Section 3.2 to deduce the Euler-Lagrange equations from the above expression because η_2 cannot be varied independently of η_1:

---

[3] We can use the implicit function theorem to assert the existence of such solutions.

these functions are connected by the constraint (6.1). Suppose that we choose some smooth but arbitrary function η_1 that satisfies the boundary conditions. The implicit function theorem indicates that for $|\epsilon|$ small there is a solution η_2 to the equation $g(t, \hat{\mathbf{q}}) = 0$, that depends on ϵ and η_1. For a fixed but arbitrary η_1 we can thus regard η_2 as a function of ϵ. Moreover, the implicit function theorem implies that, for $|\epsilon|$ sufficiently small, η_2 is a smooth function of ϵ. Now $g(t, \hat{\mathbf{q}}) = 0$, and thus

$$\frac{d}{d\epsilon} g(t, \hat{\mathbf{q}}) \Big|_{\epsilon=0} = \frac{\partial g}{\partial q_1} \eta_1 + \frac{\partial g}{\partial q_2} \eta_2 = 0. \tag{6.5}$$

(The term containing $d\eta_2/d\epsilon$ vanishes at $\epsilon = 0$.) Equations (6.3) and (6.5) therefore imply that

$$\eta_2 = -\frac{\frac{\partial g}{\partial q_1}}{\frac{\partial g}{\partial q_2}}. \tag{6.6}$$

The function L is assumed to be smooth; therefore, for any smooth function $\mathbf{q}$ the term

$$E_2(L) = \frac{d}{dx} \frac{\partial L}{\partial q_2} - \frac{\partial L}{\partial q_2}$$

is a continuous function of t. Since $\partial g/\partial q_2$ is also a continuous function of t and nonzero there exists a function λ such that

$$E_2(L) = \lambda(t) \frac{\partial g}{\partial q_2}. \tag{6.7}$$

Naturally, the function λ depends on $\mathbf{q}$.

We return now to equation (6.4). The η_2 term in the integral can be eliminated using equations (6.6) and (6.7). Specifically,

$$\int_{t_0}^{t_1} \left\{ \left(\frac{\partial L}{\partial q_1} - \frac{d}{dt} \frac{\partial L}{\partial \dot{q}_1} \right) \eta_1 + \left(\frac{\partial L}{\partial q_2} - \frac{d}{dt} \frac{\partial L}{\partial \dot{q}_2} \right) \eta_2 \right\} dt$$

$$= \int_{t_0}^{t_1} \left\{ \left(\frac{\partial L}{\partial q_1} - \frac{d}{dt} \frac{\partial L}{\partial \dot{q}_1} \right) \eta_1 - \lambda(t) \frac{\partial g}{\partial q_2} \eta_2 \right\} dt$$

$$= \int_{t_0}^{t_1} \left(\frac{\partial L}{\partial q_1} - \frac{d}{dt} \frac{\partial L}{\partial \dot{q}_1} + \lambda(t) \frac{\partial g}{\partial q_1} \right) \eta_1 \, dt$$

$$= 0.$$

We can now apply Lemma 2.2.2 and thus deduce that

$$\frac{\partial L}{\partial q_1} - \frac{d}{dt} \frac{\partial L}{\partial \dot{q}_1} + \lambda(t) \frac{\partial g}{\partial q_1} = 0. \tag{6.8}$$

Equations (6.7) and (6.8) provide two differential equations for the three unknown functions q_1, q_2, and λ. The constraint (6.1) provides the third equation. The function λ is also called a Lagrange multiplier. Equations (6.7) and (6.8) can be written in the compact form

$$\frac{d}{dt}\frac{\partial F}{\partial \dot{q}_k} - \frac{\partial F}{\partial q_k} = 0, \tag{6.9}$$

where $k = 1, 2$, and $F = L - \lambda g$.

The derivation of equations (6.9) has the merit of simplicity, yet it seems disappointingly narrow. The Lagrange multiplier has a tractable geometrical interpretation in finite-dimensional problems and even in the isoperimetric problem. Here, the approach seems somewhat destitute of geometry. In fact, there is a satisfactory geometrical interpretation available, but it requires certain concepts from differential geometry such as fibre bundles that would lead us astray from an introductory account. The reader is referred to Giaquinta and Hildebrandt [32] for a geometry-based proof of the Lagrange multiplier technique.

In summary, we have the following necessary condition.

Theorem 6.1.1 *Suppose that* $\mathbf{q} = (q_1, q_2)$ *is a smooth extremal for the functional* J *subject to the holonomic constraint* $g(t, \mathbf{q}) = 0$, *and that* $\nabla g(t, \mathbf{q}) \neq \mathbf{0}$ *for* $t \in [t_0, t_1]$. *Then there exists a function* λ *of* t *such that* $\mathbf{q}$ *satisfies the Euler-Lagrange equations* (6.9).

Example 6.1.1: Let

$$J(\mathbf{q}) = \int_0^{\pi/2} \sqrt{|\dot{\mathbf{q}}|^2 + 1}\, dt,$$

and

$$g(t, \mathbf{q}) = |\mathbf{q}|^2 - 1.$$

Find the extremals for J subject to the constraint $g(t, \mathbf{q}) = 0$ and the boundary conditions $\mathbf{q}(0) = (1, 0)$ and $\mathbf{q}(\pi/2) = (0, 1)$.

For this problem

$$F = \sqrt{|\dot{\mathbf{q}}|^2 + 1} - \lambda(t)\left(|\mathbf{q}|^2 - 1\right),$$

and the Euler-Lagrange equations are

$$\frac{d}{dt}\left(\frac{\dot{q}_1}{\sqrt{|\dot{\mathbf{q}}|^2 + 1}}\right) - 2\lambda(t)q_1 = 0,$$

$$\frac{d}{dt}\left(\frac{\dot{q}_2}{\sqrt{|\dot{\mathbf{q}}|^2 + 1}}\right) - 2\lambda(t)q_2 = 0.$$

Since the constraint requires that $|\mathbf{q}|^2 = 1$, we can use the substitution

$$q_1(t) = \cos\phi(t), \quad q_2(t) = \sin\phi(t).$$

Now $\dot{q}_1 = -\dot{\phi}\sin\phi$, $\dot{q}_2 = \dot{\phi}\cos\phi$, and hence $|\dot{\mathbf{q}}|^2 = \dot{\phi}^2$. The Euler-Lagrange equations in terms of ϕ are

$$\frac{d}{dt}\left(\frac{\dot{\phi}\sin\phi}{\sqrt{\dot{\phi}^2+1}}\right) + 2\lambda(t)\cos\phi = 0,$$

$$\frac{d}{dt}\left(\frac{\dot{\phi}\cos\phi}{\sqrt{\dot{\phi}^2+1}}\right) - 2\lambda(t)\sin\phi = 0,$$

and eliminating the function λ yields the equation

$$\sin\phi\frac{d}{dt}\left(\frac{\dot{\phi}\sin\phi}{\sqrt{\dot{\phi}^2+1}}\right) + \cos\phi\frac{d}{dt}\left(\frac{\dot{\phi}\cos\phi}{\sqrt{\dot{\phi}^2+1}}\right) = 0.$$

Integrating the left-hand side of the above equation by parts gives the relation

$$\frac{\dot{\phi}}{\sqrt{\dot{\phi}^2+1}} = const.;$$

i.e.,

$$\dot{\phi} = c_0$$

for some constant c_0. We thus have that

$$\phi(t) = c_0 t + c_1,$$

where c_1 is another constant of integration. The general solution of the Euler-Lagrange equation for the extremal $\mathbf{q}$ is therefore

$$q_1(t) = \cos(c_0 t + c_1),$$
$$q_2(t) = \sin(c_0 t + c_1).$$

The boundary condition $\mathbf{q}(0) = (1,0)$ indicates that $c_1 = 2n\pi$ for some integer n. The boundary condition $\mathbf{q}(\pi/2) = (0,1)$ implies that $c_0 = 4m+1$ for some integer m. The extremal is thus given by

$$\mathbf{q}(t) = (\cos(t), \sin(t)),$$

for $t \in [0, \pi/2]$. Note that $\nabla g = (-\sin(t), \cos(t)) \neq \mathbf{0}$ for all $t \in \mathbb{R}$.

Example 6.1.2: Simple Pendulum

The parametric equations for the motion of a simple pendulum of mass m and length ℓ can be derived using Lagrange multipliers. Let $\mathbf{q}(t) = (q_1(t), q_2(t))$ denote the position of the pendulum at time t. Here we associate q_2 with the vertical component of position. The motion of the pendulum from time t_0 to time t_1 is such that the functional

$$J(\mathbf{q}) = \int_{t_0}^{t_1} \left(\frac{m}{2} |\dot{\mathbf{q}}|^2 + gq_2 \right) dt$$

is an extremum subject to the condition[4]

$$q_1^2 + (q_2 - \ell)^2 - \ell^2 = 0. \tag{6.10}$$

Here, the term $m/2 |\dot{\mathbf{q}}|^2$ is the kinetic energy and the term gq_2, where g is the gravitation constant, is the potential energy. The Euler-Lagrange equations for the motion of a pendulum are thus

$$\ddot{q}_1 + 2\lambda(t)q_1 = 0,$$
$$\ddot{q}_2 - g + 2\lambda(t)(q_2 - \ell) = 0.$$

The method outlined in this section can be generalized in some obvious ways as was done for the isoperimetric problem. For example, we could include functionals depending on higher-order derivatives, multiple holonomic constraints, or functionals depending on n dependent variables, $n > 2$. We do not pursue these generalizations. The reader is referred to the literature ([12], [21], [27], [31], [32]) for details on these generalizations.

We close this section with a derivation of the equations for geodesics on a surface defined implicitly by $g(x, y, z) = 0$.

Example 6.1.3:   **Geodesics** Let g be a smooth function of the variables x, y, z. If $\nabla g \neq \mathbf{0}$, then an equation of the form

$$g(x, y, z) = 0 \tag{6.11}$$

describes a surface implicitly. For example, if $g(x, y, z) = x^2 + y^2 + z^2 - 1$, then equation (6.11) describes a sphere of radius 1 centred at the origin.

A general space curve γ of finite length is described (at least locally) by parametric equations of the form

$$\mathbf{r}(t) = (x(t), y(t)z(t)), \tag{6.12}$$

where $t \in [t_0, t_1]$. The arclength of γ is

$$J(x, y, z) = \int_{t_0}^{t_1} |\mathbf{r}'(t)| \, dt$$

$$= \int_{t_0}^{t_1} \sqrt{x'^2 + y'^2 + z'^2} \, dt.$$

---

[4] For the connoisseur of technical terms, constraints that do not involve time explicitly are called **scleronomic** in mechanics. Constraints that involve time explicitly are called **rheonomic**. Condition (6.10) can thus be called a scleronomic holonomic constraint. Need we say more?

Let Σ denote the surface described by equation (6.11) and let $\mathbf{P}_0$ and $\mathbf{P}_1$ be two distinct points on Σ. A geodesic on Σ from $\mathbf{P}_0$ to $\mathbf{P}_1$ is a curve on Σ with endpoints $\mathbf{P}_0$ and $\mathbf{P}_1$ such that the arclength is stationary. Assuming that such a curve can be represented by a (single) parametric function of the form (6.11) with $\mathbf{r}(t_0) = \mathbf{P}_0$ and $\mathbf{r}(t_1) = \mathbf{P}_1$, a geodesic is thus a curve such that the functional J is stationary subject to the constraint (6.11). Let

$$F = \sqrt{x'^2 + y'^2 + z'^2} - \lambda(t)g(x, y, z).$$

The (smooth) geodesics on Σ must therefore satisfy the Euler-Lagrange equations

$$\frac{d}{dt}\frac{x'}{\sqrt{x'^2 + y'^2 + z'^2}} + \lambda(t)\frac{\partial g}{\partial x} = 0,$$

$$\frac{d}{dt}\frac{y'}{\sqrt{x'^2 + y'^2 + z'^2}} + \lambda(t)\frac{\partial g}{\partial y} = 0,$$

$$\frac{d}{dt}\frac{z'}{\sqrt{x'^2 + y'^2 + z'^2}} + \lambda(t)\frac{\partial g}{\partial z} = 0.$$

Exercises 6.1:

1. **Geodesics on a Cylinder:** The equation $g(x, y, z) = x^2 + y^2 - 1 = 0$ defines a right circular cylinder. Use the multiplier rule to show that the geodesics on the cylinder are helices.
2. **Catenary on a Cylinder:** Let

$$J(\mathbf{q}) = \int_{t_0}^{t_1} q_3 \sqrt{\dot{q}_1^2 + \dot{q}_2^2 + \dot{q}_3^2} \, dt,$$

and

$$g(\mathbf{q}) = q_1^2 + q_2^2 - 1.$$

Find the extremals for J subject to the constraint $g(\mathbf{q}) = 0$ and boundary conditions of the form $\mathbf{q}(t_0) = \mathbf{q}_0$, $\mathbf{q}(t_1) = \mathbf{q}_1$.

6.2 Nonholonomic Constraints

In this section we discuss variational problems that have nonholonomic constraints. These problems are also called **Lagrange problems**. The Lagrange problem thus consists of determining the extrema for functionals of the form

$$J(\mathbf{q}) = \int_{t_0}^{t_1} L(t, \mathbf{q}, \dot{\mathbf{q}}) \, dt, \tag{6.13}$$

subject to the boundary conditions

$$\mathbf{q}(t_0) = \mathbf{q}_0, \quad \mathbf{q}(t_1) = \mathbf{q}_1, \tag{6.14}$$

and a condition of the form

$$g(t, \mathbf{q}, \dot{\mathbf{q}}) = 0. \tag{6.15}$$

It is clear that Lagrange problems include problems with holonomic constraints as a special case, but not every constraint of the form (6.15) can be integrated to yield a holonomic constraint. For example, suppose $n = 3$ and

$$g(t, \mathbf{q}, \dot{\mathbf{q}}) = P(\mathbf{q})\dot{q}_1 + Q(\mathbf{q})\dot{q}_2 + R(\mathbf{q})\dot{q}_3 = 0. \tag{6.16}$$

Then it is well known that this equation is integrable only if

$$P\left(\frac{\partial Q}{\partial q_3} - \frac{\partial R}{\partial q_2}\right) + Q\left(\frac{\partial R}{\partial q_1} - \frac{\partial P}{\partial q_3}\right) + R\left(\frac{\partial P}{\partial q_2} - \frac{\partial Q}{\partial q_1}\right) = 0 \tag{6.17}$$

(cf. [61], p. 140), and hence the constraint (6.16) can be converted into a holonomic one only for certain functions P, Q, R. For quasilinear nonholonomic constraints such as (6.16) the dimension n is crucial. If $n = 2$ and the constraint is of the form

$$P(\mathbf{q})\dot{q}_1 + Q(\mathbf{q})\dot{q}_2 = 0$$

then, in principle, this constraint can be reduced to a holonomic condition. For example, assuming $Q(\mathbf{q}) \neq 0$, we could recast the above constraint as an ordinary differential equation

$$\frac{dq_2}{dq_1} = -\frac{P(\mathbf{q})}{Q(\mathbf{q})},$$

and appeal to Picard's theorem to assert the existence of a solution $q_2(q_1)$ to this differential equation. If $n > 2$, then condition (6.17) is not generically satisfied for P, Q, R and hence the condition is not integrable.

Isoperimetric problems can also be converted into Lagrange problems. Suppose that the isoperimetric condition

$$I(\mathbf{q}) = \int_{t_0}^{t_1} g(t, \mathbf{q}, \dot{\mathbf{q}}) \, dt = \ell \tag{6.18}$$

is prescribed, where $\mathbf{q} = (q_1, \ldots, q_n)$. We can introduce a new variable q_{n+1} defined by

$$\dot{q}_{n+1} = g(t, \mathbf{q}, \dot{\mathbf{q}}) = 0, \tag{6.19}$$

along with any boundary conditions $q_{n+1}(t_0)$, $q_{n+1}(t_1)$ such that

$$q_{n+1}(t_1) - q_{n+1}(t_0) = \ell. \tag{6.20}$$

In this manner we can recast the isoperimetric problem as a Lagrange problem.

Problems that contain derivatives of order two or higher in the integrand can also be regarded as Lagrange problems. For instance, consider a basic variational problem that involves a functional of the form

$$J(y) = \int_{x_0}^{x_1} f(x, y, y', y'') \, dx.$$

In the notation of this section, let $t = x$, $q_1 = y$, $q_2 = y'$, and introduce the constraint

$$\dot{q}_1 - q_2 = 0.$$

This reformulation leads to a functional of the form (6.13) along with the above nonholonomic constraint.

The theory behind the Lagrange problem is well developed for problems involving one independent variable,[5] but the proof of the Lagrange multiplier rule for nonholonomic constraints is more complicated than that for isoperimetric or holonomic constraints. In addition, the application of the rule itself is awkward owing to the condition for an extremal to be normal. Some of the difficulties that surround the Lagrange problem concern the possibility of rigid extremals. Consider, for instance, the problem of finding extremals for the functional J defined by

$$J(y) = \int_{t_0}^{t_1} q_1 \sqrt{1 + \dot{q}_2^2} \, dt,$$

subject to the constraint

$$g(t, \mathbf{q}, \dot{\mathbf{q}}) = \dot{q}_1^2 + \dot{q}_2^2 = 0,$$

and the boundary conditions $\mathbf{q}(t_0) = \mathbf{q}_0$, $\mathbf{q}(t_1) = \mathbf{q}_1$. The only (real) solution to the constraint equation is $\dot{q}_1 = \dot{q}_2 = 0$, so that q_1 and q_2 are constant functions. If $\mathbf{q}(t_0) \neq \mathbf{q}(t_1)$ then there are no solutions that meet the constraint and the boundary conditions. If $\mathbf{q}(t_0) = \mathbf{q}(t_1)$, then the only solution to the constraint equation that satisfies the boundary conditions is $\mathbf{q} = \mathbf{q}(t_0)$, and in this case $J(\mathbf{q}) = q_1(0)(t_1 - t_0)$. In short, there are no arbitrary variations available for this problem because only one function satisfies the constraint and the boundary conditions.

In the remainder of this section we present without proof the Lagrange multiplier rule for nonholonomic constraints and limit our discussion to the

---

[5] The Lagrange problem for several independent variables is less complete (cf. Giaquinta and Hildebrandt [32]).

simplest cases and examples. Fuller accounts of the Lagrange problem resplendent with gory details can be found in [10], [12], [21], and [63].

We begin first with a general multiplier rule that includes the abnormal case.

Theorem 6.2.1   *Let J be the functional defined by (6.13), where $\mathbf{q} = (q_1, \ldots, q_n)$ and L is a smooth function of $t, \mathbf{q}$, and $\dot{\mathbf{q}}$. Suppose that J has an extremum at $\mathbf{q} \in C^2[t_0, t_1]$ subject to the boundary conditions (6.14) and the constraint (6.15), where g is a smooth function of $t, \mathbf{q}$, and $\dot{\mathbf{q}}$ such that $\partial g / \partial \dot{q}_j \neq 0$ for some j, $1 \leq j \leq n$. Then there exists a constant λ_0 and a function $\lambda_1(t)$ not both zero such that for*

$$K(t, \mathbf{q}, \dot{\mathbf{q}}) = \lambda_0 L(t, \mathbf{q}, \dot{\mathbf{q}}) - \lambda_1(t) g(t, \mathbf{q}, \dot{\mathbf{q}}),$$

$\mathbf{q}$ *is a solution to the system*

$$\frac{d}{dt} \frac{\partial K}{\partial \dot{q}_k} - \frac{\partial K}{\partial q_k} = 0, \qquad (6.21)$$

where $k = 1, \ldots, n$.

The above result includes the abnormal case, which corresponds to $\lambda_0 = 0$. In this case the function λ_1 is not identically zero on the interval $[t_0, t_1]$, and equation (6.21) implies that

$$\frac{d}{dt} \left(\lambda_1 \frac{\partial g}{\partial \dot{q}_k} \right) - \lambda_1 \frac{\partial g}{\partial q_k} = 0, \qquad (6.22)$$

for $k = 1, \ldots, n$. We thus see that if $\mathbf{q}$ is an extremal for the problem with $\lambda_0 = 0$ then λ_1 must be a *nontrivial* solution to (6.22). The existence of a nontrivial solution λ_1 thus characterizes the abnormal case. A smooth extremal $\mathbf{q}$ is thus called **abnormal** if there exists a nontrivial solution to system (6.22); otherwise, it is called **normal**. We have the following result for normal extremals.

Theorem 6.2.2 *Let J, q, L, and g be as in Theorem 6.2.1. If $\mathbf{q}$ is a normal extremal then there exists a function λ_1 such that $\mathbf{q}$ is a solution to the system*

$$\frac{d}{dt} \frac{\partial F}{\partial \dot{q}_k} - \frac{\partial F}{\partial q_k} = 0, \qquad (6.23)$$

where

$$F(t, \mathbf{q}, \dot{\mathbf{q}}) = L(t, \mathbf{q}, \dot{\mathbf{q}}) - \lambda_1(t) g(t, \mathbf{q}, \dot{\mathbf{q}}).$$

Moreover, λ_1 is uniquely determined by $\mathbf{q}$.

Note that, unlike the other constrained problems, the differential equations (6.23) will contain the term $\dot{\lambda}_1$; moreover, the condition (6.15) is a differential equation, so that solving problems with nonholonomic constraints entails solving a system of $n + 1$ differential equations.

Example 6.2.1:   Let

$$J(\mathbf{q}) = \int_{t_0}^{t_1} \left(\dot{q}_1^2 + \dot{q}_2^2 \right) \, dt.$$

Find the extremals for J subject to the boundary conditions (6.14) and the constraint

$$g(t, \mathbf{q}, \dot{\mathbf{q}}) = \dot{q}_1 + q_1 + q_2 = 0. \tag{6.24}$$

Usually, the practical approach to constrained problems is to first identify the candidates for extremals and then study whether the problem is in fact normal, i.e., proceed under the assumption that the problem is normal. For this simple problem, however, we can deduce readily that any extremals to the problem must be normal. Specifically, equation (6.22) for $k = 2$ gives

$$\lambda_1(t) \frac{\partial g}{\partial q_2} = \lambda_1(t) = 0,$$

which has only the trivial solution. We thus know in advance that we have only normal extremals.

Let

$$F(t, \mathbf{q}, \dot{\mathbf{q}}) = \dot{q}_1^2 + \dot{q}_2^2 - \lambda_1(t) \left(\dot{q}_1 + q_1 + q_2 \right).$$

Theorem 6.2.2 shows that if $\mathbf{q}$ is an extremal then it must satisfy the following system,

$$\dot{\lambda}_1 - \lambda_1 + 2q_1 = 0$$
$$\dot{\lambda}_1 - 2q_2 = 0.$$

These equations and the constraint (6.24) imply

$$\ddot{q}_1 - 2q_1 = 0;$$

i.e.,

$$q_1 = k_1 \sinh\left(\sqrt{2}t \right) + k_2 \cosh\left(\sqrt{2}t \right),$$

where k_1 and k_2 are constants. Hence,

$$q_2 = - \left(k_1 + k_2\sqrt{2} \right) \sinh\left(\sqrt{2}t \right) - \left(k_1\sqrt{2} + k_2 \right) \cosh\left(\sqrt{2}t \right),$$

and

$$\lambda_1 = -2 \left(\left(k_1 + k_2\sqrt{2} \right) \sinh\left(\sqrt{2}t \right) - \left(k_1\sqrt{2} + k_2 \right) \cosh\left(\sqrt{2}t \right) \right).$$

Example 6.2.2:   Catenary

Let us revisit the catenary problem, but this time as a Lagrange problem. Suppose that the length of the cable is ℓ and the endpoints are given by (x_0, y_0) and (x_1, y_1), where

$$(x - x_0)^2 + (y - y_0)^2 < \ell^2. \tag{6.25}$$

The potential energy functional is given by

$$J(y) = \int_0^\ell y \, ds,$$

where s denotes arclength. In order to ensure that s is arclength we need to add the constraint

$$x'^2 + y'^2 - 1 = 0$$

(cf. Section 1.2). Using the notation of this section, let $q_1 = x$, $q_2 = y$, and $s = t$. We thus seek an extremum for the functional

$$J(\mathbf{q}) = \int_0^\ell q_2 \, dt, \tag{6.26}$$

subject to the constraint

$$g(\mathbf{q}) = \dot{q}_1^2 + \dot{q}_2^2 - 1 = 0, \tag{6.27}$$

and the boundary conditions

$$\mathbf{q}(0) = (x_0, y_0), \quad \mathbf{q}(\ell) = (x_1, y_1). \tag{6.28}$$

We can show directly that any abnormal extremals to this problem must be lines. Suppose that there is a nontrivial solution to equation (6.22). Then there are constants c_1 and c_2 such that

$$\lambda_1 \dot{q}_1 = c_1,$$
$$\lambda_1 \dot{q}_2 = c_2.$$

The constraint (6.27) implies that

$$\lambda_1^2 = c_1^2 + c_2^2,$$

and consequently q_1 and q_2 must be linear functions of t. The boundary conditions (6.28) and the inequality (6.25), however, preclude linear solutions. Let

$$F = q_2 - \lambda_1(\dot{q}_1^2 + \dot{q}_2^2 - 1).$$

Equations (6.23) give

$$2\lambda_1 \dot{q}_1 = k_1, \tag{6.29}$$
$$2\lambda_1 \dot{q}_2 = t + k_2, \tag{6.30}$$

where k_1 and k_2 are constants. Equations (6.29), (6.30), and the constraint (6.27) thus give

$$\lambda_1 = \frac{1}{2}\sqrt{k_1^2 + (t + k_2)^2};$$

hence,

$$q_1 = \sinh^{-1}\left(\frac{t + k_2}{k_1}\right) + k_3$$

$$q_2 = \sqrt{k_1^2 + (t + k_2)^2} + k_4,$$

where k_3 and k_4 are constants. The familiar parametrization of the catenary in terms of the hyperbolic cosine can thus be recovered from the above expressions. Note that this problem is merely a reformulation of the isoperimetric problem so that the comments concerning the satisfaction of the boundary conditions (Example 4.2.1) still apply.

Exercises 6.2:

1. Let

$$J(\mathbf{q}) = \int_{t_0}^{t_1} \left(\dot{q}_1^2 + \dot{q}_2^2 + \dot{q}_3^2\right) dt,$$

and

$$g(t, \mathbf{q}, \dot{\mathbf{q}}) = t\dot{q}_1 + \dot{q}_2 + q_3 - 1.$$

Find the extremals for J subject to the constraint $g(t, \mathbf{q}, \dot{\mathbf{q}}) = 0$ and boundary conditions of the form $\mathbf{q}(t_0) = \mathbf{q}_0$, $\mathbf{q}(t_1) = \mathbf{q}_1$.

2. Let J be a functional of the form (6.13), where $n = 3$ and let g be of the form (6.16). Suppose that there exists a function $\mu(t, \mathbf{q})$ such that the nonconstant extremals for the constrained problem satisfy

$$\frac{d}{dt}\frac{\partial L}{\partial \dot{q}_k} - \frac{\partial L}{\partial q_k} = \mu(t, \mathbf{q})\frac{\partial g}{\partial \dot{q}_k},$$

for $k = 1, 2, 3$. Show that g must be integrable: i.e., g must satisfy equation (6.17).

6.3 Nonholonomic Constraints in Mechanics*

We digress briefly here to discuss the ticklish subject of nonholonomic constraints that occur in problems from classical mechanics. We must first set the record straight concerning our use of the term "nonholonomic." The mechanics connoisseur is doubtless affronted by our slovenly use of this term as a label for any constraint given as a differential equation. From our perspective, it is a handy catch-all term for such constraints, and it has the pleasing merit that we need not continually distinguish nonintegrable from integrable constraints. Mathematically, Theorem 6.2.1 is valid for integrable and nonintegrable constraints alike, so the distinction is not important. From a mechanics perspective, however, the term is always used in its pure sense: a nonholonomic condition is a differential equation (or system of differential equations) that is *not* integrable. One cannot, even in principle, convert such a constraint to a holonomic one without essentially solving the problem first. The distinction in mechanics is important not so much for mathematical reasons, but for physical reasons: a more general variational principle is needed to derive the equations of motion for problems with nonholonomic constraints in mechanics.

Typically, nonholonomic conditions in mechanics are of the form

$$g_k(t, \mathbf{q}, \dot{\mathbf{q}}) = \sum_{j=1}^{n} a_{jk}(t, \mathbf{q})\dot{q}_j, \tag{6.31}$$

i.e., linear in the generalized velocities. Such constraints arise, for example, as "no slip" conditions for rolling objects. For instance, the problem of a penny rolling on the horizontal xy-plane such that the disc is always vertical has constraints of the form

$$\dot{x} - a\sin\theta\dot{\phi} = 0,$$
$$\dot{y} + a\cos\theta\dot{\phi} = 0,$$

where a is a constant, θ is the angle between the axis of the disc and the x-axis, and ϕ is the angle of rotation about the disc axis. A constraint of this form cannot be reduced to a holonomic one.

Given the prominence of nonholonomic constraints in mechanics, the reader might wonder why we have studiously avoided them in the previous section. The direct answer is that the "no frills" version of Hamilton's Principle given in Section 1.3 is generally not applicable to these problems. The appropriate principle for these problems is **d'Alembert's Principle**, which states that the total virtual work of the forces is zero for all (reversible) variations that satisfy the given kinematical conditions. Here, the forces include impressed forces along with inertial forces (forces resulting from a mass in accelerated motion).

Loosely speaking, we can think of d'Alembert's Principle as the condition $\delta L = 0$. Hamilton's Principle comes from d'Alembert's Principle by integration with respect to time. For holonomic problems we have

$$\int_{t_0}^{t_1} \delta L\, dt = \delta \int_{t_0}^{t_1} L\, dt,$$

but, as Pars [59] (p. 528) points out, for nonholonomic problems

$$\int_{t_0}^{t_1} \delta L \, dt \neq \delta \int_{t_0}^{t_1} L \, dt, \tag{6.32}$$

in general.

For nonholonomic problems with m constraints of the form (6.31) d'Alembert's Principle yields equations of the form

$$\frac{d}{dt} \frac{\partial L}{\partial \dot{q}_j} - \frac{\partial L}{\partial q_j} = \sum_{k=1}^{m} \mu_k(\mathbf{q}) a_{kj}(t, \mathbf{q}), \tag{6.33}$$

where L is the (unmodified) Lagrangian, i.e., $T - V$, and the functions μ_k are multipliers to be determined along with $\mathbf{q}$ using the n Euler-Lagrange equations and the m differential equations (6.31). In general, the system (6.33) is not equivalent to the Euler-Lagrange equations of Theorem 6.2.1 using the modified Lagrangian $L - (\lambda_1 g_1 + \cdots + \lambda_m g_m)$, because this approach assumes that Hamilton's Principle is valid. Pars (*loc. cit.*) gives an insightful discussion of Hamilton's Principle as it relates to nonholonomic problems and gives a simple concrete example to illustrate relation (6.32). The rolling penny and other nonholonomic problems are treated in detail by Pars (*op. cit.*), Webster [72], and Whittaker [73].

There appear to be divergent streams of thought regarding the rôle of Hamilton's Principle in nonholonomic problems. Goldstein [35] and others maintain that Hamilton's Principle can be extended to cover nonholonomic problems; Rund [63] states that such a principle is not applicable to nonholonomic problems. The confusion of opinion on this matter is in no small part due to different interpretations of Hamilton's Principle and the use of the term nonholonomic. We end this section with the following quote from Goldstein (*op. cit.*, p. 49) that perhaps brings the real issue into perspective.

In view of the difficulties in formulating a variational principle for nonholonomic systems, and the relative ease with which the equations of motion can be obtained directly, it is natural to question the usefulness of the variational approach in this case.

7

Problems with Variable Endpoints

7.1 Natural Boundary Conditions

The fixed endpoint variational problem entails finding the extremals for a functional subject to a given set of boundary conditions. For a functional J of the form

$$J(y) = \int_{x_0}^{x_1} f(x, y, y') \, dx \qquad (7.1)$$

these boundary conditions take the form $y(x_0) = y_0$, $y(x_1) = y_1$, where y_0 and y_1 are specified numbers. If the functional contains higher order derivatives, then more boundary conditions are required. Variational problems arising in physics and geometry, however, are not always accompanied by the appropriate number of boundary conditions. For example, the shape of a cantilever beam is such that the potential energy is minimum. At the clamped end of the beam we have boundary conditions of the form $y(0) = 0$ and $y'(0) = 0$ reflecting the nature of the support. At the free end, however, there are no conditions imposed on y. Indeed, it is part of the problem to determine y and y' at this end. Now, the differential equation describing the shape of the beam is of fourth order, and four boundary conditions are thus required for uniqueness. We expect a unique solution to the problem and hence there must be some boundary data implicit in the variational formulation of the problem. We discuss this problem further in Example 7.1.3.

One of the striking features of calculus of variations is that the methods always supply exactly the right number of boundary conditions. There are essentially two types of boundary conditions. There are boundary conditions that are imposed on the problem such as those at the clamped end of the beam, and there are boundary conditions that arise from the variational process in lieu of imposed conditions. The latter type of boundary condition is called a natural boundary condition. Even if no boundary conditions are imposed, the process takes care of itself and, as we show, the condition that the functional be stationary leads to precisely the correct number of boundary conditions for the problem.

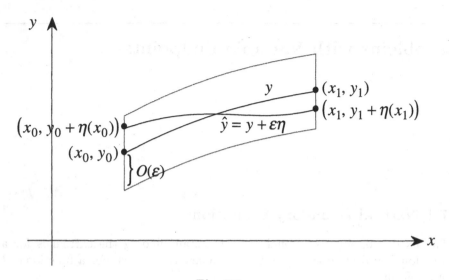

Fig. 7.1.

Let $J : C^2[x_0, x_1] \to \mathbb{R}$ be a functional of the form (7.1), where f is a smooth function. We consider the problem of determining the functions $y \in C^2[x_0, x_1]$ such that J has an extremum. No boundary conditions will be imposed on y. In this section, we derive a necessary condition for J to have an extremum at y.

Suppose that J has an extremum at y. We can proceed as in Section 2.2 by considering the value of J at a "nearby" function $\hat{y}$. Let

$$\hat{y} = y + \epsilon\eta,$$

where ϵ is a small parameter and $\eta \in C^2[x_0, x_1]$. Since no boundary conditions are imposed, we do not require η to vanish at the endpoints (Figure 7.1). Following the analysis of Section 2.2, the condition that $J(\hat{y}) - J(y)$ be of order ϵ^2 as $\epsilon \to 0$ leads to the condition

$$\int_{x_0}^{x_1} \left(\eta \frac{\partial f}{\partial y} + \eta' \frac{\partial f}{\partial y'} \right) dx = 0$$

(cf. equation (2.6)), and integrating the term containing η' by parts gives the condition

$$\eta \frac{\partial f}{\partial y'} \bigg|_{x_0}^{x_1} + \int_{x_0}^{x_1} \eta \left(\frac{\partial f}{\partial y} - \frac{d}{dx} \frac{\partial f}{\partial y'} \right) dx = 0. \tag{7.2}$$

For the fixed endpoint problem the term $\eta \, \partial f/\partial y'$ vanished at the endpoints because $\eta(x_0) = \eta(x_1) = 0$. For the present problem this term does not vanish for all η under consideration. Nonetheless, equation (7.2) must be satisfied for all $\eta \in C^2[x_0, x_1]$, and in particular the subclass H of functions that do vanish at the endpoints. Since equation (7.2) must be satisfied for all $\eta \in H$ the arguments of Section 2.2 apply and therefore

$$\frac{d}{dx}\frac{\partial f}{\partial y'} - \frac{\partial f}{\partial y} = 0, \tag{7.3}$$

for any y at which J has an extremum. Equation (7.2) must be satisfied for all $\eta \in C^2[x_0, x_1]$, however, and this includes functions that do not vanish at the endpoints; consequently, equations (7.2) and (7.3) imply that

$$\eta \frac{\partial f}{\partial y'}\bigg|_{x_1} - \eta \frac{\partial f}{\partial y'}\bigg|_{x_0} = 0, \tag{7.4}$$

for all $\eta \in C^2[x_0, x_1]$. Now we can always find functions in $C^2[x_0, x_1]$ that vanish at x_0 but not at x_1. We must therefore have that

$$\frac{\partial f}{\partial y'}\bigg|_{x_1} = 0. \tag{7.5}$$

Similarly, we can find functions that vanish at x_1 but not at x_0. This observation leads to the condition

$$\frac{\partial f}{\partial y'}\bigg|_{x_0} = 0. \tag{7.6}$$

In summary, if J has an extremum at $y \in C^2[x_0, x_1]$, and there are no imposed boundary conditions, then y must satisfy the Euler-Lagrange equation (7.3) along with equations (7.5) and (7.6). Equations (7.5) and (7.6) are relations involving y and its derivatives at the endpoints; i.e., they are boundary conditions. Because these conditions arise in the variational formulation of the problem and not from considerations outside the functional, they are called **natural boundary conditions**.

The above process is completely "modular" in the sense that if boundary conditions are imposed at each end, then the variational formulation requires η to vanish at the endpoints, and thus there are no natural boundary conditions. If only one boundary condition is imposed at say x_0, then η is required to vanish at x_0 but not at x_1; hence, the problem is supplemented by the natural boundary condition (7.5). If no boundary conditions are imposed, then we have both natural boundary conditions.

Example 7.1.1: Determine a function y such that the functional

$$J(y) = \int_0^1 \sqrt{1 + y'^2}\, dx$$

is an extremum.

Geometrically, the above problem corresponds to the problem of finding a curve γ with one endpoint on the line $x = 0$ and the other on the line $x = 1$ such that the arclength of γ is an extremum. Intuitively, we see that any function of the form $y = const.$ will produce a curve of minimum arclength. Let us see if the natural boundary conditions lead us to this conclusion.

Any extremal to the problem must satisfy the Euler-Lagrange equation (7.3). From Example 2.2.1 we know that solutions of this equation are of the form $y = mx + b$, i.e., line segments. No boundary conditions are imposed on the problem and hence the natural boundary conditions (7.5) and (7.6) must be satisfied. Now, for $y = mx + b$,

$$\frac{\partial f}{\partial y'} = \frac{y'}{\sqrt{1 + y'^2}} = \frac{m}{\sqrt{1 + m^2}},$$

so that the natural boundary conditions are satisfied only if $m = 0$. This means that $y = b$, where there is no restriction on the value of the constant b.

Example 7.1.2:   Catenary
Suppose that we revisit the catenary problem but impose only one boundary condition. We thus seek to find a function y such that the functional

$$J(y) = \int_0^1 y\sqrt{1 + y'^2}\, dx$$

is an extremum subject to the boundary condition $y(0) = h > 0$.

The general solution to the Euler-Lagrange equation for this functional was determined in Example 2.3.3. Hence we know that y is of the form

$$y(x) = \kappa_1 \cosh\left(\frac{x}{\kappa_1} + \kappa_2\right),$$

where κ_1 and κ_2 are constants. The boundary condition $y(0) = h$ implies that

$$h = \kappa_1 \cosh(\kappa_2).$$

No boundary condition has been imposed at $x = 1$; consequently, y must satisfy the natural boundary condition (7.5). Therefore,

$$\frac{\partial f}{\partial y'}\bigg|_{x=1} = \frac{y(1)y'(1)}{\sqrt{1 + y'^2(1)}} = 0.$$

Since $h > 0$, $\kappa_1 \neq 0$ and consequently $y(1) \neq 0$. We must therefore have that $y'(1) = 0$; i.e.,

$$\sinh\left(\frac{1}{\kappa_1} + \kappa_2\right) = 0,$$

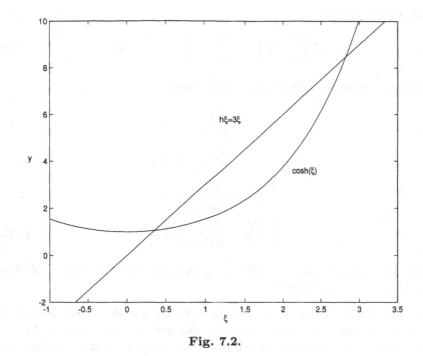

y

$h\xi = 3\xi$

$\cosh(\xi)$

ξ

Fig. 7.2.

so that $\kappa_2 = -1/\kappa_1$. The extremal is thus

$$y(x) = \kappa_1 \cosh(\frac{x - 1}{\kappa_1}).$$

Now the condition $y(0) = h$ implies that

$$h\xi = \cosh \xi,$$

where $\xi = 1/\kappa_1$. The above equation has two solutions provided h is sufficiently large (cf. Figure 7.2).

The natural boundary conditions for a functional involving higher-order
derivatives can be obtained using arguments similar to those used in the
derivation of equations (7.5) and (7.6). For example, if J is of the form

$$J(y) = \int_{x_0}^{x_1} f(x, y, y', y'') \, dx,$$

and no boundary conditions are imposed at x_0 and x_1, then it can be shown
that any smooth extremal y must satisfy the Euler-Lagrange equation

$$\frac{d^2}{dx^2} \frac{\partial f}{\partial y''} - \frac{d}{dx} \frac{\partial f}{\partial y'} + \frac{\partial f}{\partial y} = 0, \tag{7.7}$$

along with the condition

$$\eta' \frac{\partial f}{\partial y''} - \eta \left(\frac{d}{dx} \frac{\partial f}{\partial y''} - \frac{\partial f}{\partial y'} \right) \Big|_{x_0}^{x_1} = 0. \tag{7.8}$$

Equation (7.8) spawns the four boundary conditions

$$\frac{\partial f}{\partial y''} \Big|_{x_0} = 0, \tag{7.9}$$

$$\frac{\partial f}{\partial y''} \Big|_{x_1} = 0, \tag{7.10}$$

$$\frac{d}{dx} \frac{\partial f}{\partial y''} - \frac{\partial f}{\partial y'} \Big|_{x_0} = 0, \tag{7.11}$$

$$\frac{d}{dx} \frac{\partial f}{\partial y''} - \frac{\partial f}{\partial y'} \Big|_{x_1} = 0, \tag{7.12}$$

for the fourth-order ordinary differential equation (7.7). The proof of these assertions is left as an exercise.

Example 7.1.3:   We apply the above results to the study of small deflections of a beam of length ℓ having uniform cross section under a load.[1] Let y : $[0, \ell] \to \mathbb{R}$ describe the shape of the beam and $\rho : [0, \ell] \to \mathbb{R}$ be the load per unit length on the beam. Assuming small deflections, the potential energy from elastic forces is

$$V_1 = \frac{\kappa}{2} \int_0^\ell y''^2 \, dx,$$

where κ is a nonzero constant. The potential energy from gravitational forces is

$$V_2 = - \int_0^\ell \rho y \, dx.$$

The total potential energy is thus

$$J(y) = \int_0^\ell \left(\frac{\kappa y''^2}{2} - \rho y \right) dx.$$

The shape of the beam is such that J is a minimum; therefore, y must satisfy the Euler-Lagrange equation (7.7), and this produces the equation

$$y^{(iv)}(x) = \frac{\rho(x)}{\kappa}. \tag{7.13}$$

Equation (7.13) has the general solution

$$y(x) = P(x) + c_3 x^3 + c_2 x^2 + c_1 x + c_0, \tag{7.14}$$

---

[1] This example is based on one given in Lanczos [48], p. 70.

where $P^{(iv)}(x) = \rho(x)/\kappa$, and the c_ks are constants of integration. The total impressed force on the beam is

$$F = \int_0^\ell \rho(x)\, dx,$$

and the differential equation (7.13) implies that

$$\kappa\left(y'''(\ell) - y'''(0)\right) = F. \tag{7.15}$$

The terms $y'''(0)$ and $y'''(\ell)$ can be interpreted as the reaction forces at $x = 0$ and $x = \ell$, respectively, to keep the beam in equilibrium. The moment (torque) produced by the impressed force is

$$M = \int_0^\ell x\rho(x)\, dx,$$

and hence

$$M = \kappa \int_0^\ell xy^{(iv)}(x)\, dx = \kappa\left(\ell y'''(\ell) - y''(\ell) + y''(0)\right). \tag{7.16}$$

If we sum the moments at the $x = 0$ end of the beam, the term $\ell y'''(\ell)$ is the moment produced by the reaction force at $x = \ell$, and the terms $\kappa y''(0)$ and $\kappa y''(\ell)$ can be interpreted as the reaction moments at $x = 0$ and $x = \ell$, respectively.

Having made a physical interpretation of the higher derivatives of y at the endpoints, we now examine the problem under a variety of boundary conditions corresponding to support systems for the beam.

Case I: Double Clamped Beam
Suppose that the beam is clamped at each end (figure 7.3). The beam is "fixed in the wall" at each end so that at $x = 0$ we have $y(0) = y'(0) = 0$, and at $x = \ell$ we have $y(\ell) = y'(\ell) = 0$. Here, there are four imposed boundary conditions. All the allowable variations in this problem require that $\eta(0) = \eta'(0) = 0$ and $\eta(\ell) = \eta'(\ell) = 0$ so that no natural boundary conditions arise.

Case II: Cantilever Beam
Suppose that the beam is clamped at $x = 0$ (figure 7.4). Then the boundary conditions $y(0) = y'(0) = 0$ are imposed. No boundary conditions are imposed at the other end of the beam and, consequently, the natural boundary conditions (7.10) and (7.12) must be satisfied.[2] Equation (7.10) yields the relation

$$\left.\frac{\partial f}{\partial y''}\right|_{x=\ell} = \kappa y''(\ell) = 0,$$

---

[2] The assumption here is made that the unclamped endpoint of the beam still lies on the line $x = \ell$ (small deflections).

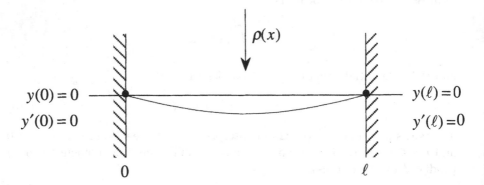

Fig. 7.3.

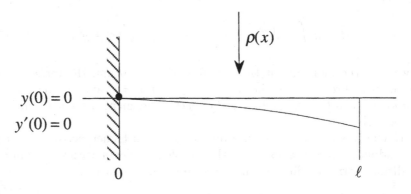

Fig. 7.4.

which states that the reaction moment at $x = \ell$ is zero. Equation (7.12) gives the condition

$$\frac{d}{dx}\frac{\partial f}{\partial y''} - \frac{\partial f}{\partial y'} = \kappa y'''(\ell) = 0,$$

which states that the reaction force at $x = \ell$ is zero. In view of the nature of the cantilever support, the natural boundary conditions reflect the physically evident situation at $x = \ell$ required for equilibrium.

Case III: Simply Supported Beam
Suppose that the beam is pinned at the ends, but no restrictions are made concerning the derivatives of y at the endpoints (figure 7.5). The imposed boundary conditions are $y(0) = 0$ and $y(\ell) = 0$. No restrictions are made on the values of η' at the endpoints, and hence we have the natural boundary

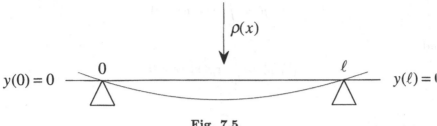

$$\rho(x)$$

$$y(0) = 0 \qquad\qquad\qquad\qquad\qquad\qquad\qquad y(\ell) = ($$

Fig. 7.5.

conditions (7.9) and (7.10). The natural boundary conditions are $y''(0) = 0$ and $y''(\ell) = 0$; i.e., the reaction moments at each endpoint are zero.

Case IV: Unsupported Beam

Suppose that the beam is unsupported. Then there are no imposed boundary conditions and we need all four natural boundary conditions. The natural boundary conditions are $y''(0) = 0$, $y'''(0) = 0$, $y''(\ell) = 0$, and $y'''(\ell) = 0$. These conditions state that the reaction force and moment at each end of the beam are zero. Note that the boundary conditions also imply

$$F = \int_0^\ell \rho(x)\, dx = 0, \tag{7.17}$$

by equation (7.15), and

$$M = \int_0^\ell x\rho(x)\, dx = 0, \tag{7.18}$$

by equation (7.16). But the function $\rho(x)$ is prescribed and may or may not satisfy equations (7.17) and (7.18). The natural boundary conditions thus tell us that the problem has a solution only if ρ is such that the total impressed force and total impressed moment is zero. Again, the natural boundary conditions lead us to physically sensible requirements.

The unsupported beam affords a glimpse of a result known as the **Fredholm alternative**. The Fredholm alternative is usually encountered in the context of integral equations, but it is a general result applicable to linear operators. Briefly, the major mathematical difference between Case IV and the earlier cases is that the homogeneous equation $y^{(iv)}(x) = 0$ with the given boundary conditions has only the trivial solution in the first three cases, whereas, in the fourth case, there are nontrivial solutions of the form $y(x) = c_1 x + c_0$ available.[3] In particular, we have the nontrivial solutions $Y_1 = x$ and $Y_0 = 1$. The Fredholm alternative states that the original boundary-value problem will have solutions only if ρ is orthogonal to both Y_0 and Y_1; i.e.,

---

[3] In general, the Fredholm alternative is concerned with solutions to the *adjoint* of the homogeneous equation. Here, the linear operator is self-adjoint. See Hochstadt [39] or Kreyszig [46] for more details on the Fredholm alternative.

$$\langle Y_0, \rho \rangle = \int_0^\ell \rho(x)\, dx = 0$$

and

$$\langle Y_1, \rho \rangle = \int_0^\ell x\rho(x)\, dx = 0.$$

Exercises 7.1:

1. For the brachystochrone problem (Example 2.3.4), let $x_0 = 0$, $x_1 = 1$. Given the condition $y(0) = 1$ show that the extremal satisfies the condition $y'(1) = 0$ and find an implicit equation for $y(1)$.

2. A simplified version of the Ramsey growth model in economics concerns a functional of the form

$$J(M) = \int_0^T c_1 \left(c_2 M(t) - M'(t) - c_3 \right)^2 dt.$$

   Here, J corresponds to the "total product," M is the capital, and the c_k are positive constants. The problem is to find the best use of capital such that J is minimized in a given planning period $[0, T]$. Now, the initial capital $M(0) = M_0$ is known, but the final capital $M(T)$ is not prescribed. Use the natural boundary conditions to find the extremal for J and the final capital $M(T)$.

3. Derive the natural boundary conditions (7.9) to (7.12) for functionals that involve second-order derivatives.

4. Let $\mathbf{q} = (q_1, \ldots, q_n)$ and

$$J(\mathbf{q}) = \int_{t_0}^{t_1} L(t, \mathbf{q}, \dot{\mathbf{q}})\, dt.$$

   Derive the natural boundary conditions that an extremal must satisfy if neither $\mathbf{q}(t_0)$ nor $\mathbf{q}(t_1)$ are prescribed.

7.2 The General Case

In the last section we considered problems where perhaps no boundary conditions are prescribed. Although the variations need not satisfy the same conditions at the endpoints, the x coördinates of the endpoints remained fixed (cf. figure 7.1). Even this restriction is not suitable for certain variational problems. In this section we consider the general case where both the independent and the dependent endpoint coördinates may be variable.

Let $y : [x_0, x_1] \to \mathbb{R}$ be a smooth function that describes a curve γ with endpoints $\mathbf{P}_0 = (x_0, y_0)$ and $\mathbf{P}_1 = (x_1, y_1)$, and let $\hat{y} : [\hat{x}_0, \hat{x}_1] \to \mathbb{R}$ be

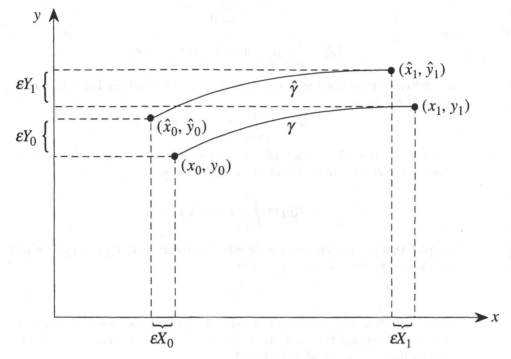

Fig. 7.6.

a smooth function that describes a curve $\hat{\gamma}$ with endpoints $\hat{\mathbf{P}}_0 = (\hat{x}_0, \hat{y}_0)$ and $\hat{\mathbf{P}}_1 = (\hat{x}_1, \hat{y}_1)$ (figure 7.6). For the ensuing analysis we wish to compare curves that are "close" to each other; however, the functions y and $\hat{y}$ are not necessarily defined on the same interval, and the norms discussed in Section 2.2 are not suitable. We can nonetheless extend the definitions of y and $\hat{y}$ so that they are defined over a common interval. Let $\tilde{x}_0 = \min\{x_0, \hat{x}_0\}$ and $\tilde{x}_1 = \max\{x_1, \hat{x}_1\}$. As we are interested in small variations on y, we can extend the functions y and $\hat{y}$ to the common interval $[\tilde{x}_0, \tilde{x}_1]$ using a Taylor series approximation where necessary. For example, if $\tilde{x}_0 = x_0$ and $x_1 < \tilde{x}_1$, then we can extend the definition of y as follows,

$$
y^*(x) = \begin{cases} y, & \text{if } x \in [x_0, x_1] \\ y(x_1) + (x - x_1)y'(x_1) + \frac{(x-x_1)^2}{2}y''(x_1), & \text{if } x \in (x_1, \tilde{x}_1], \end{cases}
$$

to get a function $y^* \in C^2[\tilde{x}_0, \tilde{x}_1]$. We assume that all such extensions have been made and retain the symbols y and $\hat{y}$. We define the distance between y and $\hat{y}$ as

$$
d(y, \hat{y}) = \|y - \hat{y}\| + |\mathbf{P}_0 - \hat{\mathbf{P}}_0| + |\mathbf{P}_1 - \hat{\mathbf{P}}_1|,
$$

where $|\mathbf{P}_k - \hat{\mathbf{P}}_k| = \sqrt{(x_k - \hat{x}_k)^2 + (y_k - \hat{y}_k)^2}$. Here, $\|\cdot\|$ is the norm defined by

$$\|y\| = \sup_{x \in [\tilde{x}_0, \tilde{x}_1]} |y(x)|,$$

or

$$\|y\| = \sup_{x \in [\tilde{x}_0, \tilde{x}_1]} |y(x)| + \sup_{x \in [\tilde{x}_0, \tilde{x}_1]} |y'(x)|,$$

whichever is appropriate to the problem under consideration. Let J be a functional of the form

$$J(y) = \int_{x_0}^{x_1} f(x, y, y') \, dx,$$

where f is a smooth function of x, y, and y'. The integration limits for this functional now depend on the choice of function; e.g.,

$$J(\hat{y}) = \int_{\hat{x}_0}^{\hat{x}_1} f(x, \hat{y}, \hat{y}') \, dx.$$

Suppose that J is stationary at y. Briefly, this means that $J(\hat{y}) - J(y) = O(\epsilon^2)$ whenever $d(\hat{y}, y) = O(\epsilon)$ as $\epsilon \to 0$. Let

$$\hat{y} = y + \epsilon \eta,$$

where $\eta \in C^2[\tilde{x}_0, \tilde{x}_1]$. No conditions aside from this smoothness condition are prescribed on η, but the condition $d(\hat{y}, y) = O(\epsilon)$ requires that the quantities $\hat{x}_k - x_k$ and $\hat{y}_k - y_k$ be of order ϵ. Let

$$\hat{x}_k = x_k + \epsilon X_k,$$
$$\hat{y}_k = y_k + \epsilon Y_k,$$

for $k = 0, 1$. Then,

$$J(\hat{y}) - J(y) = \int_{\hat{x}_0}^{\hat{x}_1} f(x, \hat{y}, \hat{y}') \, dx - \int_{x_0}^{x_1} f(x, y, y') \, dx$$

$$= \int_{x_0 + \epsilon X_0}^{x_1 + \epsilon X_1} f(x, \hat{y}, \hat{y}') \, dx - \int_{x_0}^{x_1} f(x, y, y') \, dx$$

$$= \int_{x_0}^{x_1} (f(x, \hat{y}, \hat{y}') - f(x, y, y')) \, dx + \int_{x_1}^{x_1 + \epsilon X_1} f(x, \hat{y}, \hat{y}') \, dx$$

$$- \int_{x_0}^{x_0 + \epsilon X_0} f(x, \hat{y}, \hat{y}') \, dx.$$

Using the arguments of Section 7.1 we have that

$$\int_{x_0}^{x_1} (f(x, \hat{y}, \hat{y}') - f(x, y, y')) \, dx = \epsilon \left\{ \eta \frac{\partial f}{\partial y'} \Big|_{x_0}^{x_1} + \int_{x_0}^{x_1} \eta \left(\frac{\partial f}{\partial y} - \frac{d}{dx} \frac{\partial f}{\partial y'} \right) dx \right\}$$
$$+ O(\epsilon^2),$$

and since ϵ is small

$$\int_{x_1}^{x_1+\epsilon X_1} f(x,\hat{y},\hat{y}')\,dx = \epsilon X_1 f(x,y,y')\Big|_{x_1} + O(\epsilon^2)$$

$$\int_{x_0}^{x_1+\epsilon X_0} f(x,\hat{y},\hat{y}')\,dx = \epsilon X_0 f(x,y,y')\Big|_{x_0} + O(\epsilon^2).$$

We therefore have

$$J(\hat{y}) - J(y) = \epsilon \left\{ \eta \frac{\partial f}{\partial y'}\Big|_{x_0}^{x_1} + \int_{x_0}^{x_1} \eta \left(\frac{\partial f}{\partial y} - \frac{d}{dx}\frac{\partial f}{\partial y'} \right) dx \right.$$

$$\left. + X_1 f(x,y,y')\Big|_{x_1} - X_0 f(x,y,y')\Big|_{x_0} \right\}$$

$$+ O(\epsilon^2). \tag{7.19}$$

The variations at the endpoint (x_0, y_0) must satisfy the compatibility condition

$$\hat{y} = y(\hat{x}_0) = y(x_0 + \epsilon X_0) + \epsilon\eta(x_0 + \epsilon X_0)$$
$$= y_0 + \epsilon Y_0.$$

Since

$$y(x_0 + \epsilon X_0) + \epsilon\eta(x_0 + \epsilon X_0) = y(x_0) + \epsilon X_0 y'(x_0)$$
$$+ \epsilon\eta(x_0) + O(\epsilon^2),$$

we have

$$\eta(x_0) = Y_0 - X_0 y'(x_0) + O(\epsilon). \tag{7.20}$$

Similarly at the other endpoint

$$\eta(x_1) = Y_1 - X_1 y'(x_1) + O(\epsilon). \tag{7.21}$$

Substituting relations (7.20) and (7.21) into equation (7.19) yields

$$J(\hat{y}) - J(y) = \epsilon \left\{ \eta \frac{\partial f}{\partial y'}\Big|_{x_0}^{x_1} + \int_{x_0}^{x_1} \eta \left(\frac{\partial f}{\partial y} - \frac{d}{dx}\frac{\partial f}{\partial y'} \right) dx \right.$$

$$+ Y_1 \frac{\partial f}{\partial y'}\Big|_{x_1} - Y_0 \frac{\partial f}{\partial y'}\Big|_{x_0}$$

$$\left. + X_1 \left(f - y'\frac{\partial f}{\partial y'} \right)\Big|_{x_1} - X_0 \left(f - y'\frac{\partial f}{\partial y'} \right)\Big|_{x_0} \right\}$$

$$+ O(\epsilon^2).$$

The functional J is stationary at y and therefore the terms of order ϵ must be zero for all variations in the above expression. We can always choose variations such that $X_k = Y_k = 0$ (i.e., fixed endpoint variations). Arguing as in Section 7.1 we therefore deduce that y must satisfy the equation

$$\frac{d}{dx}\frac{\partial f}{\partial y'} - \frac{\partial f}{\partial y} = 0. \tag{7.22}$$

In addition, y must satisfy the endpoint condition

$$p\delta y - H\delta x \Big|_{x_0}^{x_1} = 0, \tag{7.23}$$

where

$$p = \frac{\partial f}{\partial y'},$$
$$H = y'p - f,$$

and for $k = 0, 1$ we define the functions δy and δx as

$$\delta y(x_k) = Y_k,$$
$$\delta x(x_k) = X_k.$$

Equation (7.23) is the starting point for more specialized problems. These problems concern variations where the endpoints satisfy relations of the form

$$g_k(x_0, y_0, x_1, y_1) = 0.$$

Evidently, no more than four such relations can be prescribed, since four equations would determine the endpoints (assuming the relations are functionally independent). The fixed endpoint problem thus corresponds to the case where four such relations are given. The natural boundary problems of the previous section correspond to three (or two) such relations imposed on the problem. For example, the case of one fixed endpoint, say (x_0, y_0), is characterized by the three conditions $x_0 = const.$, $y_0 = const.$, and $x_1 = const.$ These equations are then supplemented by the natural boundary condition at (x_1, y_1) to provide the fourth equation. In this problem only the variation δy at (x_1, y_1) is arbitrary.

Typically, variational problems come with relations of the form

$$g_k(x_j, y_j) = 0, \tag{7.24}$$

for $j = 1, 2$ so that the endpoint variations of (x_0, y_0) are not linked to those of (x_1, y_1). In this case we can always include variations that leave one endpoint fixed, and this leads to the two conditions

$$p\delta y - H\delta x \Big|_{x_0} = 0, \tag{7.25}$$

$$p\delta y - H\delta x \Big|_{x_1} = 0. \tag{7.26}$$

In the next section we focus on variational problems with endpoint relations of the form (7.24). Geometrically such relations correspond to the requirement that an endpoint (x_k, y_k) lie on the curve defined by the implicit equation $g(x_k, y_k) = 0$.

It is worth noting that, in general, some relationship must be imposed among the endpoints to get compatible boundary conditions. Otherwise, the situation is much like the unsupported beam in the previous section. In particular, suppose that no relations are imposed on the endpoints. Certainly equations (7.25) and (7.26) are satisfied, but since δx and δy are independent and arbitrary at each endpoint we have that $p = 0$ and $H = 0$ at each endpoint. Hence we have the boundary conditions

$$\frac{\partial f}{\partial y'} = 0, \tag{7.27}$$

$$f = 0, \tag{7.28}$$

that must be satisfied at each endpoint. Since any extremal must also satisfy the Euler-Lagrange equation (7.22), the boundary conditions (7.27) imply

$$\int_{x_0}^{x_1} \frac{\partial f}{\partial y} \, dx = 0. \tag{7.29}$$

In addition, we know that

$$\frac{dH}{dx} = -\frac{\partial f}{\partial x}$$

(cf. Section 2.3); hence, the boundary conditions (7.28) also give

$$\int_{x_0}^{x_1} \frac{\partial f}{\partial x} \, dx = 0. \tag{7.30}$$

Equations (7.29) and (7.30) pose additional restrictions on y that are generally not compatible with the Euler-Lagrange equations. For instance, suppose that f does not depend on x explicitly. Then we know that $H = const.$ along any extremal (Section 2.3). Since $H = 0$ at the endpoints we have that $H = 0$ for all x and hence

$$y' \frac{\partial f}{\partial y'} - f = 0.$$

The above relation implies that f must be of the form

$$f(y, y') = A(y)y'.$$

Finally, we note that the above arguments can be extended to cope with functionals that depend on several dependent variables. Let

$$J(\mathbf{q}) = \int_{t_0}^{t_1} L(t, \mathbf{q}, \dot{\mathbf{q}}) \, dt,$$

where $\mathbf{q} = (q_1, q_2, \ldots, q_n)$ and L is a smooth function. If J is stationary at $\mathbf{q}$ then it can be shown that

$$\frac{d}{dt}\frac{\partial L}{\partial \dot{q}_k} - \frac{\partial L}{\partial q_k} = 0 \tag{7.31}$$

for $k = 1, \ldots, n$, and

$$\sum_{k=1}^{n} p_k \delta q_k - H \delta t = 0 \tag{7.32}$$

at the endpoints t_0 and t_1. Here, the quantities p_k and H are defined as

$$p_k = \frac{\partial L}{\partial \dot{q}_k}, \quad H = \sum_{k=1}^{n} \dot{q}_k p_k - L, \tag{7.33}$$

and δq_k, δt are defined in a manner analogous to δy and δx.

Exercises 7.2:

1. Derive the endpoint compatibility condition analogous to (7.25) and (7.26) for a functional of the form

$$J(y) = \int_{x_0}^{x_1} f(x, y, y', y'') \, dx.$$

7.3 Transversality Conditions

Let J be a functional of the form

$$J(y) = \int_{x_0}^{x_1} f(x, y, y') \, dx,$$

and consider the problem of finding smooth functions y such that J is stationary, at one end $y(x_0) = y_0$, and at the other end y is required to lie on a curve Γ described parametrically by

$$\mathbf{r}(\xi) = (x_\Gamma(\xi), y_\Gamma(\xi)), \tag{7.34}$$

for $\xi \in \mathbb{R}$. We know from Section 7.2 that any candidate for a solution to this problem must be a solution to the Euler-Lagrange equation (7.22) that passes through the point (x_0, y_0) and intersects the curve Γ (figure 7.7). A solution to this problem, however, will also have to satisfy equation (7.26), and this may (and generally does) limit the choice of extremals. If we return to the analysis of the previous section for this problem, we know that $\hat{y}(x_1)$ and $\hat{x}(x_1)$ are related through equation (7.34); i.e., all variations must have an endpoint on the curve Γ. This means that we can associate the "virtual displacement" δy at $x = x_1$ with $dy_\Gamma/d\xi$ and the "virtual displacement" δx with $dx_\Gamma/d\xi$. Condition (7.26) thus becomes

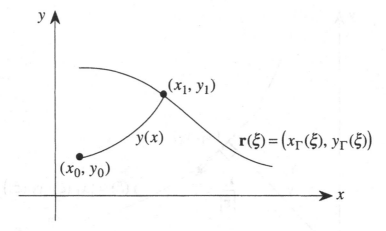

Fig. 7.7.

$$\frac{dy_\Gamma}{d\xi}p - \frac{dx_\Gamma}{d\xi}H = 0, \tag{7.35}$$

where p and H are evaluated at $x = x_1$. In this framework we do not know a priori what value to assign to x_1, but we do know that the point $(x_1, y(x_1))$ lies on the curve Γ. If we know either x_1 or $y(x_1)$, then we would also know at which value of ξ to evaluate the derivatives in equation (7.35). We can thus regard equation (7.35) as either an equation for ξ or an equation for x_1. Geometrically, the vector $(dx_\Gamma/d\xi, dy_\Gamma/d\xi)$ is a tangent vector on Γ. If $\mathbf{v} = (p, -H)$, we see that equation (7.35) corresponds to the condition that $\mathbf{v}$ be orthogonal to the tangent vector. Equation (7.35) is sometimes called a **transversality condition**.

Evidently, the above analysis can be readily extended to cope with the problem of finding extrema for J when one endpoint is required to be on a curve Γ_0 and the other endpoint on a curve Γ_1. If the curve Γ_0 is described by $(x_{\Gamma_0}(\sigma), y_{\Gamma_0}(\sigma))$, $\sigma \in [\sigma_0, \sigma_1]$ and the curve Γ_1 by $(x_{\Gamma_1}(\xi), y_{\Gamma_1}(\xi))$, $\xi \in [\xi_0, \xi_1]$, then

$$\frac{dy_{\Gamma_0}}{d\sigma}p - \frac{dx_{\Gamma_0}}{d\sigma}H = 0$$

$$\tag{7.36}$$

$$\frac{dy_{\Gamma_1}}{d\xi}p - \frac{dx_{\Gamma_1}}{d\xi}H = 0.$$

Example 7.3.1: Let

$$J(y) = \int_0^{x_1} \sqrt{1 + y'^2}\, dx,$$

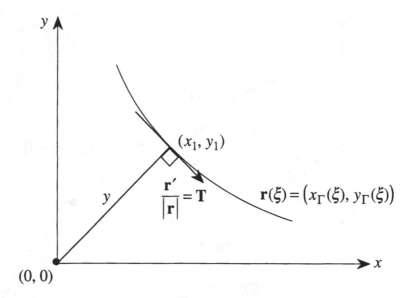

Fig. 7.8.

and consider the problem of finding the function(s) y for which J is stationary subject to the condition that $y(0) = 0$ and that $(x_1, y(x_1))$ lies on the curve described by (7.34).

Geometrically, we are finding the distance of a plane curve Γ from the origin. The extremals for this problem will be line segments through the origin, and we seek among the segments that intersect Γ the one for which the arclength is an extremum (figure 7.8).

For this problem,

$$p = \frac{\partial f}{\partial y'} = \frac{y'}{\sqrt{1 + y'^2}},$$ (7.37)

and

$$H = y' \frac{\partial f}{\partial y'} - f$$

$$= \frac{y'^2}{\sqrt{1 + y'^2}} - \sqrt{1 + y'^2}$$

$$= -\frac{1}{\sqrt{1 + y'^2}}.$$ (7.38)

We thus have

$$\frac{dy_\Gamma}{d\xi} \frac{y'}{\sqrt{1 + y'^2}} + \frac{dx_\Gamma}{d\xi} \frac{1}{\sqrt{1 + y'^2}} = 0;$$

i.e.,

$$\left(\frac{dx_\Gamma}{d\xi}, \frac{dy_\Gamma}{d\xi}\right) \cdot (1, \frac{dy}{dx}) = 0. \tag{7.39}$$

Geometrically, equation (7.39) implies that for a stationary value of J, the tangent to the extremal (i.e., the line segment) must be orthogonal to the tangent to Γ.

A bit of reflection shows that condition (7.39) can yield any number of solutions depending on the curve Γ. If for instance Γ is an arc of a circle centred at the origin then any extremal will satisfy the orthogonality condition. For illustration, let us suppose that Γ corresponds to the curve described by

$$\mathbf{r}(\xi) = (\xi - 1, \xi^2 + \frac{1}{2}),$$

for $\xi \in \mathbb{R}$. We know that the extremals for this problem are of the form $y = mx$. Now,

$$\left(\frac{dx_\Gamma}{d\xi}, \frac{dy_\Gamma}{d\xi}\right) \cdot (1, \frac{dy}{dx}) = (1, 2\xi)(1, m) = 0;$$

hence,

$$2\xi m + 1 = 0;$$

i.e.,

$$\xi = -\frac{1}{2m}.$$

The extremal and Γ have the point $(x_1, y(x_1))$ in common and therefore

$$(x_1, mx_1) = (\xi - 1, \xi^2 + \frac{1}{2})$$

$$= (-\frac{1}{2m} - 1, \frac{1}{4m^2} + \frac{1}{2}).$$

The above relation provides two equations for x_1 and m. After some algebra we see that m must satisfy the relation

$$4m^3 + 1 = 0.$$

There is only one real solution to the above equation, viz.

$$m = -\frac{1}{\sqrt[3]{4}},$$

and hence the only extremal satisfying condition (7.39) is

$$y = -\frac{1}{\sqrt[3]{4}}x.$$

Example 7.3.2: Let J be the functional of Example 7.3.1, and let Γ_0 and Γ_1 be curves described by

$$\mathbf{r}_0(\sigma) = (-\sigma^2, \sigma),$$
$$\mathbf{r}_1(\xi) = (\xi, (\xi-1)^2),$$

where $\sigma, \xi \in \mathbb{R}$, respectively. We consider the problem of finding an extremal y for J subject to the condition that (x_0, y_0) lies on Γ_0 and (x_1, y_1) lies on Γ_1. We know from Example 7.3.1 that the extremals must be of the form $y = mx + b$ for some constants m and b. The functions p and H are given by equations (7.37) and (7.38), respectively; hence,

$$p = \frac{m}{\sqrt{1+m^2}},$$

and

$$H = -\frac{1}{\sqrt{1+m^2}}.$$

The transversality conditions (7.36) thus imply

$$m - 2\sigma^* = 0, \tag{7.40}$$
$$2m(\xi^* - 1) + 1 = 0, \tag{7.41}$$

where $(x_0, y_0) = \mathbf{r}_0(\sigma^*)$ and $(x_1, y_1) = \mathbf{r}_1(\xi^*)$. Since $y_0 = mx_0 + b$, we have

$$\sigma^* = -m\sigma^{*2} + b, \tag{7.42}$$

and similarly

$$(\xi^* - 1)^2 = m\xi^* + b. \tag{7.43}$$

The four equations (7.40) to (7.43) lead to the quintic equation

$$m^5 + 6m^3 - 2m^2 - 1 = 0, \tag{7.44}$$

and the relations

$$b = \frac{m}{2}\left(1 + \frac{m^2}{2}\right),$$

$$\sigma^* = \frac{m}{2},$$

$$\xi^* = 1 - \frac{1}{2m}.$$

Transversality conditions can be derived for problems that involve several dependent variables. Consider, for example, the problem of finding smooth functions $\mathbf{q} = (q_1, q_2)$ such that the functional

$$J(\mathbf{q}) = \int_{t_0}^{t_1} L(t, \mathbf{q}, \dot{\mathbf{q}})\, dt$$

is stationary subject to the condition that $\mathbf{q}(t_0) = \mathbf{q}_0$ (fixed endpoint) and $\mathbf{q}(t_1)$ is required to lie on a surface Σ given by $t = \psi(\mathbf{q})$. Evidently, an extremal to the problem must satisfy the Euler-Lagrange equations (7.31) and the two boundary conditions given by the fixed endpoint. We can glean the appropriate boundary conditions at the other endpoint from equation (7.32).

Equation (7.32) must be satisfied for all variations near $\mathbf{q}$ with an endpoint on Σ. In particular, we can consider variations with endpoints $q_2 = const.$ on Σ. For this special class of variations equation (7.32) gives

$$p_1 \delta q_1 - H \delta t = 0,$$

and since δq_1 and δt are related by $t = \psi(\mathbf{q})$ we have $\delta t = \delta q_1 \partial \psi / \partial q_1$ for $q_2 = const.$, so that

$$\left(p_1 - H \frac{\partial \psi}{\partial q_1} \right) \delta q_1 = 0;$$

i.e.,

$$\frac{\partial L}{\partial \dot{q}_1} - H \frac{\partial \psi}{\partial q_1} = 0. \tag{7.45}$$

Similar arguments lead to the condition

$$\frac{\partial L}{\partial \dot{q}_2} - H \frac{\partial \psi}{\partial q_2} = 0. \tag{7.46}$$

Example 7.3.3:   Let

$$J(\mathbf{q}) = \int_0^{t_1} \sqrt{1 + \dot{q}_1^2 + \dot{q}_2^2}\, dt,$$

$$\mathbf{q}(0) = \mathbf{0},$$

and

$$t_1 = \psi(\mathbf{q}(t_1)) = \sqrt{(q_1(t_1) - 1)^2 + (q_2(t_1) - 1)^2}. \tag{7.47}$$

We seek an extremal for J subject to the condition that the endpoint $\mathbf{q}(t_1)$ lies on the surface defined by ψ. Geometrically, the problem amounts to finding the curve in $\mathbb{R}^3$ from the origin to the surface defined by ψ (a cone with vertex at $(1, 1, 0)$) such that arclength is minimum.

The Euler-Lagrange equations show that $\mathbf{q}$ is of the form

$$\mathbf{q} = \alpha t + \beta,$$

where $\alpha = (\alpha_1, \alpha_2)$ $\beta = (\beta_1, \beta_2)$ are constants. The boundary condition $\mathbf{q}(0) = 0$ implies that $\beta = \mathbf{0}$, and hence the extremals are of the form

$$\mathbf{q} = \alpha t.$$

Now,

$$p_k = \frac{\dot{q}_k}{\sqrt{1 + \dot{q}_1^2 + \dot{q}_2^2}},$$

and

$$H = p_1 \dot{q}_1 + p_2 \dot{q}_2 - L$$
$$= -\frac{1}{\sqrt{1 + \dot{q}_1^2 + \dot{q}_2^2}},$$

so that the transversality conditions (7.45) and (7.46) yield

$$\dot{q}_k + \frac{\partial \psi}{\partial q_k} = 0, \tag{7.48}$$

for $k = 1, 2$. Geometrically, the above condition indicates that at the t_1 endpoint the tangent vector to the extremal $(\dot{q}_1, \dot{q}_2)$ is parallel to $\nabla \psi$; i.e., the extremal is normal to the surface. The transversality condition reduces to

$$\alpha_k = -\frac{q_k(t_1) - 1}{\sqrt{(q_1(t_1) - 1)^2 + (q_2(t_1) - 1)^2}} = -\frac{\alpha_k t_1 - 1}{t_1};$$

i.e.,

$$2\alpha_k t_1 = 1;$$

hence,

$$\alpha_1 = \alpha_2.$$

Equation (7.47) implies

$$\alpha_1^2 + \alpha_2^2 = 1,$$

so that

$$\alpha_k t_1 = \frac{1}{\sqrt{2}}.$$

The extremal is thus given by the line

$$q_k(t) = \frac{t}{\sqrt{2}},$$

which intersects the cone at $(1/2, 1/2, 1/\sqrt{2})$.

Exercises 7.3:

1. The functional for the brachystochrone is

$$J(y) = \int_0^{x_1} \sqrt{\frac{1 + y'^2}{y}} \, dx.$$

Find an extremal for J subject to the condition that $y(0) = 0$ and $(x_1, y(x_1))$ lies on the curve $y = x - 1$.

2. Let

$$J(y) = \int_0^{x_1} \left(y'^2 + y^2\right) dx.$$

Find an extremal for J subject to the condition that $y(0) = 0$ and $(x_1, y(x_1))$ lies on the curve $y = 1 - x$. Determine the appropriate constants in terms of implicit relations.

3. Lagrange multipliers provide an alternative approach to deriving transversality conditions. Consider the problem where the (x_1, y_1) endpoint is required to be on the curve defined by $g(x, y) = 0$, and let

$$\Theta(\epsilon) = \int_{x_0}^{x_1(\epsilon)} f(x, \hat{y}, \hat{y}') \, dx - \lambda g(x_1(\epsilon), \hat{y}(x_1(\epsilon))),$$

where λ is a Lagrange multiplier and $\hat{y} = y + \epsilon \eta$. Derive the transversality condition

$$p \frac{\partial g}{\partial x} + H \frac{\partial g}{\partial y} = 0$$

at (x_1, y_1). (In this problem, the δx and δy variations are independent.)

4. Let $\mathbf{q} = (q_1, q_2)$ and consider a functional of the form

$$J(\mathbf{q}) = \int_{t_0}^{t_1} n(t, \mathbf{q}) \sqrt{1 + \dot{q}_1^2 + \dot{q}_2^2} \, dt,$$

along with the boundary condition $\mathbf{q}(t_0) = \mathbf{q}_0$ (fixed endpoint) and the condition that the t_1 endpoint lie on a surface Σ defined by $t = \psi(\mathbf{q})$. Show that the extremals must be orthogonal to Σ.

5. Let $\mathbf{q} = (q_1, q_2)$ and

$$J(\mathbf{q}) = \int_0^{t_1} \left(\dot{q}_1^2 + \dot{q}_2^2 + 2q_1 q_2\right) dt.$$

Given the condition that $\mathbf{q}(0) = \mathbf{0}$ and that $t_1 = q_1(t_1)$ determine the form of the extremal for J and derive the implicit equations for the integration constants and t_1.

6. Let $\mathbf{q} = (q_1, \ldots, q_n)$. Derive the general transversality conditions for a functional J to have an extremum subject to one endpoint fixed and the other endpoint on a hypersurface defined implicitly by $g(\mathbf{q}, t) = 0$.

8

The Hamiltonian Formulation

Given the existence of a certain transformation, the n Euler-Lagrange equations associated with a variational problem can be converted into an equivalent system of $2n$ first-order ordinary differential equations. These equations are called Hamilton's equations, and they have some special properties. In particular, the derivatives in this system are uncoupled, and the differential equations can be derived from a single (scalar) function called the Hamiltonian. Given a Hamiltonian system, another Hamiltonian system can be constructed by a special type of transformation called a symplectic map. It may be possible to find a symplectic map that produces a Hamiltonian system that can be solved and thereby used to solve the original problem. The search for such a map leads to a partial differential equation called the Hamilton-Jacobi equation.

In this chapter we discuss the connexions between the Euler-Lagrange equations and Hamilton's equations. We first discuss a certain transformation, the Legendre transformation, and then use it to derive Hamilton's equations. Symplectic maps are discussed briefly in the third section, and the Hamilton-Jacobi equation is then derived. The motivation in this chapter for the alternative formulation is solving the Euler-Lagrange equations. We thus focus on the use of the Hamilton-Jacobi equation as a tool for solving certain variational problems. In particular, we discuss the method of additive separation for solving the Hamilton-Jacobi equation. This method has many limitations, but there is a paucity of analytical techniques for solving variational problems, and the Hamilton-Jacobi equation provides one additional (albeit specialized) tool that has applications to problems of interest. Beyond being simply a tool for solving the Euler-Lagrange equations, the Hamilton-Jacobi equation is important in its own right. It plays a central rôle in the theory underlying the calculus of variations.

8.1 The Legendre Transformation

The reader has doubtless encountered point transformations and used them to solve differential equations or evaluate integrals. A **point transformation** from one pair $(x, y(x))$ to another pair $(X, Y(X))$ consists of relations of the form

$$X = X(x, y)$$
$$Y = Y(x, y).$$

Another type of transformation that plays an important part in differential equations and geometry is called a **contact transformation**. Contact transformations differ from point transformations in that the functions defining the transformation depend on the derivatives of the dependent variable. One of the simplest and most useful contact transformations is called the **Legendre transformation**. This transformation has some remarkable properties and provides the link between the Euler-Lagrange equations and Hamilton's equations. We consider first the simplest Legendre transformation involving one independent variable.

Let $y : [x_0, x_1] \to \mathbb{R}$ be a smooth function, and define the new variable p by

$$p = y'(x). \tag{8.1}$$

Equation (8.1) can be used to define the variable x in terms of p provided $y''(x) \neq 0$. For definiteness, let us suppose that

$$y''(x) > 0 \tag{8.2}$$

for all $x \in [x_0, x_1]$. Inequality (8.2) implies that the curve γ described by $\mathbf{r}(x) = (x, y(x))$, $x \in [x_0, x_1]$ is strictly convex upwards in shape. The new variable p corresponds to the slope of the tangent line (figure 8.1). Geometrically, one can see that under these conditions any point on γ is determined uniquely by the slope of its tangent line. Suppose now that we introduce the function

$$H(p) = -y(x) + px. \tag{8.3}$$

Here, we regard x as a function of p. Equations (8.1) and (8.3) provide a transformation from the pair $(x, y(x))$ to the pair $(p, H(p))$. This is an example of a Legendre transformation. A remarkable property of this transformation is that it is an **involution**; i.e., the transformation is its own inverse. To see this, note that

$$\frac{dH}{dp} = -\frac{d}{dp} y(x) + \frac{d}{dp}(px)$$

$$= -\frac{dy}{dx}\frac{dx}{dp} + p\frac{dx}{dp} + x$$

$$= (-y'(x) + p)\frac{dx}{dp} + x$$

$$= x,$$

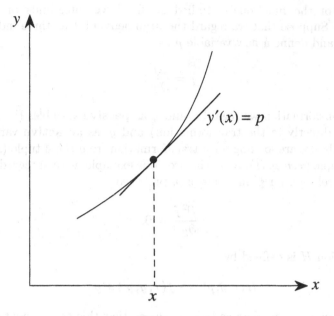

Fig. 8.1.

where we have used equation (8.1). Note also that

$$-H(p) + xp = -(-y(x) + px) + px = y(x).$$

These calculations show that if we apply the Legendre transformation to the pair $(p, H(p))$ we recover the original pair $(x, y(x))$.

Example 8.1.1: Let $y(x) = x^4/4$. Then

$$p = y'(x) = x^3$$

so that $x = p^{1/3}$. The function H is given by

$$H(p) = -\frac{x^4}{4} + px = \frac{3}{4}p^{4/3}.$$

Note that

$$H'(p) = \frac{4}{3}\left(\frac{3}{4}p^{1/3}\right) = p^{1/3} = x,$$

and that

$$-H(p) + xp = -\frac{3}{4}p^{4/3} + xp$$

$$= -\frac{3}{4}x^4 + x^4$$

$$= y.$$

Many of the functionals studied so far have integrands of the form $f(x, y, y')$. Suppose that we regard the arguments of f as three independent variables, and define a new variable p as

$$p = \frac{\partial f}{\partial y'}. \tag{8.4}$$

In this transformation we regard x and y as **passive** variables (i.e., not participating directly in the transformation) and y' as an **active** variable. In other words, we are looking for a transformation from the 4-tuple (x, y, y', f) to the 4-tuple (x, y, p, H). As in the previous example, we can regard equation (8.4) as a relation for y' in terms of p, provided

$$\frac{\partial^2 f}{\partial y'^2} \neq 0. \tag{8.5}$$

The function H is defined by

$$H(x, y, p) = -f(x, y, y') + y'p. \tag{8.6}$$

Using the same arguments as before, we see that this transformation is also an involution.

Example 8.1.2:   Let $f(x, y, y') = \sqrt{1 + y'^2}$. Then

$$p = \frac{\partial f}{\partial y'} = \frac{y'}{\sqrt{1 + y'^2}};$$

hence,

$$y' = \frac{p}{\sqrt{1 - p^2}},$$

since y' and p must be of the same sign. The function H is thus

$$H(x, y, p) = -\sqrt{1 + y'^2} + y'p$$
$$= -\frac{1}{\sqrt{1 - p^2}} + \frac{p^2}{\sqrt{1 - p^2}}$$
$$= -\sqrt{1 - p^2}.$$

The quantities p and H defined by the Legendre transformation have already come into prominence in the theory. For example, it is precisely the quantity H that is constant along extremals when f does not contain x explicitly (Section 2.3). Moreover, H appears as a term in the general endpoint condition derived in Section 7.2. Note that, for the passive variables in the transformation,

$$\frac{\partial H}{\partial x} = -\frac{\partial f}{\partial x},$$
$$\frac{\partial H}{\partial y} = -\frac{\partial f}{\partial y}.$$

Let us now consider a Legendre transformation involving a function $L(t, \mathbf{q}, \dot{\mathbf{q}})$, where L is a smooth function and $\mathbf{q} = (q_1, q_2, \ldots, q_n)$. In this transformation the variables t and $\mathbf{q}$ are regarded as passive. Let

$$p_k = \frac{\partial L}{\partial \dot{q}_k}, \tag{8.7}$$

for $k = 1, 2, \ldots, n$. Equations (8.7) connect the active variables $\dot{q}_k$ and p_k. The implicit function theorem can be invoked to show that equations (8.7) can (in principle) be solved for the $\dot{q}_k$ provided the $n \times n$ Hessian matrix

$$\mathbf{M}_L = \begin{pmatrix} \frac{\partial^2 L}{\partial \dot{q}_1 \partial \dot{q}_1} & \frac{\partial^2 L}{\partial \dot{q}_1 \partial \dot{q}_2} & \cdots & \frac{\partial^2 L}{\partial \dot{q}_1 \partial \dot{q}_n} \\ \frac{\partial^2 L}{\partial \dot{q}_2 \partial \dot{q}_1} & \frac{\partial^2 L}{\partial \dot{q}_2 \partial \dot{q}_2} & \cdots & \frac{\partial^2 L}{\partial \dot{q}_2 \partial \dot{q}_n} \\ \vdots & \vdots & & \vdots \\ \frac{\partial^2 L}{\partial \dot{q}_n \partial \dot{q}_1} & \frac{\partial^2 L}{\partial \dot{q}_n \partial \dot{q}_2} & \cdots & \frac{\partial^2 L}{\partial \dot{q}_n \partial \dot{q}_n} \end{pmatrix}$$

is nonsingular, i.e., satisfies the Jacobian condition

$$\frac{\partial (p_1, \ldots, p_n)}{\partial (\dot{q}_1, \ldots, \dot{q}_n)} = \det \mathbf{M}_L \neq 0. \tag{8.8}$$

The n-dimensional analogue of equation (8.6) is

$$H(t, \mathbf{q}, \mathbf{p}) = -L(t, \mathbf{q}, \dot{\mathbf{q}}) + \sum_{k=1}^{n} \dot{q}_k p_k, \tag{8.9}$$

where the variables $\dot{\mathbf{q}}$ are regarded as functions of $\mathbf{p} = (p_1, p_2, \ldots, p_n)$, $\mathbf{q}$ and t.

The Legendre transformation defined by equations (8.7) and (8.9) is also an involution. In particular,

$$\frac{\partial H}{\partial p_k} = \sum_{j=1}^{n} \left(-\frac{\partial L}{\partial \dot{q}_j} + p_j \right) \frac{\partial \dot{q}_j}{\partial p_k} + \dot{q}_k$$
$$= \dot{q}_k, \tag{8.10}$$

and

$$-H(t, \mathbf{q}, \mathbf{p}) + \sum_{k=1}^{n} \dot{q}_k p_k = L(t, \mathbf{q}, \dot{\mathbf{q}}). \tag{8.11}$$

The function H in the above transformation is called a **Hamiltonian function** and the function L is called a **Lagrangian**. The "new" coördinates

$(t, \mathbf{q}, \mathbf{p})$ are sometimes called **generalized coördinates**. The set of points defined by the pairs $(\mathbf{q}, \mathbf{p})$ is called the **phase space**.

In mechanics the variables p_k are called the **generalized momenta**. The name stems from functionals modelling the motion of particles (Section 1.3). The integrand in this case is of the form

$$L(t, \mathbf{q}, \dot{\mathbf{q}}) = T(t, \mathbf{q}, \dot{\mathbf{q}}) - V(t, \mathbf{q}),$$

where T is the kinetic energy, V is the potential energy, and $\mathbf{q}$ represents the positions of the particles at time t. For a single particle of mass m in space, if $\mathbf{q} = (q_1, q_2, q_3)$ are the Cartesian coördinates of the particle, then

$$T(t, \mathbf{q}, \dot{\mathbf{q}}) = \frac{1}{2} m |\dot{\mathbf{q}}|^2 = \frac{1}{2} m (\dot{q}_1^2 + \dot{q}_2^2 + \dot{q}_3^2);$$

hence,

$$p_k = \frac{\partial L}{\partial \dot{q}_k} = \frac{\partial T}{\partial \dot{q}_k} = m \dot{q}_k,$$

and p_k is thus a component of the momentum vector. For j particles in space, we have $n = 3j$ and each p_k is a component of a momentum vector.

Exercises 8.1:

1. Find the Hamiltonian H for the functionals with integrands:
   (a) $f(x, y, y') = y\sqrt{1 + y'^2}$ (the catenary);
   (b) $f(x, y, y') = \sqrt{\frac{1 + y'^2}{y}}$ (the brachystochrone).
2. Show that the Hamiltonian for the motion of a single particle of mass m is:
   (a) $H = \frac{1}{2m} \left(p_1^2 + p_2^2 + p_3^2 \right) + V(t, \mathbf{q})$, in Cartesian coördinates;
   (b) $H = \frac{1}{2m} \left(p_1^2 + \frac{p_2^2}{q_1^2} + p_3^2 \right) + V(t, \mathbf{q})$, in cylindrical coördinates $\mathbf{q} = (r, \phi, z)$;
   (c) $H = \frac{1}{2m} \left(p_1^2 + \frac{p_2^2}{q_1^2} + \frac{p_3^2}{q_1^2 \sin^2 q_2} \right) + V(t, \mathbf{q})$, in spherical coördinates $\mathbf{q} = (r, \theta, \phi)$.

8.2 Hamilton's Equations

Let J be a functional of the form

$$J(\mathbf{q}) = \int_{t_0}^{t_1} L(t, \mathbf{q}, \dot{\mathbf{q}}) \, dt, \qquad (8.12)$$

where $\mathbf{q} = (q_1, q_2, \ldots, q_n)$ and L is a smooth function satisfying condition (8.8). If $\mathbf{q}$ is a smooth extremal for J then $\mathbf{q}$ satisfies the equations

$$\frac{d}{dt}\frac{\partial L}{\partial \dot{q}_k} - \frac{\partial L}{\partial q_k} = 0 \tag{8.13}$$

for $k = 1, 2, \ldots, n$. Applying the Legendre transformation defined by equations (8.7) and (8.9) we have

$$p_k = \frac{\partial L}{\partial \dot{q}_k}, \tag{8.14}$$

and

$$\dot{q}_k = \frac{\partial H}{\partial p_k}, \tag{8.15}$$

where H is defined by equation (8.9). Now t and $\mathbf{q}$ are passive variables in this transformation so that

$$\frac{\partial H}{\partial t} = -\frac{\partial L}{\partial t}, \tag{8.16}$$

and

$$\frac{\partial H}{\partial q_k} = -\frac{\partial L}{\partial q_k}. \tag{8.17}$$

Since $\mathbf{q}$ is an extremal we have from equations (8.13) that

$$\frac{\partial L}{\partial q_k} = \frac{d}{dt}\frac{\partial L}{\partial \dot{q}_k} = \dot{p}_k,$$

and therefore equation (8.17) implies that

$$\dot{p}_k = -\frac{\partial H}{\partial q_k}. \tag{8.18}$$

The solutions $\mathbf{q}$ to the Euler-Lagrange equations (8.13) are thus mapped to solutions $(\mathbf{q}, \mathbf{p})$ to the equations (8.15) and (8.18) under the Legendre transformation. Conversely, suppose that $(\mathbf{q}, \mathbf{p})$ is a solution to equations (8.15) and (8.18), and that the $n \times n$ matrix with elements $\{\frac{\partial^2 H}{\partial p_j \partial q_k}\}$ is nonsingular;[1] i.e.,

$$\frac{\partial(\dot{q}_1, \dot{q}_2, \ldots, \dot{q}_n)}{\partial(p_1, p_2, \ldots, p_n)} = \det\mathbf{M}_H \neq 0. \tag{8.19}$$

Then equations (8.15) define a Legendre transformation from $(t, \mathbf{q}, \mathbf{p}, H)$ to $(t, \mathbf{q}, \dot{\mathbf{q}}, L)$ with L as defined by equation (8.11). We thus have that equations (8.14) and (8.17) are satisfied, and hence equation (8.18) implies that

$$\frac{\partial H}{\partial q_k} + \frac{\partial L}{\partial q_k} = -\frac{dp_k}{dt} + \frac{\partial L}{\partial q_k} = -\frac{d}{dt}\frac{\partial L}{\partial \dot{q}_k} + \frac{\partial L}{\partial q_k} = 0,$$

---

[1] Note that

$$\frac{\partial(\dot{q}_1, \dot{q}_2, \ldots, \dot{q}_n)}{\partial(p_1, p_2, \ldots, p_n)}\frac{\partial(p_1, p_2, \ldots, p_n)}{\partial(\dot{q}_1, \dot{q}_2, \ldots, \dot{q}_n)} = 1$$

so that condition (8.19) is satisfied if and only if condition (8.8) is satisfied.

so that $\mathbf{q}$ is a solution to the Euler-Lagrange equations (8.13). The involutive character of the Legendre transformation thus indicates that the problem of solving the system of n Euler-Lagrange equations is equivalent to the problem of solving the system of $2n$ equations (8.15) and (8.18).

Prima facie, it seems that we have gained little by exchanging n second-order differential equations for $2n$ first-order differential equations, but the new system of equations has some attractive features. The Euler-Lagrange equations are second order and generically nonlinear in the first derivatives. Moreover, the first derivatives in this system are generally coupled. The new equations are of first order. The derivatives are uncoupled, and the system can be derived from a single generating function, the Hamiltonian. The system of of $2n$ equations (8.15) and (8.18) is called a **Hamiltonian system**, and the equations are called **Hamilton's equations**.

In the transition from the Euler-Lagrange equations to Hamilton's equations n new variables, the generalized momenta, are introduced. In Hamilton's equations, the position variables $\mathbf{q}$ and the generalized momenta variables $\mathbf{p}$ are on the same footing and regarded as independent. This approach can be somewhat confusing at first encounter given that $\mathbf{q}$ and $\dot{\mathbf{q}}$ (hence $\mathbf{q}$ and $\mathbf{p}$) are dependent in the original problem.[2] The concern here is that $\mathbf{q}$ and $\dot{\mathbf{q}}$ are not independent and therefore cannot be varied independently. In contrast, $\mathbf{q}$ and $\mathbf{p}$ are independent in Hamilton's equations and can thus be varied independently. In fact, the Legendre transformation that defines the new variables $\mathbf{p}$ also ensures that these variables can be varied independent of $\mathbf{q}$. To see this, we introduce the functional $\tilde{J}$ defined by

$$\tilde{J}(\mathbf{q},\mathbf{p}) = \int_{t_0}^{t_1} \left(\sum_{j=1}^{n} p_j \dot{q}_j - H(t,\mathbf{q},\mathbf{p}) \right) dt, \tag{8.20}$$

where $\mathbf{q}$ and $\mathbf{p}$ are regarded as independent variables and $\dot{q}_k$ is the derivative of q_k (i.e., not regarded as a function of $\mathbf{p}$). Evidently the integrands defining J and $\tilde{J}$ are equivalent under the Legendre transformation and hence $\tilde{J} = J$. Suppose now that we vary the p_k but leave the q_k fixed. Let

$$\hat{\mathbf{p}} = \mathbf{p} + \epsilon\rho,$$

where $\rho = (\rho_1, \rho_2, \ldots, \rho_n)$ and ϵ is a small parameter. Then

$$\tilde{J}(\mathbf{q},\hat{\mathbf{p}}) - \tilde{J}(\mathbf{q},\mathbf{p}) = \int_{t_0}^{t_1} \left\{ H(t,\mathbf{q},\mathbf{p}) - H(t,\mathbf{q},\hat{\mathbf{p}}) + \sum_{j=1}^{n} \dot{q}_j(\hat{p}_j - p_j) \right\} dt$$

$$= \epsilon \int_{t_0}^{t_1} \left\{ \sum_{j=1}^{n} \rho_j \left(-\frac{\partial H}{\partial p_j} + \dot{q}_j \right) \right\} dt + O(\epsilon^2).$$

---

[2] We are not alone. In his book *Applied Differential Geometry* Burke [20] addresses the dedication as follows: "To all those who, like me, have wondered how in the hell you can change $\dot{q}$ without changing q."

Now, the p_k are defined through the Legendre transformation (8.14), and this implies equations (8.15). We therefore have that

$$\tilde{J}(\mathbf{q}, \hat{\mathbf{p}}) - \tilde{J}(\mathbf{q}, \mathbf{p}) = O(\epsilon^2).$$

This calculation shows that variations on $\mathbf{p}$ do not affect variations on $\tilde{J}$. Although $\mathbf{q}$ and $\mathbf{p}$ are independent variables in $\tilde{J}$, only the variations of $\mathbf{q}$ affect the variation of $\tilde{J}$. This situation is also reflected in the derivation of Hamilton's equations. Specifically, equation (8.15) is valid for any pair $(\dot{\mathbf{q}}, \mathbf{p})$ as it is a property of the Legendre transformation; in contrast, equation (8.18) comes directly from the Euler-Lagrange equation. Only functions $\mathbf{q}$ and $\mathbf{p}$ corresponding to an extremal for J will satisfy these equations.

Example 8.2.1: Simple Pendulum
Consider the pendulum of Example 1.3.1. The kinetic energy is

$$T = \frac{1}{2}m\left(\dot{x}^2(t) + \dot{y}^2(t)\right) = \frac{1}{2}m\ell^2\dot{\phi}^2(t),$$

and the potential energy is

$$V = gy(t) = mg\ell(1 - \cos\phi(t)).$$

Hamilton's Principle implies that the motion of the pendulum is such that the functional

$$J(\phi) = \int_{t_0}^{t_1}\left\{\frac{1}{2}m\ell^2\dot{\phi}^2(t) - mg\ell(1 - \cos\phi(t))\right\}dt$$

is an extremum. Let $q = \phi$ and

$$L(t, q, \dot{q}) = \frac{1}{2}m\ell^2\dot{q}^2 - mg\ell(1 - \cos q).$$

Then

$$p = \frac{\partial L}{\partial \dot{q}} = m\ell^2\dot{q},$$

so that

$$\dot{q} = \frac{p}{m\ell^2}.$$

The Hamiltonian H is given by

$$\begin{aligned}
H(t, q, p) &= p\dot{q} - L \\
&= p\frac{p}{m\ell^2} - \frac{1}{2}m\ell^2\dot{q}^2 + mg\ell(1 - \cos q) \\
&= \frac{p^2}{2m\ell^2} + mg\ell(1 - \cos q).
\end{aligned}$$

Hamilton's equations are thus

$$\dot{q} = \frac{\partial H}{\partial p} = \frac{p}{m\ell^2}$$

$$\dot{p} = -\frac{\partial H}{\partial q} = -mg\ell \sin q.$$

For this example H corresponds to the total energy of the pendulum. The Euler-Lagrange equation is

$$\frac{d}{dt}\frac{\partial L}{\partial \dot{q}} - \frac{\partial L}{\partial q} = m\ell^2 \ddot{q} + mg\ell \sin q = 0;$$

i.e.,

$$\ddot{q} + \frac{g}{\ell} \sin q = 0.$$

Note that, according to Hamilton's equations,

$$\ddot{q} = \frac{\dot{p}}{m\ell^2} = \frac{mg\ell \sin q}{m\ell^2} = -\frac{g}{\ell} \sin q,$$

in agreement with the Euler-Lagrange equation.

Example 8.2.2:   Geometrical Optics

Let $(x(z), y(z), z)$, $z \in [z_0, z_1]$ describe a space curve γ. The optical path length of γ in a medium with refractive index $n(x, y, z)$ is given by

$$J(x, y) = \int_{z_0}^{z_1} n(x, y, z)\sqrt{1 + x'^2 + y'^2}\, dz.$$

Fermat's Principle implies that a necessary condition for γ to be a light ray is that J be an extremum. For notational convenience, let $q_1 = x$, $q_2 = y$, $z = t$, $z_0 = t_0$, and $z_1 = t_1$. The quantity

$$L(t, \mathbf{q}, \dot{\mathbf{q}}) = n(t, \mathbf{q})\sqrt{1 + |\dot{\mathbf{q}}|^2}$$

is called the **optical Lagrangian**. Here $\mathbf{q} = (q_1, q_2)$, $n(t, \mathbf{q}) = n(x, y, z)$, and $\cdot$ denotes d/dt. The generalized momenta are given by

$$p_k = \frac{\partial L}{\partial q_k} = \frac{n\dot{q}_k}{\sqrt{1 + |\dot{\mathbf{q}}|^2}}.$$

Geometrically, the quantities p_k/n correspond to the direction cosines of the curve with respect to the q_k axis. For this reason the p_k are generally called the **optical cosines** in geometrical optics. Now,

$$p_1^2 + p_2^2 - n^2 = -\frac{n^2}{1 + |\dot{\mathbf{q}}|^2},$$

and hence

$$\dot{q}_k = \frac{p_k}{n}\sqrt{1 + |\dot{\mathbf{q}}|^2}$$

$$= \frac{p_k}{n}\sqrt{\frac{n^2}{n^2 - p_1^2 - p_2^2}}$$

$$= \frac{p_k}{\sqrt{n^2 - p_1^2 - p_2^2}}.$$

The Hamiltonian is therefore

$$H(t, \mathbf{q}, \mathbf{p}) = \sum_{k=1}^{2} p_k \dot{q}_k - L(t, \mathbf{q}, \dot{\mathbf{q}})$$

$$= \frac{p_1^2 + p_2^2}{\sqrt{n^2 - p_1^2 - p_2^2}} - \frac{n^2}{\sqrt{n^2 - p_1^2 - p_2^2}}$$

$$= -\sqrt{n^2 - p_1^2 - p_2^2},$$

and Hamilton's equations are

$$\dot{q}_k = \frac{\partial H}{\partial p_k} = \frac{p_k}{\sqrt{n^2 - p_1^2 - p_2^2}}$$

$$\dot{p}_k = -\frac{\partial H}{\partial q_k} = \frac{1}{\sqrt{n^2 - p_1^2 - p_2^2}}\frac{\partial n}{\partial q_k}.$$

The quantity H has a tractable geometrical interpretation. Since

$$H(t, \mathbf{q}, \mathbf{p}) = -n\sqrt{1 - \frac{p_1^2}{n^2} - \frac{p_2^2}{n^2}},$$

H corresponds to the negative of the optical cosine with respect to the t (z) axis.

We have already encountered the Hamiltonian (in a slightly different guise) in Section 2.3. We know from Chapter 2 that if L does not contain the variable t explicitly, then H is constant along any extremal. It is clear from the Legendre transformation that L contains t explicitly if and only if H contains t explicitly. Hamiltonian systems that do not depend on t explicitly are called **conservative**. The pendulum in Example 8.2.1 is an example of a conservative system. In this example $H = const.$ corresponds to the condition that the total energy of the system is conserved. The Hamiltonian system derived in Example 8.2.2 is not conservative unless the refractive index is independent of z. Note that a nonconservative system such as this one may still have conservation laws. Note also that a nonconservative system can be converted into a conservative one by the introduction of a new "position" variable corresponding to t in the original formulation and using a new variable for "time." For solving specific problems, however, this observation is of limited value because the first integral afforded by a conservative system is offset by the introduction of a new dependent variable. Nonetheless, it is a useful observation because, when convenient, we can always reformulate a problem to get a conservative system and thus use general results for conservative systems.

Exercises 8.2:

1. The Lagrangian for a linear harmonic oscillator is

$$L(t, q, \dot{q}) = \frac{1}{2m} m\dot{q}^2 - \frac{1}{2} kq,$$

where m is mass and k is a restoring force coefficient. Show that

$$H(t, q, p) = \frac{1}{2m} \left(p^2 + \omega^2 q^2 \right),$$

where

$$\omega = \sqrt{\frac{k}{m}}.$$

Derive and solve the Euler-Lagrange equation assuming m and k are constants. Derive Hamilton's equations and verify that the solution obtained for the Euler-Lagrange equation is also a solution for Hamilton's equations.

2. Derive Hamilton's equations for the catenary (Exercises 8.1-1). Verify that the solution found in Example 2.3.3 is also a solution of Hamilton's equations.

3. For any smooth functions $\Phi(t, \mathbf{q}, \mathbf{p})$ and $\Theta(t, \mathbf{q}, \mathbf{p})$, the **Poisson bracket** is defined by

$$[\Phi, \Theta] = \sum_{k=1}^{n} \left(\frac{\partial \Phi}{\partial q_k} \frac{\partial \Theta}{\partial p_k} - \frac{\partial \Phi}{\partial p_k} \frac{\partial \Theta}{\partial q_k} \right).$$

Let H be the Hamiltonian function associated with a functional J, and suppose that along the extremals for J

$$\Phi(\mathbf{q}, \mathbf{p}) = const.$$

The function Φ is then called a **first integral** of the system. Show that

$$[\Phi, H] = 0.$$

Prove the converse: if $[\Phi, H] = 0$, then Φ is a first integral of the system.

4. Show that $\dot{p}_k = [p_k, H]$ and $\dot{q}_k = [q_k, H]$.

8.3 Symplectic Maps

Let $\mathbf{q}$ and $\mathbf{p}$ be generalized coördinates and H be a function of t, $\mathbf{q}$, $\mathbf{p}$ such that

$$\dot{q}_k = \frac{\partial H}{\partial p_k},$$

$$\dot{p}_k = -\frac{\partial H}{\partial q_k},$$

(8.21)

for $k = 1, 2, \ldots, n$. A **symplectic map**[3] is a transformation of the form

$$Q_k = Q_k(t, \mathbf{q}, \mathbf{p}),$$

$$P_k = P_k(t, \mathbf{q}, \mathbf{p}),$$

(8.22)

such that the Hamiltonian system (8.21) transforms into another Hamiltonian system

$$\dot{Q}_k = \frac{\partial \hat{H}}{\partial P_k},$$

$$\dot{P}_k = -\frac{\partial \hat{H}}{\partial Q_k},$$

(8.23)

where $\hat{H}$ is a function of t, $\mathbf{Q}$, and $\mathbf{P}$. In short, a symplectic map is a transformation on the generalized coördinates that preserves the Hamiltonian structure. Symplectic maps are also called **canonical transformations**. These maps loom large in the classical mechanics lore. The reader is directed to Abraham and Marsden [1], Arnold [6], Goldstein [35], Lanczos [48], and Whittaker [73] among numerous other works on classical mechanics. In this section we briefly discuss symplectic maps primarily as a herald to the Hamilton-Jacobi equation.

We know from the previous section that Hamiltonian systems such as (8.21) and (8.23) can be associated with the extremals to the functionals

$$J(\mathbf{q}) = \int_{t_0}^{t_1} L(t, \mathbf{q}, \dot{\mathbf{q}}) \, dt,$$

$$\hat{J}(\mathbf{Q}) = \int_{t_0}^{t_1} \hat{L}(t, \mathbf{Q}, \dot{\mathbf{Q}}) \, dt,$$

---

[3] The word **symplectic** comes from the Greek word *sumplektikos* meaning "intertwined." There is also a bone in the skull of a fish by this name.

respectively, where

$$L(t, \mathbf{q}, \dot{\mathbf{q}}) = \sum_{k=1}^{n} p_k \dot{q}_k - H(t, \mathbf{q}, \mathbf{p})$$

$$\hat{L}(t, \mathbf{Q}, \dot{\mathbf{Q}}) = \sum_{k=1}^{n} P_k \dot{Q}_k - \hat{H}(t, \mathbf{Q}, \mathbf{P}).$$

If we regard $\mathbf{q}$ and $\mathbf{p}$ as independent variables and $\dot{\mathbf{q}}$ as the derivative of $\mathbf{q}$, then the Euler-Lagrange equations for the functional

$$J(\mathbf{q}, \mathbf{p}) = \int_{t_0}^{t_1} \left(\sum_{k=1}^{n} p_k \dot{q}_k - H(t, \mathbf{q}, \mathbf{p}) \right) dt$$

are precisely Hamilton's equations (8.21), and the solutions to equations (8.21) correspond to the extremals for J. A similar remark holds for the other functional $\hat{J}$.

We say that two functionals J and $\hat{J}$ are **variationally equivalent** if they produce the same set of extremals. A symplectic map is essentially a transformation from the $(\mathbf{q}, \mathbf{p})$ phase space to the $(\mathbf{Q}, \mathbf{P})$ phase space such that the resulting functionals J and $\hat{J}$ are variationally equivalent.

In Section 2.5 we showed that any nonsingular coördinate transformation leads to a variationally equivalent functional. This result can be extended to functionals involving several dependent variables. Transformations that involve only position coördinates lead to variationally equivalent functionals and hence this class of transformations is symplectic. But transformations of this type are too restrictive, and, in the spirit of the Hamiltonian approach, we should let the momenta variables participate in transformations as independent variables. The problem is, if the p_k transform, the resulting transformation need not be symplectic.

One method for constructing symplectic maps involves the introduction of a generating function. The method is based on the observation that two functionals are variationally equivalent if their integrands differ by a perfect differential (cf. Exercises 3.2-4). Suppose that there is a smooth function Φ such that

$$\sum_{k=1}^{n} p_k \dot{q}_k - H(t, \mathbf{q}, \mathbf{p}) = \sum_{k=1}^{n} P_k \dot{Q}_k - \hat{H}(t, \mathbf{Q}, \mathbf{P}) + \frac{d}{dt} \Phi(t, \mathbf{q}, \mathbf{p}). \qquad (8.24)$$

Then the corresponding functionals J and $\hat{J}$ are variationally equivalent and the transformation (8.22) is symplectic. We can use the transformation (8.22) to convert Φ to a function of t, $\mathbf{q}$, and $\mathbf{Q}$, and equation (8.24) can thus be recast in the form

$$\frac{d}{dt} \Phi(t, \mathbf{q}, \mathbf{Q}) = \sum_{k=1}^{n} p_k \dot{q}_k - \sum_{k=1}^{n} P_k \dot{Q}_k + \hat{H}(t, \mathbf{Q}, \mathbf{P}) - H(t, \mathbf{q}, \mathbf{p}).$$

Since

$$\frac{d\Phi}{dt} = \sum_{k=1}^{n} \left(\frac{\partial \Phi}{\partial q_k} \dot{q}_k + \frac{\partial \Phi}{\partial Q_k} \dot{Q}_k \right) + \frac{\partial \Phi}{\partial t},$$

we have

$$p_k = \frac{\partial \Phi}{\partial q_k}, \quad P_k = -\frac{\partial \Phi}{\partial Q_k}, \tag{8.25}$$

and

$$\hat{H}(t, \mathbf{Q}, \mathbf{P}) = H(t, \mathbf{q}, \mathbf{p}) + \frac{\partial \Phi}{\partial t}. \tag{8.26}$$

Equations (8.25) provide relations for the symplectic map. Equation (8.26) provides an expression for the transformed Hamiltonian function.

Example 8.3.1: Harmonic Oscillator

The Hamiltonian for a linear harmonic oscillator in one dimension is of the form

$$H = \frac{1}{2m}(p^2 + \omega^2 q^2),$$

where q corresponds to the position of a particle of mass m at time t, p is the momentum, and ω is a constant relating to the restoring force (see Exercises 8.2-1). The Hamiltonian system for the equation of motion is

$$\dot{q} = \frac{\partial H}{\partial p} = \frac{p}{m}$$

$$\dot{p} = -\frac{\partial H}{\partial q} = -\frac{\omega^2 q}{m}.$$

Let

$$\Phi(q, Q) = \frac{\omega q^2}{2} \cot Q.$$

This peculiar generating function is chosen so that the resulting Hamiltonian system is particularly simple.[4] The momenta coördinates are

$$p = \frac{\partial \Phi}{\partial q} = \omega q \cot Q, \tag{8.27}$$

and

$$P = -\frac{\partial \Phi}{\partial Q} = \frac{\omega q^2}{2 \sin^2 Q}. \tag{8.28}$$

Equations (8.27) and (8.28) can be used to determine the symplectic map (8.22). It is more convenient, however, to give the inverse transformation equations

$$q = \sqrt{\frac{2P}{\omega}} \sin Q, \tag{8.29}$$

$$p = \sqrt{2\omega P} \cos Q. \tag{8.30}$$

---

[4] Goldstein [35], p. 389 provides a derivation of the transformation.

The transformed Hamiltonian is

$$\hat{H} = H + \frac{\partial \Phi}{\partial t} = H(q(Q,P), p(Q,P))$$

$$= \frac{1}{2m}\left(\left(\sqrt{2\omega P}\cos Q\right)^2 + \omega^2\left(\sqrt{\frac{2P}{\omega}}\sin Q\right)^2\right)$$

$$= \frac{\omega P}{m},$$

and the associated Hamiltonian system is

$$\dot{Q} = \frac{\partial \hat{H}}{\partial P} = \frac{\omega}{m},$$

$$\dot{P} = -\frac{\partial \hat{H}}{\partial Q} = 0.$$

This is a particularly simple system to solve.[5] We have

$$Q = \frac{\omega}{m}t + c_1,$$

$$P = c_2,$$

where c_1 and c_2 are constants, which gives

$$q = \sqrt{\frac{2c_2}{\omega}}\sin\left(\frac{\omega}{m}t + c_1\right),$$

$$p = \sqrt{2\omega c_2}\cos\left(\frac{\omega}{m}t + c_1\right).$$

Exercises 8.3:

1. Suppose that we regard Φ as a function of $\mathbf{q}$ and $\mathbf{P}$. Show that

$$p_k = \frac{\partial \Phi}{\partial q_k}, \quad Q_k = \frac{\partial \Phi}{\partial P_k}, \quad \hat{H} = H + \frac{\partial \Phi}{\partial t}.$$

   Derive similar equations for a symplectic map if Φ is regarded as a function of $\mathbf{p}$ and $\mathbf{Q}$.

2. Let $\Phi = \sum_{k=1}^{n} q_k Q_k$. Show that this generating function leads to a symplectic map that essentially interchanges the spatial variables with the momenta variables. This further shows that these variables are on the same footing in the Hamiltonian framework.

---

[5] Of course, it is even easier to solve the Euler-Lagrange equation directly, but this example gives a simple illustration of how a symplectic map can be used to reduce Hamilton's equations to a simple form.

3. Let $\Phi = \sum_{k=1}^{n} q_k P_k$. Show that this function merely generates the identity transformation.

4. Let $\Phi = \sum_{k=1}^{n} g_k(t, \mathbf{q}) P_k$. Show that this generating function leads to point transformations; i.e., the Q_k depend only on $\mathbf{q}$ and t.

8.4 The Hamilton-Jacobi Equation

Although symplectic maps are of intrinsic interest, they are also of practical interest because they may lead to simpler Hamiltonian systems. In this section we target a particularly simple Hamiltonian system that can be readily solved. The problem is to derive a generating function that produces a symplectic map leading to the simpler system. It turns out that the generating function must satisfy a first-order (generally nonlinear) partial differential equation called the **Hamilton-Jacobi equation**. Once a general solution is found to the Hamilton-Jacobi equation, the solution to the Hamiltonian system can be derived by solving a set of implicit equations. The problem of solving a Hamiltonian system can thus be exchanged for the problem of solving a single partial differential equation. From a practical standpoint, a single partial differential equation is generally at least as difficult to solve as a system of ordinary differential equations, and in this sense the victory may seem Pyrrhic. Nonetheless, there are special cases of interest when the Hamilton-Jacobi equation can be solved. We discuss some of these cases in the next section. Although our motivation here is to solve Hamilton's equations, it turns out that the Hamilton-Jacobi equation plays a pivotal rôle in the theory. The real profit from this reformulation is a deeper understanding of variational processes.

8.4.1 The General Problem

Suppose that a generating function Φ can be found such that the transformed Hamiltonian is a constant, say $\hat{H} = 0$. The symplectic map produced by Φ then yields the Hamiltonian system

$$\dot{Q}_k = \frac{\partial \hat{H}}{\partial P_k} = 0,$$

$$\dot{P}_k = -\frac{\partial \hat{H}}{\partial Q_k} = 0,$$

that can easily be solved to get

$$Q_k = \alpha_k,$$
$$P_k = \beta_k,$$

where the α_k and β_k are constants. Since $\hat{H} = 0$, equation (8.26) implies that Φ must satisfy

$$H(t, \mathbf{q}, \mathbf{p}) + \frac{\partial \Phi}{\partial t} = 0.$$

In the above equation, the function Φ is regarded as a function of $\mathbf{q}$ and $\mathbf{Q}$. To eliminate the p_k variables in this expression we can use equation (8.25), and thus get

$$H\left(t, q_1, \ldots, q_n, \frac{\partial \Phi}{\partial q_1}, \ldots, \frac{\partial \Phi}{\partial q_n}\right) + \frac{\partial \Phi}{\partial t} = 0. \tag{8.31}$$

Equation (8.31) is a first-order partial differential equation for the generating function Φ called the **Hamilton-Jacobi equation**. Hamilton derived the equation in 1834; in 1837 Jacobi made a precise connexion between solutions of the differential equation (8.31) and the corresponding Hamiltonian system (Theorem 8.4.1).

Example 8.4.1: Geometrical Optics

The Hamiltonian derived for the path of a light ray in Example 8.2.2 is

$$H(t, \mathbf{q}, \mathbf{p}) = -\sqrt{n^2 - p_1^2 - p_2^2}.$$

The Hamilton-Jacobi equation for this problem is

$$-\sqrt{n^2 - \frac{\partial \Phi}{\partial q_1}^2 - \frac{\partial \Phi}{\partial q_2}^2} + \frac{\partial \Phi}{\partial t} = 0;$$

i.e.,

$$\left(\frac{\partial \Phi}{\partial q_1}\right)^2 + \left(\frac{\partial \Phi}{\partial q_2}\right)^2 + \left(\frac{\partial \Phi}{\partial t}\right)^2 = n^2.$$

In the original (x, y, z) notation this equation can be written in the compact form

$$|\nabla \Phi|^2 = n^2, \tag{8.32}$$

where $\nabla = (\frac{\partial}{\partial x}, \frac{\partial}{\partial y}, \frac{\partial}{\partial z})$. Equation (8.32) is called the **eikonal equation** of geometrical optics.

The Hamilton-Jacobi equation has two notable features. Firstly, the function Φ does not appear explicitly in the differential equation. Only the partial derivatives of Φ are present in the equation. Secondly, the differential equation does not depend on any of the Q_k variables or partial derivatives of Φ with respect to the Q_k. In essence, this means that if Φ is a solution to equation (8.31) then so is any function of the form $\Phi + f(\mathbf{Q})$, where f is an arbitrary function. The function Φ depends on the Q_k, and one might rightfully query exactly how these variables enter into the problem given no Q_k dependence in the differential equation. The answer is that the Q_k enter into the problem as *initial data* for the differential equation. Typically partial differential equations such as (8.31) are solved subject to a condition that Φ take prescribed values along a curve in the $\mathbf{q}$ space. Our problem is to find a generating function that produces the simplified Hamiltonian system, and this amounts to finding a general solution to (8.31) containing n arbitrary functions of $\mathbf{Q}$. No uniqueness is expected for solutions to this problem; we need the arbitrary functions in order to invert the transformation to solve for the q_k. We explain this more precisely after we introduce the concept of a complete solution.

Although we speak of arbitrary functions of $\mathbf{Q}$ entering into solutions of the Hamilton-Jacobi equation, we know by construction that the Q_k are in fact constants and hence the arbitrary functions of $\mathbf{Q}$ are also constants. We can thus regard a general solution to equation (8.31) as a function of the form $\Phi(t, \mathbf{q}, \alpha)$, where $\alpha = (\alpha_1, \ldots, \alpha_n)$, and the α_k are parameters that can be thought of as the Q_k when convenient. A solution $\Phi = \Phi(t, \mathbf{q}, \alpha)$ is called **complete**[6] if Φ has continuous second derivatives with respect to the q_k, the α_k and t variables, and the matrix $\mathbf{M}$ defined by

$$
\mathbf{M} = \begin{pmatrix}
\dfrac{\partial^2 \Phi}{\partial q_1 \partial \alpha_1} & \dfrac{\partial^2 \Phi}{\partial q_1 \partial \alpha_2} & \cdots & \dfrac{\partial^2 \Phi}{\partial q_1 \partial \alpha_n} \\
\vdots & \vdots & & \vdots \\
\dfrac{\partial^2 \Phi}{\partial q_n \partial \alpha_1} & \dfrac{\partial^2 \Phi}{\partial q_n \partial \alpha_2} & \cdots & \dfrac{\partial^2 \Phi}{\partial q_n \partial \alpha_n}
\end{pmatrix}
$$

is nonsingular; i.e.,

$$\det \mathbf{M} \neq 0, \tag{8.33}$$

in the relevant $\mathbf{q}$, α domain of the problem. The condition (8.33) is a Jacobian condition for the solvability of the q_k given the functions $\frac{\partial \Phi}{\partial \alpha_k}$. The next result is fundamental to the theory: it connects a complete solution to the Hamilton-Jacobi equation with the general solution to Hamilton's equations.

Theorem 8.4.1 (Hamilton-Jacobi) *Suppose that $\Phi(t, \mathbf{q}, \alpha)$ is a complete solution to the Hamilton-Jacobi equation (8.31). Then the general solution to the Hamiltonian system*

$$\dot{q}_k = \frac{\partial H}{\partial p_k}, \quad \dot{p}_k = -\frac{\partial H}{\partial q_k} \tag{8.34}$$

is given by the equations

$$\frac{\partial \Phi}{\partial \alpha_k} = -\beta_k, \tag{8.35}$$

$$\frac{\partial \Phi}{\partial q_k} = p_k, \tag{8.36}$$

where the β_k are n arbitrary constants.

Proof: Suppose that $\Phi(t, \mathbf{q}, \alpha)$ is a complete solution to the Hamilton-Jacobi equation (8.31). Then Φ satisfies condition (8.33), and the implicit function

---

[6] Some authors call these solutions **complete integrals**.

theorem implies that equations (8.35) can be solved for the q_k in terms of t, the α_k, and the β_k. Once this is accomplished equations (8.36) define the p_k in terms of these variables as well. Hence, equations (8.35) and (8.36) define the functions

$$q_k = q_k(t, \alpha, \beta),$$
$$p_k = p_k(t, \alpha, \beta),$$

for $k = 1, 2, \ldots, n$. Here, $\beta = (\beta_1, \ldots, \beta_n)$. To establish the result we need to show that the q_k and p_k defined by equations (8.35) and (8.36) satisfy the Hamiltonian system (8.34).

Substituting the solution Φ into the Hamilton-Jacobi equation and differentiating with respect to α_1 yields

$$\frac{\partial}{\partial \alpha_1}\left(H + \frac{\partial \Phi}{\partial t}\right) = \frac{\partial^2 \Phi}{\partial \alpha_1 \partial t} + \sum_{k=1}^{n} \frac{\partial^2 \Phi}{\partial \alpha_1 \partial q_k} \frac{\partial H}{\partial p_k} = 0. \qquad (8.37)$$

Now the equation

$$\frac{\partial \Phi}{\partial \alpha_1} = -\beta_1$$

must be satisfied identically, and therefore differentiating with respect to t yields[7]

$$\frac{d}{dt}\frac{\partial \Phi}{\partial \alpha_1} = \frac{\partial^2 \Phi}{\partial t \partial \alpha_1} + \sum_{k=1}^{n} \frac{\partial^2 \Phi}{\partial q_k \partial \alpha_1} \frac{dq_k}{dt} = 0. \qquad (8.38)$$

By hypothesis, the solution is complete and thus all the second-order derivatives of Φ are continuous; hence,

$$\frac{\partial^2 \Phi}{\partial t \partial \alpha_1} = \frac{\partial^2 \Phi}{\partial \alpha_1 \partial t}, \qquad \frac{\partial^2 \Phi}{\partial q_k \partial \alpha_1} = \frac{\partial^2 \Phi}{\partial \alpha_1 \partial q_k}.$$

Equations (8.37) and (8.38) thus imply that

$$\sum_{k=1}^{n} \frac{\partial^2 \Phi}{\partial q_k \partial \alpha_1}\left(\dot{q}_k - \frac{\partial H}{\partial p_k}\right) = 0. \qquad (8.39)$$

We can, of course, repeat the above arguments to derive n equations similar to equation (8.39), viz.,

$$\sum_{k=1}^{n} \frac{\partial^2 \Phi}{\partial q_k \partial \alpha_j}\left(\dot{q}_k - \frac{\partial H}{\partial p_k}\right) = 0$$

---

[7] Strictly speaking, we should use $\partial/\partial t$ instead of d/dt to denote partial differentiation with respect to t holding the α_k and β_k constant. We nonetheless use d/dt or $\dot{}$ to denote this differentiation to avoid confusion with the operator $\partial/\partial t$ in the Hamilton-Jacobi equation, which denotes differentiation with respect to t holding the q_k (as well as the α_k and β_k) constant.

for $j = 2, 3, \ldots, n$. The above equations can be written in the more compact form

$$\begin{pmatrix} \dot{q}_1 - \frac{\partial H}{\partial p_1} \\ \vdots \\ \dot{q}_n - \frac{\partial H}{\partial p_n} \end{pmatrix}^T \mathbf{M} = \mathbf{0}.$$

The completeness of the solution implies that $\det \mathbf{M} \neq 0$; consequently,

$$\dot{q}_k = \frac{\partial H}{\partial p_k} \tag{8.40}$$

for all $k = 1, 2, \ldots, n$.

To get the second set of Hamilton equations we again substitute the solution Φ into the Hamilton-Jacobi equation but now differentiate with respect to q_j. For $j = 1$,

$$\frac{\partial^2 \Phi}{\partial q_1 \partial t} + \frac{\partial H}{\partial q_1} + \sum_{k=1}^{n} \frac{\partial^2 \Phi}{\partial q_1 \partial q_k} \frac{\partial H}{\partial p_k} = 0. \tag{8.41}$$

The equation

$$p_1 = \frac{\partial \Phi}{\partial q_1}$$

must be satisfied identically and therefore

$$\begin{aligned} \dot{p}_1 &= \frac{d}{dt} \frac{\partial \Phi}{\partial q_1} = \frac{\partial^2 \Phi}{\partial t \partial q_1} + \sum_{k=1}^{n} \frac{\partial^2 \Phi}{\partial q_k \partial q_1} \dot{q}_k \\ &= \frac{\partial^2 \Phi}{\partial q_1 \partial t} + \sum_{k=1}^{n} \frac{\partial^2 \Phi}{\partial q_1 \partial q_k} \frac{\partial H}{\partial p_k}, \end{aligned} \tag{8.42}$$

where we have used equations (8.40) and the relations

$$\frac{\partial^2 \Phi}{\partial t \partial q_k} = \frac{\partial^2 \Phi}{\partial q_k \partial t}, \qquad \frac{\partial^2 \Phi}{\partial q_k \partial q_1} = \frac{\partial^2 \Phi}{\partial q_1 \partial q_k},$$

that follow from the continuity of the second derivatives. Subtracting equation (8.41) from equation (8.42) gives

$$\dot{p}_1 = -\frac{\partial H}{\partial q_1}.$$

Similar arguments can be used to show that

$$\dot{p}_k = -\frac{\partial H}{\partial q_k},$$

for $k = 2, 3, \ldots, n$.   □

From the standpoint of solving Hamilton's equations, the above theorem shows that we need not be concerned with the absence of initial data for equation (8.31) and the resulting nonuniqueness. *Any* complete solution to the Hamilton-Jacobi equation suffices to enable us to construct a solution to Hamilton's equations involving $2n$ arbitrary constants and hence a general solution to the underlying Euler-Lagrange equations. Given a variational problem we can thus outline a procedure to get the solution based on the Hamilton-Jacobi equation as follows:

(a) determine the Hamiltonian H for the given problem;
(b) form the Hamilton-Jacobi equation;
(c) find a complete solution Φ to the Hamilton-Jacobi equation;
(d) form the equations

$$\beta_k = -\frac{\partial \Phi}{\partial \alpha_k},$$

where the β_k are constants; and

(e) solve the n equations in part (d) for the q_k to get a general solution $\mathbf{q}(t, \alpha, \beta)$.

Example 8.4.2:   Geometrical Optics

Suppose that the optical medium in Example 8.4.1 has a refractive index $n = \mu\sqrt{q_1}$, where $\mu > 1$ is a constant. The relevant domain for this problem is $q_1 \geq 1$ (so that $n \geq 1$).[8] The Hamilton-Jacobi equation for this problem is

$$\left(\frac{\partial \Phi}{\partial q_1}\right)^2 + \left(\frac{\partial \Phi}{\partial q_2}\right)^2 + \left(\frac{\partial \Phi}{\partial t}\right)^2 = \mu^2 q_1. \tag{8.43}$$

The reader may verify directly that

$$\Phi(t, \mathbf{q}, \alpha) = \frac{2}{3\mu^2}\left(\mu^2 q_1 - (\alpha_1^2 + \alpha_2^2)\right)^{3/2} + \alpha_1 q_2 + \alpha_2 t \tag{8.44}$$

is a solution to equation (8.43). The matrix $\mathbf{M}$ is given by

$$\mathbf{M} = \begin{pmatrix} \frac{-\alpha_1}{A} & \frac{-\alpha_2}{A} \\ 1 & 0 \end{pmatrix},$$

where $A = \sqrt{\mu^2 q_1 - (\alpha_1^2 + \alpha_2^2)}$, and hence

$$\det \mathbf{M} = \frac{\alpha_2}{A}.$$

The solution is complete in the domain defined by $\alpha_1^2 + \alpha_2^2 < \mu^2 q_1$, $\alpha_2 \neq 0$. To get q_1 and q_2 we form the equations

---

[8] Recall that the refractive index is the ratio of the speed of light *in vacuo* to the speed of light in the medium.

$$\beta_1 = -\frac{\partial \Phi}{\partial \alpha_1} = \frac{\alpha_1 A}{\mu^2} - q_2$$

$$\beta_2 = -\frac{\partial \Phi}{\partial \alpha_2} = \frac{\alpha_2 A}{\mu^2} - t,$$

which are readily solved to get

$$q_1(t, \alpha, \beta) = \frac{\mu^2}{\alpha_2^2}(\beta_2 + t)^2 + \frac{\alpha_1^2 + \alpha_2^2}{\mu^2}$$

$$q_2(t, \alpha, \beta) = \frac{\alpha_1}{\alpha_2} t + \beta_2 - \beta_1.$$

Although part (d) in the above plan may in itself be a formidable task to accomplish in practice, the crux is part (c). First-order nonlinear partial differential equations are generally harder to solve than systems of ordinary differential equations. The only general solution technique for these partial differential equations involves the use of characteristics, which are defined by a system of ordinary differential equations. It turns out that the system of differential equations defining the characteristics is equivalent to the original Hamiltonian system. In general, to implement part (c) of the above plan we first have to solve Hamilton's equations in order to solve the Hamilton-Jacobi equation. This rather defeats the purpose of using the Hamilton-Jacobi equation to solve the original problem. Generically, the Hamilton-Jacobi formulation does not actually help to find solutions. There are cases, however, when the solution to the Hamilton-Jacobi equation can be found without resorting to characteristics. Solution techniques such as separation of variables do not rely on knowledge of the characteristics and therefore circumvent the problem of integrating Hamilton's equations first. The success of the solution technique depends crucially on the type of Hamiltonian, but it turns out that a number of problems of interest have Hamiltonians that allow a separation of variables. We discuss this technique in the next section.

8.4.2 Conservative Systems

A special but important case concerns conservative Hamiltonian systems. The Hamiltonian does not depend on explicitly t for such systems, and we know from Section 3.2 that H is constant along any extremal. We can exploit this situation because we know that the variable t can be separated out in the complete solution to the Hamilton-Jacobi equation. In other words, we know that there is a complete solution of the form

$$\Phi(t, \mathbf{q}, \alpha) = \Psi(\mathbf{q}, \alpha) - f(\alpha)t,$$

where $H = f(\alpha) = const.$ along the extremal $\mathbf{q}(t, \alpha, \beta)$. We can simplify matters further by identifying one of the coördinates, say $Q_n = \alpha_n$, with $f(\alpha)$. This approach produces the partial differential equation

$$H\left(q_1, \ldots, q_n, \frac{\partial \Phi}{\partial q_1}, \ldots, \frac{\partial \Phi}{\partial q_n}\right) = \alpha_n, \tag{8.45}$$

which we call the **reduced Hamilton-Jacobi equation**.

The function Ψ in the solution to the Hamilton-Jacobi equation is evidently a solution to the reduced Hamilton-Jacobi equation. Moreover, it is clear that Φ is a complete solution if Ψ is a complete solution; i.e., Ψ has continuous derivatives of second order and the matrix

$$\mathbf{M} = \begin{pmatrix} \frac{\partial^2 \Psi}{\partial q_1 \partial \alpha_1} & \cdots & \frac{\partial^2 \Psi}{\partial q_1 \partial \alpha_n} \\ \vdots & & \vdots \\ \frac{\partial^2 \Psi}{\partial q_n \partial \alpha_1} & \cdots & \frac{\partial^2 \Psi}{\partial q_n \partial \alpha_n} \end{pmatrix}$$

is nonsingular.

The function Ψ is of interest as a generating function for a symplectic map in its own right. The symplectic map $S : (\mathbf{q}, \mathbf{p}) \to (\mathbf{Q}, \mathbf{P})$ produced by Ψ transforms the Hamiltonian $H(\mathbf{q}, \mathbf{p})$ to the Hamiltonian $\hat{H}(\mathbf{Q}, \mathbf{P}) = Q_n$. The new position and momenta coördinates must satisfy the equations

$$\dot{Q}_k = \frac{\partial \hat{H}}{\partial P_k} = 0,$$

$$\dot{P}_k = -\frac{\partial \hat{H}}{\partial Q_k} = \begin{cases} 0, & \text{if } 1 \le k \le n-1 \\ -1, & \text{if } k = n, \end{cases}$$

and hence the Q_k are constants for $k = 1, 2, \ldots, n$, and the P_k are constants for $k = 1, 2, \ldots, n-1$. The anomalous coördinate is P_n because $P_n = \beta_n - t$, where β_n is a constant.

The symplectic map S has an interesting geometrical interpretation. For a given constant E, the condition $H(\mathbf{q}, \mathbf{p}) = E$ produces a hypersurface in the $2n$-dimensional phase space and a (hyper)cylinder in the $2n + 1$-dimensional $(t, \mathbf{q}, \mathbf{p})$ space. The extremals correspond to a family of curves that lie on the cylinder. The symplectic map S transforms the picture dramatically. In the $(t, \mathbf{q}, \mathbf{p})$ space, the cylinder is transformed to a hyperplane and, even more remarkable, the family of extremals in the original space is transformed into a family of extremals in the new space, where each extremal is a straight line inclined at an angle $\pi/4$ to the t-axis. Roughly speaking, the symplectic map S "flattens out" the cylinders $H = E$ and "straightens up" the extremals.

If the Hamiltonian system is conservative, we generally start with the reduced Hamilton-Jacobi equation. Once a complete solution Ψ is determined we can return to the solution $\Phi = \Psi - \alpha_n t$ and proceed as before. This amounts to solving the equations

$$\beta_k = -\frac{\partial \Psi}{\partial \alpha_k}, \quad k = 1, 2, \ldots, n-1,$$

$$\beta_n = -\frac{\partial \Psi}{\partial \alpha_n} - t,$$

for the q_k in terms of t and the constants α, β. The absence of t in the Hamiltonian simplifies the problem slightly, but all the comments about the difficulty of solving the Hamilton-Jacobi equation still apply to the reduced equation.

Example 8.4.3: Harmonic Oscillator

The linear harmonic oscillator of Example 8.3.1 has the Hamiltonian

$$H = \frac{1}{2m} \left(p^2 + \omega^2 q^2 \right).$$

The reduced Hamilton-Jacobi equation is thus

$$\frac{1}{2m} \left(\left(\frac{\partial \Psi}{\partial q} \right)^2 + \omega^2 q^2 \right) = \alpha;$$

i.e.,

$$\frac{\partial \Psi}{\partial q} = \sqrt{2m\alpha - \omega^2 q^2},$$

where $\alpha > 0$ is a constant. The generating function is therefore of the form

$$\Psi = \int \sqrt{2m\alpha - \omega^2 \xi^2} \, d\xi + const.$$

We need only one arbitrary constant for a complete solution so we can ignore the integration constant. A solution to the reduced Hamilton-Jacobi equation is thus

$$\Psi(q, \alpha) = \frac{\omega}{2} \left(q\sqrt{a^2 - q^2} + a^2 \sin^{-1} \left(\frac{q}{a} \right) \right),$$

where

$$a = \frac{\sqrt{2m\alpha}}{\omega}.$$

To get the position q, we form the equation

$$\beta = \frac{\partial \Psi}{\partial \alpha} - t$$

$$= -\frac{m}{\omega} \sin^{-1} \left(\frac{q}{a} \right) - t,$$

where β is another constant. We thus arrive at the solution

$$q(t) = -\frac{\sqrt{2m\alpha}}{\omega} \sin \left(\frac{\omega}{m} (\beta + t) \right),$$

which is equivalent to the solution found in Example 8.3.1.

Exercises 8.4:

1. Let

$$J(y) = \int_{x_0}^{x_1} y'^2 \, dx.$$

Derive the Hamilton-Jacobi equation corresponding to this functional. Solve the Euler-Lagrange equation for this functional and construct a solution to the Hamilton-Jacobi equation.

2. Let $n = n(y)$ denote the refractive index in an optical medium. Fermat's Principle implies that the path of a light ray from a point (x_0, y_0) to a point (x_1, y_1) is an extremal of the functional

$$J(y) = \int_{x_0}^{x_1} n(y)\sqrt{1 + y'^2} \, dx.$$

Derive the Hamilton-Jacobi equation for this functional and verify that a solution to this equation is

$$\Phi = \alpha x + \int_{y_0}^{y} \sqrt{n^2(\xi) - \alpha^2} \, d\xi + \beta,$$

where α and β are constants. Use this solution to find the corresponding extremals implicitly.

3. The Lagrangian for the motion of a particle of unit mass in the plane under the action of a uniform field is

$$L(t, \mathbf{q}, \dot{\mathbf{q}}) = \frac{1}{2}(\dot{q}_1^2 + \dot{q}_2^2) - gq_2,$$

where $\mathbf{q}$ denotes the Cartesian coördinates of the particle and g is a constant (Example 3.2.2). Derive the Hamilton-Jacobi equation and show that a solution to this equation is

$$\Phi = -\left(\frac{1}{2}\alpha^2 + gk\right)t + \alpha x + \sqrt{2g}\frac{2}{3}(k - y)^{3/2},$$

where α and k are arbitrary constants.

8.5 Separation of Variables

The only chance we have of solving a problem using the Hamilton-Jacobi equation without essentially solving the Hamiltonian system first is if a solution to the partial differential equation can be obtained without resorting to characteristics. One solution technique that avoids characteristics is called the

method of additive separation or simply **separation of variables**.[9] In
this section we present the method and give some examples. We also discuss
conditions under which we know that a separable solution exists. We limit our
discussion to conservative systems.

8.5.1 The Method of Additive Separation

Consider the reduced Hamilton-Jacobi equation for a conservative system

$$H\left(q_1, \ldots, q_n, \frac{\partial \Psi}{\partial q_1}, \ldots, \frac{\partial \Psi}{\partial q_n}\right) - E = 0, \tag{8.46}$$

where E is a constant. Suppose that the terms q_1 and $\frac{\partial \Psi}{\partial q_1}$ appear in this
equation only through the combination $g_1(q_1, \frac{\partial \Psi}{\partial q_1})$, where g_1 is some known
function. Equation (8.46) could be then recast in the form

$$F\left(g_1(q_1, \frac{\partial \Psi}{\partial q_1}), q_2, \ldots, q_n, \frac{\partial \Psi}{\partial q_2}, \ldots, \frac{\partial \Psi}{\partial q_n}\right) - E = 0, \tag{8.47}$$

and this motivates us to seek a solution of the form

$$\Psi = \Psi_1(q_1, \mathbf{Q}) + R_1(q_2, \ldots, q_n, \mathbf{Q}). \tag{8.48}$$

Substituting the above expression for Ψ into equation (8.47) gives

$$F\left(g_1(q_1, \frac{\partial \Psi_1}{\partial q_1}), q_2, \ldots, q_n, \frac{\partial R_1}{\partial q_2}, \ldots, \frac{\partial R_1}{\partial q_n}\right) - E = 0, \tag{8.49}$$

and this equation must be satisfied for a continuum of q_1 values. Assuming g_1
is a differentiable function, this means that

$$\frac{\partial F}{\partial q_1} = \frac{\partial F}{\partial g_1} \frac{\partial g_1}{\partial q_1} = 0;$$

i.e., $\partial g_1 / \partial q_1 = 0$. Now Ψ_1 by construction depends only on q_1 and $\frac{\partial \Psi_1}{\partial q_1}$, and
hence the above observation leads to the equation

$$g_1(q_1, \Psi_1'(q_1)) = C_1(\mathbf{Q}), \tag{8.50}$$

where C_1 is an arbitrary function of $\mathbf{Q}$ and $'$ denotes d/dq_1. Equation (8.50)
is a first-order ordinary differential equation for Ψ_1.

The best scenario is if each pair q_k, $\partial \Psi / \partial q_k$ enter into equation (8.46) only
through a combination $g_k(q_k, \partial \Psi / \partial q_k)$. In this case, the partial differential
equation can be written in the form

---

[9] The latter term also includes the method of multiplicative separation, which we
   do not use.

$$F\left(g_1(q_1, \frac{\partial \Psi}{\partial q_1}), \ldots, g_n(q_n, \frac{\partial \Psi}{\partial q_n})\right) - E = 0,$$

and we seek a solution of the form

$$\Psi = \Psi_1(q_1, \mathbf{Q}) + \Psi_2(q_2, \mathbf{Q}) + \cdots + \Psi_n(q_n, \mathbf{Q}). \tag{8.51}$$

The arguments leading to equation (8.50) can be used to show that the Ψ_k must satisfy the n (uncoupled) first-order ordinary differential equations

$$g_k(q_k, \Psi'_k(q_k)) = C_k(\mathbf{Q}). \tag{8.52}$$

Here, the C_k are functions of $\mathbf{Q}$ satisfying the equation

$$F(C_1(\mathbf{Q}), \ldots, C_n(\mathbf{Q})) - E = 0, \tag{8.53}$$

but are otherwise arbitrary. A reduced Hamilton-Jacobi equation is called **separable** if there exists a complete solution Ψ of the form (8.51). The function Ψ is called a **separable solution**.

The method of additive separation amounts to the steps:

(a) assume a solution of the form (8.51) and substitute it into equation (8.46);
(b) identify the g_k and form the differential equations (8.52); and
(c) solve the ordinary differential equations for the Ψ_k.

Once the separable solution Ψ is determined, a complete solution $\Phi = \Psi - Et$ can be obtained for the Hamilton-Jacobi equation and we can proceed to find the position functions q_k as discussed in the previous section.

The obvious weakness with the above method is that a separable solution need not exist: there is no guarantee that the requisite g_k can be found. The existence of separable solutions depends on the Hamiltonian and even a simple coördinate transformation of the position variables can affect whether a separable solution is available. Conditions under which we can predict the existence of a separable solution are discussed in the next subsection.

Example 8.5.1:   In Cartesian coördinates, the motion of a particle in space under the action of gravity acting in the q_3 direction produces the Hamiltonian

$$H = \frac{1}{2m} \left(p_1^2 + p_2^2 + p_3^2 \right) + mgq_3,$$

where m is the mass of the particle and g is a gravitational constant. The reduced Hamilton-Jacobi equation is

$$\left(\frac{\partial \Psi}{\partial q_1}\right)^2 + \left(\frac{\partial \Psi}{\partial q_2}\right)^2 + \left(\frac{\partial \Psi}{\partial q_3}\right)^2 + 2m^2 gq_3 - 2mE = 0, \tag{8.54}$$

where E, the total energy of the particle, is a constant. Suppose that equation (8.54) has a solution of the form

$$\Psi = \Psi_1(q_1, \mathbf{Q}) + \Psi_2(q_2, \mathbf{Q}) + \Psi_3(q_3, \mathbf{Q}).$$

Substituting this expression into equation (8.54) yields

$$\left(\frac{\partial \Psi_1}{\partial q_1}\right)^2 + \left(\frac{\partial \Psi_2}{\partial q_2}\right)^2 + \left(\frac{\partial \Psi_3}{\partial q_3}\right)^2 + 2m^2 g q_3 - 2mE = 0.$$

We can thus take

$$g_k(q_k, \frac{\partial \Psi_k}{\partial q_k}) = \frac{\partial \Psi_k}{\partial q_k},$$

for $k = 1, 2,$ and

$$g_3(q_3, \frac{\partial \Psi_3}{\partial q_3}) = \left(\frac{\partial \Psi}{\partial q_3}\right)^2 + 2m^2 g q_3.$$

We have that

$$\frac{\partial \Psi_k}{\partial q_k} = C_k(\mathbf{Q}),$$

for $k = 1, 2,$ so that

$$\Psi_k = C_k(\mathbf{Q})q_k + K_k(\mathbf{Q}),$$

where the K_k are arbitrary functions. The differential equation involving g_3 yields

$$\frac{\partial \Psi}{\partial q_3} = \sqrt{C_3(\mathbf{Q}) - 2m^2 g q_3}$$

$$= \sqrt{2mE - C_1^2(\mathbf{Q}) - C_2^2(\mathbf{Q}) - 2m^2 g q_3},$$

where we have used equation (8.53) to eliminate $C_3(\mathbf{Q})$. The function Ψ_3 is thus of the form

$$\Psi_3 = -\frac{1}{3m^2 g} \left\{ 2mE - C_1^2(\mathbf{Q}) - C_2^2(\mathbf{Q}) - 2m^2 g q_3 \right\}^{3/2} + K_3(\mathbf{Q}),$$

where K_3 is another arbitrary function. A separable solution to equation (8.54) is therefore

$$\Psi = C_1(\mathbf{Q})q_1 + C_2(\mathbf{Q})q_2 - \frac{1}{3m^2 g}\tilde{\Lambda}^3 + K(\mathbf{Q}),$$

where C_1, C_2, and K are arbitrary functions of $\mathbf{Q}$, and

$$\tilde{\Lambda} = \sqrt{2mE - C_1^2(\mathbf{Q}) - C_2^2(\mathbf{Q}) - 2m^2 g q_3}.$$

For a complete solution we need only three arbitrary constants present in Ψ. Let $C_1(\mathbf{Q}) = Q_1 = \alpha_1$ and $C_2(\mathbf{Q}) = Q_2 = \alpha_2$. The constant E is also arbitrary in the above solution so we may take $E = Q_3 = \alpha_3$ and let $K(\mathbf{Q}) = 0$. The solution Ψ is then of the form

$$\Psi(\mathbf{q}, \alpha) = \alpha_1 q_1 + \alpha_2 q_2 - \frac{1}{3m^2 g}\Lambda^3,$$

where

$$\Lambda = \sqrt{2m\alpha_3 - \alpha_1^2 - \alpha_2^2 - 2m^2 g q_3}\,.$$

The matrix $\mathbf{M}$ with entries $\{\frac{\partial^2 \Psi}{\partial q_j \partial \alpha_k}\}$ is given by

$$\mathbf{M} = \begin{pmatrix} 1 & 0 & 0 \\ 0 & 1 & 0 \\ 0 & 0 & -\frac{mg}{\Lambda} \end{pmatrix}$$

so that $\det \mathbf{M} \neq 0$, and the solution is thus complete.

Equipped with a complete solution to the reduced Hamilton-Jacobi equation, we can proceed to determine the position functions $\mathbf{q}$ from the equations

$$\beta_k = -\frac{\partial \Psi}{\partial \alpha_k}, \quad k = 1, 2,$$

$$\beta_3 = -\frac{\partial \Psi}{\partial \alpha_3} - t,$$

in terms of the constants α and β. We find that

$$q_k = -\beta_k + \frac{\alpha_k}{m}(\beta_3 + t), \quad k = 1, 2,$$

$$q_3 = -g(t - \beta_3)^2 + \frac{2}{mg}\alpha_3 - \frac{1}{2m^2 g}(\alpha_1^2 + \alpha_2^2).$$

The above example is a somewhat complicated method for obtaining a solution that can readily be obtained from the Euler-Lagrange equations. The next problem is not quite so easy to solve using the Euler-Lagrange equations.

Example 8.5.2:   The motion of a particle in a plane under a central force field whose potential per unit mass is V leads to a Hamiltonian of the form

$$H(\mathbf{q}, \mathbf{p}) = \frac{1}{2}\left(p_1^2 + \left(\frac{p_2}{q_1}\right)^2\right) + V(q_1).$$

Here, $q_1 = r$ and $q_2 = \theta$ are polar coördinates. The reduced Hamilton-Jacobi equation leads to the differential equation

$$q_1^2 \left\{ \left(\frac{\partial \Psi}{\partial q_1}\right)^2 + 2(V(q_1) - E) \right\} + \left(\frac{\partial \Psi}{\partial q_2}\right)^2 = 0,$$

and hence we may take

$$g_1\left(q_1, \frac{\partial\Psi}{\partial q_1}\right) = q_1^2\left\{\left(\frac{\partial\Psi}{\partial q_1}\right)^2 + 2(V(q_1) - E)\right\}$$

$$g_2\left(q_2, \frac{\partial\Psi}{\partial q_2}\right) = \frac{\partial\Psi}{\partial q_2}.$$

We thus seek a solution of the form

$$\Psi(\mathbf{q}, \alpha) = \Psi_1(q_1, \alpha) + \Psi_2(q_2, \alpha),$$

where $\alpha = (\alpha_1, \alpha_2)$ is a constant. Let $\alpha_1 = 2E$. The differential equations for the Ψ_k are

$$\frac{\partial\Psi_2}{\partial q_2} = \alpha_2,$$

and

$$q_1^2\left\{\left(\frac{\partial\Psi_1}{\partial q_1}\right)^2 + 2V(q_1) - \alpha_1\right\} + \alpha_2^2 = 0,$$

where we have used relation (8.53). Ignoring integration constants we therefore have

$$\Psi_1(q_1, \alpha) = \int \sqrt{\alpha_1 - 2V(q_1) - \frac{\alpha_2^2}{q_1^2}}\, dq_1\,,$$

$$\Psi_2(q_2, \alpha) = \alpha_2 q_2.$$

A solution to the reduced Hamilton-Jacobi equation is thus

$$\Psi = \int \sqrt{f(q_1, \alpha)}\, dq_1 + \alpha_2 q_2,$$

where

$$f(q_1, \alpha) = \alpha_1 - 2V(q_1) - \frac{\alpha_2^2}{q_1^2}.$$

Now,

$$\mathbf{M} = \begin{pmatrix} \frac{1}{2\sqrt{f}} & -\frac{\alpha_2}{q_1^2\sqrt{f}} \\ 0 & 1 \end{pmatrix},$$

so that

$$\det\mathbf{M} = \frac{1}{2\sqrt{f}},$$

and hence the solution is complete provided $f > 0$. The position functions q_1 and q_2 are determined from the equations

$$\beta_1 = -\frac{\partial\Psi}{\partial\alpha_1} - t = -\frac{1}{2}\int \frac{dq_1}{\sqrt{f(q_1, \alpha)}} - t$$

$$\beta_2 = -\frac{\partial\Psi}{\partial\alpha_2} = \alpha_2\int \frac{dq_1}{q_1^2\sqrt{f(q_1, \alpha)}} - q_2,$$

where the β_k are constants. The solution of the problem is thus reduced to quadratures. Note, that by equation (8.36),

$$p_2 = \frac{\partial \Psi}{\partial q_2} = \alpha_2 = const.,$$

and hence the angular momentum of the particle is conserved.

8.5.2 Conditions for Separable Solutions*

The method of additive separation enables us to circumvent the problem of integrating Hamilton's equations or the equivalent system of Euler-Lagrange equations. Unfortunately, the method is not generally applicable to Hamilton-Jacobi equations and its success depends largely on the form of the Hamiltonian. As remarked earlier, even coördinate transformations affect the process. We saw in Example 8.5.2 that the reduced Hamilton-Jacobi equation for the central force problem is separable in polar coördinates. The same problem, however, is not separable when it is formulated in Cartesian coördinates (see Example 8.5.3). Some problems have more than one coördinate system in which the Hamilton-Jacobi equation is separable; others have no coördinate systems that lead to separable solutions.[10]

The importance of identifying Hamilton-Jacobi equations that are separable was recognized soon after the equation was first derived. Liouville (ca. 1846) studied the problem for the case $n = 2$. For a special but important class of Hamiltonians he established necessary and sufficient conditions for separability. Later, Stäckel (ca. 1890) generalized the results of Liouville for systems where $n \geq 3$. Both Liouville and Stäckel were concerned with Hamiltonians where the underlying coördinate system is orthogonal. Levi-Civita (ca. 1904) generalized the results for nonorthogonal coördinate systems. There are still many unanswered questions concerning the separability of the Hamilton-Jacobi equation. The monograph by Kalnins [43] details some of the newer, more specialized results in this field. Kalnins also discusses some of the basic questions and provides a number of key references on the subject. Here, we limit our discussion to a few elementary results with examples.

A significant class of problems in mechanics has a Hamiltonian of the form

$$H(\mathbf{q}, \mathbf{p}) = T(\mathbf{q}, \mathbf{p}) + V(\mathbf{q}),$$

where V is a potential energy term, and T is a kinetic energy term of the form

$$T(\mathbf{q}, \mathbf{p}) = \frac{1}{2} \sum_{k=1}^{n} C_k(\mathbf{q}) p_k^2,$$

---

[10] The famous "three body problem" is among these.

where the C_k are positive functions. The feature to note in the above form for H is that the p_k appear only in the combination p_k^2 (this indicates that the underlying coördinate system is orthogonal). Hamiltonians of this form lead to a reduced Hamilton-Jacobi equation of the form

$$\frac{1}{2} \sum_{k=1}^{n} C_k \left(\frac{\partial \Psi}{\partial q_k} \right)^2 + V = \alpha_1, \tag{8.55}$$

where α_1 is a constant. The results of Liouville and Stäckel concern essentially equations of the form (8.55).

Theorem 8.5.1 (Liouville) *A necessary and sufficient condition for the Hamilton-Jacobi equation*

$$\frac{1}{2} \left\{ C_1 \left(\frac{\partial \Psi}{\partial q_1} \right)^2 + C_2 \left(\frac{\partial \Psi}{\partial q_2} \right)^2 \right\} + V(\mathbf{q}) = \alpha_1, \tag{8.56}$$

where the C_k are positive functions of $\mathbf{q}$, to have a separable solution is that there exist functions ν_1, μ_1, σ_1 depending only on q_1 and functions ν_2, μ_2, σ_2 depending only on q_2 such that

$$C_1 = \frac{\mu_1}{\sigma_1 + \sigma_2},$$

$$C_2 = \frac{\mu_2}{\sigma_1 + \sigma_2},$$

and

$$V = \frac{\nu_1 + \nu_2}{\sigma_1 + \sigma_2}.$$

Proof: We first show that the equations for C_1, C_2, and V are necessary conditions for separability. Suppose that equation (8.56) is separable. Then there exists a complete solution Ψ of the form

$$\Psi(\mathbf{q}, \alpha) = \Psi_1(q_1, \alpha) + \Psi_2(q_2, \alpha),$$

and substituting this solution into equation (8.56) gives

$$\frac{1}{2} \left\{ C_1 \left(\frac{\partial \Psi_1}{\partial q_1} \right)^2 + C_2 \left(\frac{\partial \Psi_2}{\partial q_2} \right)^2 \right\} = \alpha_1 - V. \tag{8.57}$$

For simplicity, let

$$A = \frac{1}{2} \left(\frac{\partial \Psi}{\partial q_1} \right)^2,$$

$$B = \frac{1}{2} \left(\frac{\partial \Psi}{\partial q_2} \right)^2.$$

Note that A depends only on q_1 and B depends only on q_2. Equation (8.57) is satisfied for a continuum of α_1 and α_2 values, and differentiating this equation with respect to α_1 yields

$$C_1 A_1 + C_2 B_1 = 1, \tag{8.58}$$

where $A_k = \partial A/\partial \alpha_k$, $B_k = \partial B/\partial \alpha_k$. Similarly, differentiation with respect to α_2 gives

$$C_1 A_2 + C_2 B_2 = 0. \tag{8.59}$$

Equations (8.58) and (8.59) can be viewed as a system of linear equations for the C_k. Now,

$$A_1 B_2 - A_2 B_1 = \frac{\partial \Psi_1}{\partial q_1} \frac{\partial \Psi_2}{\partial q_2} \frac{\partial \left(\frac{\partial \Psi_1}{\partial q_1}, \frac{\partial \Psi_2}{\partial q_2} \right)}{\partial (\alpha_1, \alpha_2)}, \tag{8.60}$$

where the final factor on the right-hand side is a Jacobian term. The solution Ψ is complete and therefore the terms $\partial \Psi_j/\partial q_k$ cannot vanish identically. Moreover, $\det \mathbf{M} \neq 0$, so that the Jacobian term cannot vanish identically. We may thus choose a particular set of values α_1, α_2 such that these terms are nonzero and solve equations (8.58) and (8.59) to get

$$C_1 = \frac{\frac{1}{A_2}}{\frac{A_1}{A_2} - \frac{B_1}{B_2}}, \quad C_2 = \frac{-\frac{1}{B_2}}{\frac{A_1}{A_2} - \frac{B_1}{B_2}}.$$

Since A does not depend on q_2, neither A_1 nor A_2 depend on q_2; a similar statement can be made regarding B_1 and B_2. We may thus take $\mu_1 = 1/A_2$, $\mu_2 = -1/B_2$, $\sigma_1 = A_1/A_2$, and $\sigma_2 = -B_1/B_2$. We can use equation (8.56) to show that

$$V = \frac{\frac{\alpha_1 A_1 - A}{A_2} - \frac{\alpha_1 B_1 - B}{B_2}}{\frac{A_1}{A_2} - \frac{B_1}{B_2}}$$
$$= \frac{\nu_1 + \nu_2}{\sigma_1 + \sigma_2};$$

hence, the equations for the C_k and V are necessary conditions for separability. To establish sufficiency, consider a Hamilton-Jacobi equation of the form

$$\frac{1}{2(\sigma_1 + \sigma_2)} \left\{ \mu_1 \left(\frac{\partial \Psi}{\partial q_1} \right)^2 + \mu_2 \left(\frac{\partial \Psi}{\partial q_2} \right)^2 \right\} + \frac{\nu_1 + \nu_2}{\sigma_1 + \sigma_2} = \alpha_1, \tag{8.61}$$

where $\sigma_1 + \sigma_2 \neq 0$. We can determine a solution of the form $\Psi(\mathbf{q}, \alpha) = \Psi_1(q_1, \alpha) + \Psi_2(q_2, \alpha)$ from the equations

$$\frac{1}{2} \mu_1 \left(\frac{\partial \Psi_1}{\partial q_1} \right)^2 + \nu_1 - \alpha_1 \sigma_1 = \alpha_2,$$

$$\frac{1}{2} \mu_2 \left(\frac{\partial \Psi_2}{\partial q_2} \right)^2 + \nu_2 - \alpha_1 \sigma_2 = -\alpha_2,$$

where α_2 is a constant. Hence,

$$\Psi_1 = \int \sqrt{\frac{2}{\mu_1}(\alpha_2 + \alpha_1\sigma_1 - \nu_1)}\, dq_1,$$

and

$$\Psi_2 = \int \sqrt{\frac{2}{\mu_2}(-\alpha_2 + \alpha_1\sigma_1 - \nu_2)}\, dq_2.$$

(Note that $\mu_1, \mu_2 > 0$ since $C_1, C_2 > 0$ by hypothesis.) It remains only to show that the solution Ψ determined in this manner is complete. The solution Ψ has continuous second-order derivatives provided

$$D = \frac{2}{\mu_1}(\alpha_2 + \alpha_1\sigma_1 - \nu_1) > 0,$$

$$E = \frac{2}{\mu_2}(-\alpha_2 + \alpha_1\sigma_1 - \nu_2) > 0.$$

Moreover,

$$\mathbf{M} = \begin{pmatrix} \frac{-\sigma_1}{\mu_1\sqrt{D}} & \frac{1}{\mu_1\sqrt{D}} \\ \frac{\sigma_2}{\mu_2\sqrt{E}} & \frac{1}{\mu_2\sqrt{E}} \end{pmatrix},$$

and thus

$$\det\mathbf{M} = \frac{-(\sigma_1 + \sigma_2)}{\mu_1\mu_2\sqrt{D}} \neq 0.$$

The solution is thus complete in the domain defined by $D > 0, E > 0$. $\square$

Example 8.5.3: The reduced Hamilton-Jacobi equation of Example 8.5.2 can be readily put in the form (8.61). The reduced Hamilton-Jacobi equation from this example is equivalent to

$$\frac{1}{2q_1^2}\left\{ q_1^2\left(\frac{\partial\Psi}{\partial q_1}\right)^2 + \left(\frac{\partial\Psi}{\partial q_2}\right)^2 \right\} + \frac{q_1^2 V(q_1)}{q_1^2} = \alpha_1,$$

and we may take $\mu_1 = q_1^2$, $\mu_2 = 1$, $\sigma_1 = q_1^2$, $\sigma_2 = 0$, $\nu_1 q_1^2 V(q_1)$, and $\nu_2 = 0$. We could have thus concluded that a separable solution exists before we embarked on finding it.

Suppose, however, that the problem was initially posed in Cartesian coördinates (x, y). The reduced Hamilton-Jacobi equation in this coördinate system is

$$\frac{1}{2}\left\{ \left(\frac{\partial\Psi}{\partial x}\right)^2 + \left(\frac{\partial\Psi}{\partial y}\right)^2 \right\} + V(\sqrt{x^2 + y^2}) = \alpha_1.$$

For the central force problem, the potential function V must depend on x and y only through the combination $\sqrt{x^2 + y^2}$. This means that we cannot get V in the separated form required by Liouville's theorem (unless V is constant) and hence no separable solution exists for this equation.

The sufficiency part of the above theorem can be easily extended to higher dimensions. Specifically, it can be shown that a Hamilton-Jacobi equation of the form

$$\frac{1}{2(\sigma_1 + \cdots + \sigma_n)} \left\{ \mu_1 \left(\frac{\partial \Psi}{\partial q_1} \right)^2 + \cdots + \mu_n \left(\frac{\partial \Psi}{\partial q_n} \right)^2 \right\} + \frac{\nu_1 + \cdots + \nu_n}{\sigma_1 + \cdots + \sigma_n} = \alpha_1,$$

(8.62)

where the functions ν_k, μ_k, and σ_k depend only on q_k, $\sum_{k=1}^{n} \sigma_k > 0$, and $\mu_k > 0$, admits a complete solution of the form

$$\Psi(\mathbf{q}, \alpha) = \Psi_1(q_1, \alpha) + \cdots + \Psi_n(q_n, \alpha).$$

In fact, Ψ is given by

$$\Psi_1 = \int \sqrt{\frac{2}{\mu_1} \left(\alpha_1 \sigma_1 - \nu_1 + \sum_{k=2}^{n} \alpha_k \right)} \, dq_1,$$

and

$$\Psi_k = \int \sqrt{\frac{2}{\mu_k} \left(\alpha_1 \sigma_k - \nu_k - \alpha_k \right)} \, dq_k,$$

for $k = 2, \ldots, n$. Reduced Hamilton-Jacobi equations of the type (8.62) are said to be in **Liouville form**.

If $n = 2$, then a Hamilton-Jacobi equation must be reducible to Liouville form for a separable solution to exist. If $n \geq 3$, however, there are equations that are not reducible to Liouville form that are nonetheless separable. Stäckel studied this problem and arrived at the following characterization.

Theorem 8.5.2 (Stäckel) *A necessary and sufficient condition for the reduced Hamilton-Jacobi equation (8.55) to be separable is that there exists a nonsingular matrix $\mathbf{U}$ with entries u_{kj}, where for $j = 1, \ldots, n$, u_{kj} is a function of q_k only, and a column matrix $\mathbf{w} = (w_1, \ldots, w_n)^T$, where w_k is a function of q_k only, such that*

$$\sum_{k=1}^{n} C_k u_{k1} = 1,$$

(8.63)

$$\sum_{k=1}^{n} C_k u_{kj} = 0, \quad j = 2, \ldots, n,$$

(8.64)

$$\sum_{k=1}^{n} C_k w_k = V.$$

(8.65)

Proof: The proof of Stäckel's theorem is similar to that given for Liouville's theorem. We give only a sketch of the proof here. We first establish that equations (8.63) to (8.65) are necessary for a separable solution. Suppose that the Hamilton-Jacobi equation (8.55) is separable. Then there exists a complete solution of the form

$$\Psi = \Psi_1(q_1, \alpha) + \cdots + \Psi_n(q_n, \alpha). \tag{8.66}$$

Substituting the above solution into equation (8.55) and differentiating with respect to the α_j gives the equations

$$\sum_{k=1}^{n} C_k \frac{\partial \Psi_k}{\partial q_k} \frac{\partial^2 \Psi_k}{\partial \alpha_1 \partial q_k} = 1 \tag{8.67}$$

and

$$\sum_{k=1}^{n} C_k \frac{\partial \Psi_k}{\partial q_k} \frac{\partial^2 \Psi_k}{\partial \alpha_j \partial q_k} = 1, \quad j = 2, \ldots, n. \tag{8.68}$$

The coefficients of the C_k in the above linear equations are functions of q_k only. Moreover, the determinant of the coefficients is

$$\Delta = \frac{\partial \Psi_1}{\partial q_1} \cdots \frac{\partial \Psi_n}{\partial q_n} \det \mathbf{M},$$

where $\mathbf{M}$ is the matrix with entries $\{\partial^2 \Psi / \partial \alpha_j \partial q_k\}$. Since Ψ is a complete solution we have that Δ cannot vanish identically and therefore we may choose a particular set of α such that $\Delta \neq 0$. We can thus take

$$u_{kj} = \frac{\partial \Psi_k}{\partial q_k} \frac{\partial^2 \Psi}{\partial \alpha_j \partial q_k},$$

and substituting these expressions into equations (8.67) and (8.68) yields equations (8.63) and (8.64). Note that the matrix $\mathbf{U}$ thus defined is nonsingular since $\Delta \neq 0$ for our choice of α. The Hamilton-Jacobi equation (8.55) implies that the potential term V can be written in the form

$$V = \alpha_1 - \frac{1}{2} \sum_{k=1}^{n} C_k \left(\frac{\partial \Psi_k}{\partial q_k} \right)^2,$$

and using equation (8.63) this equation is equivalent to

$$V = \sum_{k=1}^{n} C_k \left(\alpha_1 u_{k1} - \frac{1}{2} \left(\frac{\partial \Psi_k}{\partial q_k} \right)^2 \right),$$

so that equation (8.65) is satisfied with

$$w_k = \alpha_1 u_{k1} - \frac{1}{2} \left(\frac{\partial \Psi_k}{\partial q_k} \right)^2.$$

Suppose now that the reduced Hamilton-Jacobi equation (8.55) satisfies equations (8.63) to (8.65). We can therefore recast equation (8.55) as

$$\frac{1}{2} \sum_{k=1}^{n} C_k \left(\frac{\partial \Psi_k}{\partial q_k} \right)^2 + \sum_{k=1}^{n} C_k w_k = \alpha_1 \sum_{k=1}^{n} C_k u_{k1} + \cdots$$

$$+ \alpha_n \sum_{k=1}^{n} C_k u_{kn},$$

where $\alpha_2, \ldots, \alpha_n$ are arbitrary constants. The above equation can be reorganized in the form

$$\sum_{k=1}^{n} C_k \left\{ \frac{1}{2} \left(\frac{\partial \Psi_k}{\partial q_k} \right)^2 - (\alpha_1 u_{k1} + \cdots + \alpha_n u_{kn} - w_k) \right\} = 0.$$

Although the C_k may depend on $q_1, \ldots, q_n$, the coefficient of each C_k in the above equation involves only $\partial \Psi / \partial q_k$ and q_k. We can thus construct a solution of the form (8.66) from the equations

$$\left(\frac{\partial \Psi_k}{\partial q_k} \right)^2 = 2(\alpha_1 u_{k1} + \cdots + \alpha_n u_{kn} - w_k) \equiv h_k(q_k, \alpha); \qquad (8.69)$$

i.e.,

$$\Psi_k = \int \sqrt{h_k(q_k, \alpha)} \, dq_k, \qquad (8.70)$$

for $k = 1, \ldots, n$, and it can be shown that such a solution is complete provided $h_k(q_k, \alpha) > 0$ for $k = 1, \ldots, n$. $\qquad \Box$

The equations (8.63) to (8.65) are sometimes called the **Stäckel conditions**. The matrix $\mathbf{U}$ is nonsingular and hence equations (8.63) and (8.64) can be solved for the C_k. The inverse matrix $\mathbf{S} = \mathbf{U}^{-1}$ is called a **Stäckel matrix**.

If the reduced Hamilton-Jacobi equation is separable, then the underlying Hamiltonian and Lagrangian systems can be solved by quadratures. It is interesting to note that in this case the Hamilton-Jacobi formulation and the Stäckel matrix can also be used to derive n conservation laws (first integrals) for the system. In detail, if there exists a complete solution to equation (8.55) of the form (8.66) then the Ψ_k satisfy equations (8.70). Now

$$\dot{q}_k = \frac{\partial H}{\partial p_k} = C_k p_k,$$

and therefore

$$\dot{q}_k = C_k \frac{\partial \Psi}{\partial q_k} = C_k \sqrt{h_k(q_k, \alpha)}. \qquad (8.71)$$

(Note that the C_k depend on $q_1, \ldots, q_n$ and so the $\dot{q}_k$ depend on these variables.) Rearranging equation (8.71) and using the definition of h_k gives

$$\frac{1}{2}\left(\frac{\dot{q}_k}{C_k}\right)^2 = \alpha_1 u_{k1} + \cdots + \alpha_n u_{kn} - w_k,$$

and using the inverse matrix $\mathbf{S}$, this expression yields the conservation laws

$$\sum_{j=1}^{n} s_{kj}\left(\frac{1}{2}\left(\frac{\dot{q}_j}{C_j}\right)^2 + w_j\right) = \alpha_k, \qquad (8.72)$$

for $k = 1, \ldots, n$. Here, s_{kj} denotes the kth row, jth column entry in the Stäckel matrix.

Example 8.5.4:   The central force problem in three dimensions using spherical coördinates leads to a Hamiltonian of the form

$$H(\mathbf{q}, \mathbf{p}) = \frac{1}{2m}\left(p_1^2 + \frac{p_2^2}{q_1^2} + \frac{p_3^2}{q_1^2 \sin^2 q_2}\right) + V(q_1),$$

where, in familiar spherical coördinate notation, $q_1 = r$, $q_2 = \theta$, and $q_3 = \phi$, $(x = q_1\cos q_2 \sin q_3$, $y = q_1\sin q_2 \sin q_3$, $z = q_1\cos q_3)$. The corresponding reduced Hamilton-Jacobi equation is of the form

$$\frac{1}{2}\sum_{k=1}^{3} C_k \left(\frac{\partial \Psi}{\partial q_k}\right)^2 + V(q_1) = \alpha_1, \qquad (8.73)$$

where α_1 is a constant, and

$$C_1 = \frac{1}{m}, \quad C_2 = \frac{1}{mq_1^2}, \quad C_3 = \frac{1}{mq_1^2 \sin^2 q_2}.$$

To find a suitable matrix $\mathbf{U}$ we need to find elements u_{kj} such that each u_{kj} depends only on q_k, and

$$C_1 u_{11} + C_2 u_{21} + C_3 u_{31} = 1,$$
$$C_1 u_{12} + C_2 u_{22} + C_3 u_{32} = 0,$$
$$C_1 u_{13} + C_2 u_{23} + C_3 u_{33} = 0.$$

The first equation is satisfied if $u_{11} = m$ and $u_{21} = u_{31} = 0$. The second equation is satisfied if $u_{12} = 0$, $u_{22} = -1/\sin^2 q_2$, and $u_{32} = 1$. The third equation is satisfied if $u_{13} = -1/q_1^2$, $u_{23} = 1$, and $u_{33} = 0$. Hence we have the matrix

$$\mathbf{U} = \begin{pmatrix} m & 0 & -\frac{1}{q_1^2} \\ 0 & -\frac{1}{\sin^2 q_2} & 1 \\ 0 & 1 & 0 \end{pmatrix},$$

and since

$$\det \mathbf{U} = -m \neq 0,$$

$\mathbf{U}$ is nonsingular for $q_1 > 0$, $0 < q_2 < \pi$. The choices $w_1 = mV(q_1)$, $w_2 = w_3 = 0$ suffice to meet the Stäckel condition (8.65). We can thus conclude that the reduced Hamilton-Jacobi equation is separable. Now,

$$h_1(q_1, \alpha) = 2\left(m\alpha_1 - \frac{\alpha_3}{q_1^2} - mV(q_1)\right),$$

$$h_2(q_2, \alpha) = 2\left(-\frac{\alpha_2}{\sin^2 q_2} + \alpha_3\right),$$

$$h_3(q_3, \alpha) = 2\alpha_2,$$

and hence

$$\Psi_1 = \int \sqrt{2\left(m\alpha_1 - \frac{\alpha_3}{q_1^2} - mV(q_1)\right)}\, dq_1,$$

$$\Psi_2 = \int \sqrt{2\left(-\frac{\alpha_2}{\sin^2 q_2} + \alpha_3\right)}\, dq_2,$$

$$\Psi_2 = 2\alpha_2 q_3.$$

The Stäckel matrix $\mathbf{S}$ is given by

$$\mathbf{S} = \begin{pmatrix} \frac{1}{m} & \frac{1}{mq_1^2} & \frac{1}{mq_1^2 \sin^2 q_2} \\ 0 & 0 & 1 \\ 0 & 1 & \frac{1}{\sin^2 q_2} \end{pmatrix},$$

and it can be shown that the three conservation laws associated with the system correspond to the conservation of energy, angular momentum about the polar axis (the z-component of the angular momentum), and the angular momentum.

Finally, it is interesting to note that if we generalize the problem to allow for a general potential function $V(\mathbf{q})$, the Stäckel condition (8.65) implies that V must be of the form

$$V(\mathbf{q}) = V_1(q_1) + \frac{1}{q_1^2}V_2(q_2) + \frac{1}{q_1^2 \sin^2 q_2}V_3(q_3)$$

in order that the corresponding reduced Hamilton-Jacobi equation be separable.

The results of Liouville and Stäckel apply to Hamiltonians where the underlying $q_1, \ldots, q_n$ coördinate system is orthogonal (there are no cross terms $\dot{q}_j \dot{q}_k$, $j \neq k$, in the kinetic energy function). It is natural to enquire whether a characterization of separable systems exists for nonorthogonal systems. The following result, which we state without proof,[11] applies to coördinate systems not necessarily orthogonal.

Theorem 8.5.3 (Levi-Civita) *A necessary and sufficient condition for the reduced Hamilton-Jacobi equation defined by*

$$H(q_1, \ldots, q_n, p_1, \ldots, p_n) = \alpha_1,$$

$$p_k = \frac{\partial \Psi}{\partial q_k},$$

to be separable is that the Hamiltonian H satisfy the $\frac{1}{2}n(n-1)$ equations

$$\frac{\partial H}{\partial p_k} \frac{\partial H}{\partial p_j} \frac{\partial^2 H}{\partial q_k \partial q_j} - \frac{\partial H}{\partial p_k} \frac{\partial H}{\partial q_j} \frac{\partial^2 H}{\partial q_k \partial p_j}$$

$$- \frac{\partial H}{\partial q_k} \frac{\partial H}{\partial p_j} \frac{\partial^2 H}{\partial p_k \partial q_j} + \frac{\partial H}{\partial q_k} \frac{\partial H}{\partial q_j} \frac{\partial^2 H}{\partial p_k \partial p_j},$$

where $j \neq k$, and $j, k = 1, \ldots, n$.

Theorem 8.5.3 is a launching point for much of the work on separable systems, particularly with the characterization of coördinate systems in which certain Hamilton-Jacobi equations can be separated. The reader is directed to Kalnins [43] for further results and references.

In closing this section (and chapter) we comment again that although the Hamiltonian formulation and the Hamilton-Jacobi equation are often of limited practical value for actually solving the Euler-Lagrange equations, they are useful in developing the underlying theory and making connexions across seemingly disparate theories such as electromagnetism and geometrical optics. In defense of the Hamilton-Jacobi equation as a tool for solving a variational problem, the sobering reality is that there is no general method for finding solutions analytically. For a limited but important class of Hamiltonians, the Hamilton-Jacobi equation is separable and produces general solutions. In its wake it also brings a wealth of byproducts such as conservation laws.

Exercises 8.5:

1. Consider an optical medium with a refractive index of the form

$$n(x, y) = \sqrt{f(x) + g(y)},$$

where f and g are functions such that $n \geq 1$. The light rays in this medium are given by the extremals $(x(t), y(t))$ to the functional

$$J(x, y) = \int_{t_0}^{t_1} n(x, y) \sqrt{x'^2 + y'^2} \, dt.$$

---

[11] The proof can be found in Kalnins [43], p. 13.

Determine the associated Hamilton-Jacobi equation and use Liouville's theorem to show that it must be separable. Reduce the problem of finding the extremals to quadratures. Note that this functional also models geodesics on a class of surfaces called **Liouville surfaces**.

2. The motion of a particle under gravity on a smooth spherical surface of radius R gives a kinetic energy term

$$T(\theta, \phi) = \frac{1}{2} m R^2 (\dot{\theta}^2 + \sin^2 \theta \dot{\phi}^2),$$

and a potential energy term

$$V(\theta) = mgR \cos \theta.$$

Here, θ and ϕ are polar angles, with θ being measured from the upward vertical. Derive the Hamilton-Jacobi equation and show that it is separable. Reduce the problem of finding the extremals to quadratures.

3. The motion of a particle of mass m in parabolic coördinates $(\xi, \eta, \phi) = (q_1, q_2, q_3)$ gives a Hamiltonian of the form

$$H = \frac{2}{m} \frac{q_1 p_1^2 + q_2 p_2^2}{q_1 + q_2} + \frac{p_3^2}{2m q_1 q_2} + V(\mathbf{q}).$$

Let

$$V(\mathbf{q}) = \frac{f(q_1) + g(q_2)}{q_1 + q_2}.$$

Use Stäckel's theorem to show that the corresponding Hamilton-Jacobi equation is separable. Find a separable solution.

9

Noether's Theorem

9.1 Conservation Laws

Let J be a functional of the form

$$J(y) = \int_{x_0}^{x_1} f(x, y, y', \dots, y^{(n)}) \, dx. \tag{9.1}$$

If there is a function $\phi(x, y, y', \dots, y^{(k)})$ such that

$$\frac{d}{dx} \phi(x, y, y', \dots, y^{(k)}) = 0 \tag{9.2}$$

for all extremals of J then relation (9.2) is called a kth order **conservation law** for J (and the associated Euler-Lagrange equation). For example,

$$H = y' \frac{\partial f}{\partial y'} - f \tag{9.3}$$

is a first-order conservation law for any functional of the form

$$J(y) = \int_{x_0}^{x_1} f(y, y') \, dx \tag{9.4}$$

(Theorem 2.3.1). The definition of a conservation law can be adapted to cope with functionals that involve several dependent variables. The definition can also be generalized for functionals that involve several independent variables. In this case ϕ is a vector function, and conservation laws are characterized by the divergence condition

$$\nabla \cdot \phi = 0.$$

We focus exclusively on the single independent variable case, but note that the results can be extended for functionals that involve several independent variables.

Conservation laws usually have an important/interesting physical interpretation (e.g., conservation of energy). In addition, they can materially simplify the problem of finding extremals when the order of the conservation law is less than that for the corresponding Euler-Lagrange equation. The Euler-Lagrange equation for the functional defined by (9.1) is of order $2n$, and equation (9.2) implies that

$$\phi(x, y, y', \ldots, y^{(k)}) = const. \tag{9.5}$$

If $k < 2n$ then the above relation is a differential equation of lower order that each extremal must satisfy. Such relations are called a **first integral** to the Euler-Lagrange equation. The right-hand side of the relation is a constant of integration that is determined by boundary conditions.

Given a functional of the form (9.1), it is not obvious how one might derive a conservation law, or for that matter, if it even has a conservation law. If the functional arises from some model, then the application itself might suggest the existence of a conservation law (e.g., conservation of energy). Some functionals may have several conservation laws; others may have no conservation laws. The problem is thus to develop a systematic method to identify functionals that have conservation laws and derive an algorithm for their construction.

A central result called Noether's theorem links conservation laws with certain invariance properties of the functional, and it provides an algorithm for finding the conservation law. In this chapter, we present a simple version of Noether's theorem that is motivated primarily by the pragmatic desire to find first integrals. We limit our discussion mostly to the simplest case when $n = 1$. A more complete study of Noether's theorem can be found in Bluman and Kumei [11] and Olver [57] especially for the case of several independent variables.

9.2 Variational Symmetries

The key to finding a conservation law for a functional lies in identifying transformations under which the functional is invariant. Let

$$J(y) = \int_{x_0}^{x_1} f(x, y, y') \, dx. \tag{9.6}$$

We consider a one-parameter family of transformations of the form

$$X = \theta(x, y; \epsilon), \quad Y = \psi(x, y; \epsilon), \tag{9.7}$$

where θ and ψ are smooth functions of x, y, and the parameter ϵ. In addition, we require

$$\theta(x, y; 0) = x, \quad \psi(x, y; 0) = y, \tag{9.8}$$

so that the parameter value $\epsilon = 0$ corresponds to the identity transformation. Examples of such families are given by the **translation transformations**

$$X = x + \epsilon, \quad Y = y, \tag{9.9}$$

$$X = x, \quad Y = y + \epsilon, \tag{9.10}$$

and a **rotation transformation**

$$X = x \cos \epsilon + y \sin \epsilon, \quad Y = -x \sin \epsilon + y \cos \epsilon. \tag{9.11}$$

The Jacobian matrix for the transformation (9.7) is

$$\frac{\partial(X, Y)}{\partial(x, y)} = \begin{pmatrix} \theta_x & \theta_y \\ \psi_x & \psi_y \end{pmatrix},$$

with determinant

$$\Delta(x, y; \epsilon) = \theta_x \psi_y - \theta_y \psi_x.$$

Now, θ and ψ are smooth functions; therefore, Δ is a smooth function. Moreover, since

$$\Delta(x, y; 0) = 1,$$

the continuity of Δ with respect to ϵ indicates that

$$\Delta(x, y; \epsilon) \neq 0 \tag{9.12}$$

for $|\epsilon|$ sufficiently small. Relation (9.12) implies that the transformation (9.7) has a unique inverse

$$x = \Theta(X, Y; \epsilon), \quad y = \Psi(X, Y; \epsilon), \tag{9.13}$$

provided $|\epsilon|$ is small (Theorem A.2.2). For example, the inverse of transformation (9.9) is

$$x = X - \epsilon, \quad y = Y,$$

and the inverse for transformation (9.11) is

$$x = X \cos \epsilon - Y \sin \epsilon, \quad y = X \sin \epsilon + Y \cos \epsilon.$$

For a given function $y(x)$ we can use relations (9.13) to eliminate x and determine Y as a function of X. In the following discussion we have occasion to consider Y as a function of X and some confusion might arise. We thus use the symbol $Y_\epsilon(X)$ to distinguish this case from $Y(x)$. Consider, for example, the transformation (9.9). Here $x = X - \epsilon$ and hence for any y

$$y(x) = Y(x) = y(X - \epsilon) = Y_\epsilon(X).$$

If, for instance, $y(x) = \cos(x)$, then $Y_\epsilon(X) = \cos(X - \epsilon)$. For another example, consider transformation (9.11) with $y(x) = x$. Then,

$$x = X \cos \epsilon - Y \sin \epsilon = y(x) = X \sin \epsilon + Y \cos \epsilon.$$

Solving the above relation for Y in terms of X and ϵ gives

$$Y_\epsilon(X) = \frac{\cos\epsilon - \sin\epsilon}{\sin\epsilon + \cos\epsilon}X.$$

We also use the notation

$$\dot{Y}_\epsilon(X) = \frac{d}{dX}Y_\epsilon(X).$$

Note that, for transformation (9.7), (9.13),

$$dx = (\Theta_X + \Theta_Y \dot{Y}_\epsilon(X))\, dX, \tag{9.14}$$
$$dy = (\Psi_X + \Psi_Y \dot{Y}_\epsilon(X))\, dX, \tag{9.15}$$

and hence

$$y'(x) = \frac{\Psi_X + \Psi_Y \dot{Y}_\epsilon(X)}{\Theta_X + \Theta_Y \dot{Y}_\epsilon(X)}. \tag{9.16}$$

We studied the effect of point transformations such as (9.7) on variational problems in Section 2.5 (for fixed values of ϵ). Theorem 2.5.1 shows that the transformed problem is variationally equivalent to the original problem. Generically, however, the integrand defining the functional changes under a transformation. Of special interest here are transformations that do not change the form of the integrand.

The integrand $f(x, y, y')$ of the functional J is said to be **variationally invariant** over the interval $[x_0, x_1]$ under the transformation (9.7) if, for all ϵ sufficiently small, in *any* subinterval $[a, b] \subseteq [x_0, x_1]$ we have

$$\int_a^b f(x, y(x), y'(x))\, dx = \int_{a_\epsilon}^{b_\epsilon} f(X, Y_\epsilon(X), \dot{Y}_\epsilon(X))\, dX \tag{9.17}$$

for *all* smooth functions y on $[a, b]$. Here,

$$a_\epsilon = \theta(a, y(a); \epsilon), \quad b_\epsilon = \theta(b, y(b); \epsilon).$$

In this case the transformation (9.7) is called a **variational symmetry** of J.

Example 9.2.1:   Let $x_0 = 0$, $x_1 = 1$,

$$f(x, y, y') = y'^2(x) + y^2(x),$$

and consider the transformation (9.9). For any ϵ we have by equation (9.14) $dx = dX$, and by equation (9.16)

$$y'(x) = \dot{Y}_\epsilon(X).$$

Therefore, for any $[a, b] \subseteq [0, 1]$ we have

$$\int_a^b \left(y'^2(x) + y^2(x) \right) dx = \int_{a+\epsilon}^{b+\epsilon} \left(\dot{Y}_\epsilon^2(X) + Y_\epsilon^2(X) \right) dX$$

$$= \int_{a+\epsilon}^{b+\epsilon} f\left(X, Y_\epsilon(X), \dot{Y}_\epsilon(X) \right) dX.$$

We thus conclude that transformation (9.9) is a variational symmetry for J.

Example 9.2.2:   Let $x_0 = 0$, $x_1 = 1$,

$$f(x, y, y') = y'^2(x) + xy^2(x),$$

and consider again the transformation (9.9). Now,

$$y'^2(x) + xy^2(x) = \dot{Y}_\epsilon^2(X) + (X - \epsilon)Y_\epsilon^2(X)$$
$$= f\left(X, Y_\epsilon(X), \dot{Y}_\epsilon(X) \right) - \epsilon Y_\epsilon^2(X),$$

and hence for any $[a, b] \subseteq [0, 1]$,

$$J(y) = \int_a^b f(x, y, y') \, dx$$

$$= \int_{a+\epsilon}^{b+\epsilon} f(X, Y, \dot{Y}_\epsilon) \, dX$$

$$- \epsilon \int_{a+\epsilon}^{b+\epsilon} Y_\epsilon^2(X) \, dX,$$

so that the transformation (9.9) is not a variational symmetry for J.

In fact, it can be shown that transformation (9.9) is a variational symmetry for any functional of the form (9.4). It can also be shown that transformation (9.10) is a variational symmetry for any functional of the form

$$J(y) = \int_{x_0}^{x_1} f(x, y') \, dx. \tag{9.18}$$

The notion of variational invariance can be extended to functionals that involve several dependent variables. Let $\mathbf{q} = (q_1, \ldots, q_n)$ and

$$J(\mathbf{q}) = \int_{t_0}^{t_1} L(t, \mathbf{q}, \dot{\mathbf{q}}) \, dt.$$

We consider transformations of the form

$$T = \theta(t, \mathbf{q}; \epsilon), \quad Q_k = \psi_k(t, \mathbf{q}; \epsilon), \tag{9.19}$$

where $k = 1, \ldots, n$. Here, θ and ψ_k are smooth functions that satisfy

$$\theta(t, \mathbf{q}; 0) = t, \quad \psi_k(t, \mathbf{q}; 0) = q_k.$$

Similar arguments to those used for the previous case can be used to show that the transformation (9.19) is invertible. The integrand $L(t, \mathbf{q}, \dot{\mathbf{q}})$ is **variationally invariant** over $[t_0, t_1]$ under the transformation (9.19) if, for all $|\epsilon|$ small, in any subinterval $[\alpha, \beta] \subseteq [t_0, t_1]$, we have

$$\int_{\alpha}^{\beta} L(t, \mathbf{q}, \dot{\mathbf{q}}) \, dt = \int_{\alpha_\epsilon}^{\beta_\epsilon} L(T, \mathbf{Q}_\epsilon(T), \mathbf{Q}'_\epsilon(T)) \, dT, \tag{9.20}$$

for all smooth functions $\mathbf{q}$ on $[\alpha, \beta]$. Here, $\alpha_\epsilon = \theta(\alpha, \mathbf{q}(\alpha); \epsilon)$, $\beta_\epsilon = \theta(\beta, \mathbf{q}(\beta); \epsilon)$, and $'$ denotes d/dT.

Example 9.2.3:   Let $n = 2$, $t_0 = 0$, $t_1 = 1$, and

$$L(t, \mathbf{q}, \dot{\mathbf{q}}) = \frac{1}{2}m \left(\dot{q}_1^2 + \dot{q}_2^2 \right) + \frac{K}{\sqrt{q_1^2 + q_2^2}}, \tag{9.21}$$

where m and K are constants. A Lagrangian such as this arises in the Kepler problem modelling planetary motion (Example 1.3.2). The integrand does not contain t explicitly and we can show that the transformation

$$T = t + \epsilon, \quad Q_k = q_k \tag{9.22}$$

is a variational symmetry. For variety, however, consider the transformation

$$\begin{aligned} T &= t, \\ Q_1 &= q_1 \cos \epsilon + q_2 \sin \epsilon, \\ Q_2 &= -q_1 \sin \epsilon + q_2 \cos \epsilon, \end{aligned} \tag{9.23}$$

which represents a rotation of the space variables $\mathbf{q}$. Now, $dT = dt$, and

$$\begin{aligned} Q'_{\epsilon 1} &= \dot{q}_1 \cos \epsilon + \dot{q}_2 \sin \epsilon, \\ Q'_{\epsilon 2} &= -\dot{q}_1 \sin \epsilon + \dot{q}_2 \cos \epsilon, \end{aligned}$$

so that

$$\begin{aligned} L(T, \mathbf{Q}_\epsilon, \mathbf{Q}'_\epsilon) &= \frac{1}{2}m \left(Q'^2_{\epsilon 1} + Q'^2_{\epsilon 2} \right) + \frac{K}{\sqrt{Q^2_{\epsilon 1} + Q^2_{\epsilon 2}}} \\ &= \frac{1}{2}m \left((\dot{q}_1 \cos \epsilon + \dot{q}_2 \sin \epsilon)^2 + (-\dot{q}_1 \sin \epsilon + \dot{q}_2 \cos \epsilon)^2 \right) \\ &\quad + \frac{K}{\sqrt{(q_1 \cos \epsilon + q_2 \sin \epsilon)^2 + (-q_1 \sin \epsilon + q_2 \cos \epsilon)^2}} \\ &= \frac{1}{2}m \left(\dot{q}_1^2 + \dot{q}_2^2 \right) + \frac{K}{\sqrt{q_1^2 + q_2^2}} \\ &= L(t, \mathbf{q}, \dot{\mathbf{q}}). \end{aligned}$$

Hence, for any $[\alpha, \beta] \subseteq [0, 1]$ we have

$$\int_{\alpha}^{\beta} L(t, \mathbf{q}, \dot{\mathbf{q}}) \, dt = \int_{\alpha}^{\beta} L(T, \mathbf{Q}_\epsilon, \mathbf{Q}'_\epsilon) \, dT,$$

and consequently L is variationally invariant under transformation (9.22). We thus see that this functional has at least two variational symmetries.

Exercises 9.2:

1. Let

$$J(y) = \int_{x_0}^{x_1} xy'^2 \, dx.$$

Show that the transformation

$$X = x + \epsilon 2x \ln x, \quad Y = (1 + \epsilon)y \qquad (9.24)$$

is a variational symmetry for J.

9.3 Noether's Theorem

We know from Section 2.3 that the quantity H (the Hamiltonian) defined by equation (9.3) is constant along any extremal for functionals of the form (9.4) and from Section 9.2 that such functionals have the variational symmetry (9.9). In addition, we know (Section 2.3) that any functional of the form (9.18) has a conservation law, viz.,

$$\frac{\partial f}{\partial y'} = const.,$$

and that the transformation (9.10) is a variational symmetry for such a functional. Although this is a special selection, we may suspect that the existence of a conservation law is linked with that of a variational symmetry. In this section we present a result called Noether's theorem, which shows that each variational symmetry for a functional corresponds to a conservation law. Noether's theorem also provides the conservation law.

Before we state Noether's theorem, we need to introduce another term. Taylor's theorem shows that transformation (9.7) can be written

$$X = \theta(x, y; 0) + \epsilon \frac{\partial \theta}{\partial \epsilon}\bigg|_{(x,y;0)} + O(\epsilon^2)$$

$$Y = \psi(x, y; 0) + \epsilon \frac{\partial \psi}{\partial \epsilon}\bigg|_{(x,y;0)} + O(\epsilon^2),$$

provided $|\epsilon|$ is small. Let

$$\xi(x, y) = \left.\frac{\partial \theta}{\partial \epsilon}\right|_{(x,y;0)},$$

$$\eta(x, y) = \left.\frac{\partial \psi}{\partial \epsilon}\right|_{(x,y;0)}.$$

Then, relation (9.8) gives

$$X = x + \epsilon\xi + O(\epsilon^2),$$
$$Y = y + \epsilon\eta + O(\epsilon^2),$$

so that, for $|\epsilon|$ small, the linear approximation to the transformation is

$$X \approx x + \epsilon\xi$$
$$Y \approx y + \epsilon\eta.$$

The functions ξ and η are called the **infinitesimal generators** for the transformation (9.7). Similarly, the infinitesimal generators for a transformation of the form (9.19) are given by

$$\xi(t, \mathbf{q}) = \left.\frac{\partial \theta}{\partial \epsilon}\right|_{(t,\mathbf{q};0)},$$

$$\eta_k(t, \mathbf{q}) = \left.\frac{\partial \psi_k}{\partial \epsilon}\right|_{(t,\mathbf{q};0)}.$$

Theorem 9.3.1 (Noether) *Suppose that $f(x, y, y')$ is variationally invariant on $[x_0, x_1]$ under transformation (9.7) with infinitesimal generators ξ and η. Then*

$$\eta\frac{\partial f}{\partial y'} + \xi\left(f - \frac{\partial f}{\partial y'}y'\right) = const. \tag{9.25}$$

along any extremal of

$$J(y) = \int_{x_0}^{x_1} f(x, y, y')\, dx.$$

Remark: In the notation of Chapters 7 and 8, equation (9.25) can be written

$$\eta p - \xi H = const. \tag{9.26}$$

The left-hand side of this equation is precisely the same quantity encountered in the general variation condition (7.23). Noether's theorem can be proved using a calculation similar to that leading to this condition. The main difference is that we are no longer dealing with arbitrary variations; instead, we are restricted to the one-parameter family of functions defined by transformation (9.7).

Proof: Let

$$\tilde{J}(y) = \int_a^b f(x, y, y') \, dx,$$

where a and b are numbers such that $a < b$, $[a, b] \subseteq [x_0, x_1]$, but otherwise arbitrary. Here, we regard $\tilde{J}$ as a functional with variable limits of integration. By hypothesis f is variationally invariant and hence for any X, Y_ϵ defined by transformation (9.7) we have

$$\tilde{J}(Y_\epsilon) - \tilde{J}(y) = \int_{a_\epsilon}^{b_\epsilon} f(X, Y_\epsilon, \dot{Y}_\epsilon) \, dX - \int_a^b f(x, y, y') \, dx = 0. \qquad (9.27)$$

Suppose that y is an extremal for $\tilde{J}$. For $|\epsilon|$ small we can regard Y_ϵ along with the limits a_ϵ, b_ϵ as a special case of a variation with free endpoints, since condition (9.27) is stronger than simply requiring that $\tilde{J}(Y_\epsilon) - \tilde{J}(y) = O(\epsilon^2)$. We may thus use the calculations leading to equation (7.23). Here,

$$X = x + \epsilon\xi + O(\epsilon^2) = x + \epsilon X_0,$$
$$Y_\epsilon = y + \epsilon\eta + O(\epsilon^2) = x + \epsilon Y_0.$$

We cannot argue as in Section 7.2 that the Euler-Lagrange equation is satisfied because we are not free to choose the special class of endpoint variations that vanish. We can nonetheless assert that the Euler-Lagrange equation is satisfied by Y_ϵ because extremals map to extremals under point transformations (Theorem 2.5.1), and the invariance of f implies that the Euler-Lagrange equation is unchanged for these transformations. We are thus led to the relation

$$\left. \eta p - \xi H \right|_a^b = 0. \qquad (9.28)$$

Since $[a, b]$ is an arbitrary subinterval of $[x_0, x_1]$, it follows that $\eta p - \xi H$ must be constant along any extremal. $\qquad \square$

Example 9.3.1: We can rapidly recover Theorem 2.3.1 from Noether's theorem. Let J be a functional of the form (9.4). We know from Section 9.2 that the translational transformation (9.9) is a variational symmetry for J. Now,

$$\xi(x, y) = \left. \frac{\partial\theta}{\partial\epsilon} \right|_{\epsilon=0} = 1,$$
$$\eta(x, y) = \left. \frac{\partial\psi}{\partial\epsilon} \right|_{\epsilon=0} = 0.$$

Equation (9.25) thus implies

$$y' \frac{\partial f}{\partial y'} - f = H = const.$$

along any extremal.

Example 9.3.2: Consider the functional and transformation of Exercise 9.2-1. The infinitesimal generators for this variational symmetry are

$$\xi(x, y) = -2x \ln x,$$
$$\eta(x, y) = y.$$

Noether's theorem indicates that

$$xyy' - y'^2 x^2 \ln x = const. \tag{9.29}$$

along the extremals for J. We can verify that the above expression is satisfied for all extremals by differentiating the left-hand side of equation (9.29) and applying the Euler-Lagrange equation,

$$(xy')' = 0,$$

associated with the functional. In detail

$$\frac{d}{dx} \left((xy') (y - xy' \ln x) \right)$$
$$= (xy')' (y - xy' \ln x)$$
$$\quad + (xy') (y' - (xy')' \ln x - y')$$
$$= 0.$$

Note that the translational transformation (9.10) is also a variational symmetry for J. This symmetry leads to the conservation law

$$xy' = const.,$$

which can be deduced readily from the Euler-Lagrange equation.

Noether's theorem can be generalized to accommodate functionals that involve several dependent variables. In this case, it takes the following form.

Theorem 9.3.2 (Noether) *Suppose that $L(t, \mathbf{q}, \dot{\mathbf{q}})$ is variationally invariant on $[t_0, t_1]$ under the transformation (9.19), where $\mathbf{q} = (q_1, \ldots, q_n)$. Let ξ and η_k be the infinitesimal generators for this transformation,*

$$p_k = \frac{\partial L}{\partial \dot{q}_k},$$

and

$$H = \sum_{k=1}^{n} p_k \dot{q}_k - L$$

(the Hamiltonian). Then

$$\sum_{k=1}^{n} p_k \eta_k - H\xi = const. \tag{9.30}$$

along any extremal of

$$J(\mathbf{q}) = \int_{t_0}^{t_1} L(t, \mathbf{q}, \dot{\mathbf{q}}) \, dt.$$

Example 9.3.3: Consider the Lagrangian L of Example 9.2.3. We know that the translational transformation (9.22) is a variational symmetry. For this transformation $\xi = 1$, $\eta_k = 0$, and Noether's theorem gives

$$H = const. \tag{9.31}$$

along extremals. The rotational transformation (9.23) is also a variational symmetry. In this case $\xi = 0$, and

$$\begin{aligned}
\eta_1 &= \frac{\partial}{\partial \epsilon} \left(q_1 \cos \epsilon + q_2 \sin \epsilon \right) \Big|_{\epsilon=0} \\
&= (-q_1 \sin \epsilon + q_2 \cos \epsilon) \Big|_{\epsilon=0} \\
&= q_2,
\end{aligned}$$

and

$$\begin{aligned}
\eta_2 &= \frac{\partial}{\partial \epsilon} \left(-q_1 \sin \epsilon + q_2 \cos \epsilon \right) \Big|_{\epsilon=0} \\
&= (-q_1 \cos \epsilon - q_2 \sin \epsilon) \Big|_{\epsilon=0} \\
&= -q_1.
\end{aligned}$$

Equation (9.30) thus gives

$$\dot{q}_1 q_2 - \dot{q}_2 q_1 = const. \tag{9.32}$$

along extremals.

The Lagrangian in the above example comes from the Kepler problem (Example 1.3.2), where $\mathbf{q}$ denotes the position of the planet. Equation (9.31) indicates that energy is conserved along the orbit of the planet (see Example 3.2.2). The second conservation law (9.32) is less obvious. This equation corresponds to Kepler's second law of planetary motion, viz., the conservation of "areal velocity." In the next example we explore connexions between some well-known conservation laws from classical mechanics and the corresponding variational symmetries.

Example 9.3.4: Let $\mathbf{q}(t) = (q_1(t), q_2(t), q_3(t))$ denote the position of a particle (in Cartesian coördinates) at time t. The kinetic energy is

$$T(\dot{\mathbf{q}}) = \frac{1}{2} m \left(\dot{q}_1^2 + \dot{q}_2^2 + \dot{q}_3^2 \right),$$

where m denotes the mass of the particle. Let $V(t, \mathbf{q})$ denote the potential energy. Hamilton's Principle implies that $\mathbf{q}$ is an extremal for

$$J(\mathbf{q}) = \int_{t_0}^{t_1} L(t, \mathbf{q}, \dot{\mathbf{q}})\, dt,$$

where the Lagrangian is

$$L(t, \mathbf{q}, \dot{\mathbf{q}}) = T(\dot{\mathbf{q}}) - V(t, \mathbf{q}). \tag{9.33}$$

The well-known conservation laws of energy, momentum, and angular momentum correspond to translational or rotational variational symmetries for J.

A. Conservation of Energy

Suppose that L is variationally invariant under the transformation

$$T = t + \epsilon, \quad Q_k = q_k, \tag{9.34}$$

where $k = 1, 2, 3$. Then $\xi = 1$, and $\eta_k = 0$ for $k = 1, 2, 3$. Noether's theorem thus gives the

$$H = T + V = const.,$$

so that the total energy is conserved along an extremal.

B. Conservation of Momentum

Suppose that L is variationally invariant under the transformation

$$T = t, \quad Q_1 = q_1 + \epsilon, \quad Q_k = q_k, \tag{9.35}$$

for $k = 2, 3$. Then $\xi = \eta_2 = \eta_3 = 0$, and $\eta_1 = 1$. In this case Noether's theorem gives

$$p_1 = \frac{\partial L}{\partial \dot{q}_1} = m\dot{q}_1 = const.,$$

which indicates that the q_1 component of momentum is conserved.

C. Conservation of Angular Momentum

Suppose that L is variationally invariant under the transformation

$$T = t,$$
$$Q_1 = q_1 \cos \epsilon + q_2 \sin \epsilon,$$
$$Q_2 = -q_1 \sin \epsilon + q_2 \cos \epsilon \tag{9.36}$$
$$Q_3 = q_3. \tag{9.37}$$

Then $\xi = 0$, $\eta_1 = q_2$, $\eta_2 = -q_1$, and $\eta_3 = 0$. For this case Noether's theorem yields

$$p_1 q_2 - p_2 q_1 = const.$$

Now, the momentum vector is $\mathbf{p} = (p_1, p_2, p_3)$, and the angular momentum about the origin is $\mathbf{p} \wedge \mathbf{q}$. Evidently, the term $p_1 q_2 - p_2 q_1$ is the q_3 component of the angular momentum vector; hence, Noether's theorem implies that this component of the angular momentum is conserved.

Noether's theorem thus provides a nice mechanism for interpreting well-known conservation laws in terms of variational symmetries. For Lagrangians of the form (9.33) we can readily deduce the appropriate variational symmetries from the potential energy function V. If, for instance, V does not contain t explicitly then we have the conservation of energy. If V does not depend on one of the q_k then we know that the corresponding component of momentum is conserved. If V corresponds to a central force, i.e., $V = V(t,r)$, where say $r^2 = q_1^2 + q_2^2$, then we have that the q_3 component of the angular momentum is conserved.

9.4 Finding Variational Symmetries

The reader will appreciate at this stage that the crux with Noether's Theorem is finding the variational symmetry. In fact, it is clear from the statement of Noether's theorem that we need find only the infinitesimal generators of a variational symmetry in order to construct the corresponding conservation law. In this section we give a method for finding variational symmetries. The method is based on the following result, the proof of which we can be found in Wan [71], or in a more general form, in Giaquinta and Hildebrandt [32].

Theorem 9.4.1 *Let*

$$J(y) = \int_{x_0}^{x_1} f(x, y, y')\, dx.$$

The transformation (9.7) with infinitesimal generators ξ and η is a variational symmetry for J if and only if

$$\xi \frac{\partial f}{\partial x} + \eta \frac{\partial f}{\partial y} + (\eta' - y'\xi') \frac{\partial f}{\partial y'} + \xi' f = 0, \tag{9.38}$$

for any smooth function y on $[x_0, x_1]$. Here,

$$\eta' = \frac{\partial \eta}{\partial x} + \frac{\partial \eta}{\partial y} y',$$

and

$$\xi' = \frac{\partial \xi}{\partial x} + \frac{\partial \xi}{\partial y} y'.$$

Equation (9.38) can be used to find the infinitesimal generators η and ξ. *Prima facie*, it seems that we have one differential equation for two unknown functions, but the equation must hold for *any* y not just extremals, and it is this condition that yields additional equations. The condition (9.38) is a relation of the form

$$W(x, y, y') = 0, \tag{9.39}$$

that must hold for all y. The unknown functions η and ξ depend only on x and y, so that we do know how W depends on y' in terms of ξ and η. Now, equation (9.39) is an identity that must hold pointwise on $[x_0 x_1]$ for any choice of y. Since we may always choose $y(x^*)$ and $y'(x^*)$ independently at any point $x^* \in [x_0 x_1]$, we can regard y' as an independent variable for this identity. This means that we can supplement equation (9.39) with equations of the form

$$\frac{\partial^k W}{\partial y'^k} = 0, \tag{9.40}$$

where x and y (hence η and ξ) are held fixed for the differentiation and $k = 1, 2, \ldots$.[1] For example, suppose that

$$W(x, y, y') = Ay'^2 + By' + C,$$

where A, B, and C are functions that depend explicitly on x, y, ξ, η along with the partial derivatives of the generators. Then, equation (9.40) implies that the coefficients of y'^2, y' must vanish; i.e., $A = B = 0$, and hence $C = 0$. These three equations can then be used to determine ξ and η. Note that we expect an overdetermined system, since variational symmetries are special and not every functional has them. Moreover, if there exist ξ and η that satisfy these equations we expect these functions to be determined to within a constant of integration because no initial data are specified. The above comments are perhaps best illustrated through specific examples.

Example 9.4.1:   Consider the functional of Exercise 9.2-1. For this functional, equation (9.38) is

$$\xi y'^2 + 2xy' \left(\eta_x + y' \eta_y - y' \xi_x - y'^2 \xi_y \right) + xy'^2 \left(\xi_x + y' \xi_y \right)$$
$$= -x \xi_y y'^3 + \left(\xi + 2x\eta_y - x\xi_x \right) y'^2 + 2x\eta_x y'$$
$$= 0,$$

where $\xi_x = \partial \xi / \partial x$ etc. The coefficients of y'^3, y'^2, and y' must vanish; hence,

$$x\xi_y = 0, \tag{9.41}$$

$$\xi + 2x\eta_y - x\xi_x = 0, \tag{9.42}$$

$$x\eta_x = 0, \tag{9.43}$$

---

[1] Of course, we could argue that similar expressions can be ascertained by differentiating W with respect to the other independent variables x and y, but we do not know η or ξ at this stage (this is the purpose of studying the equation) and hence we do not know how W depends on x and y.

for all $x \in [x_0, x_1]$ and y. Equation (9.41) implies that $\xi_y = 0$; hence, $\xi = \xi(x)$. Similarly, equation (9.43) implies that $\eta = \eta_y$. Since ξ depends only on x, and η depends only on y, equation (9.42) is satisfied only if $\eta_y = const.$; therefore,

$$\eta(y) = c_1 y + c_2, \tag{9.44}$$

where c_1 and c_2 are constants. Equation (9.42) now gives the first-order differential equation

$$\xi_x - \frac{1}{x}\xi + 2c_1 = 0. \tag{9.45}$$

The differential equation (9.45) has the general solution

$$\xi = -2c_1 x \ln x + c_3 x, \tag{9.46}$$

where c_3 is a constant. Equations (9.44) and (9.46) thus define a three-parameter family of infinitesimal generators that correspond to variational symmetries. The infinitesimal generators for the transformation of Exercise 9.2-2 correspond to the choice $c_2 = c_3 = 0$, $c_1 = 1$.

Example 9.4.2:   Let J be the functional defined by

$$J(y) = \int_{x_0}^{x_1} \left(x^2 y'^2 + y^4 \right) dx.$$

For this functional, equation (9.38) leads to the relations

$$x^2 \xi_y = 0, \tag{9.47}$$

$$2x\xi + 2x^2 \eta_y - x^2 \xi_x = 0, \tag{9.48}$$

$$2x^2 \eta_x + y^4 \xi_y = 0, \tag{9.49}$$

$$4\eta y^3 + \xi_x y^4 = 0, \tag{9.50}$$

which the infinitesimal generators ξ and η must satisfy for the corresponding transformation to be a variational symmetry. Equation (9.47) implies that $\xi = \xi(x)$; hence, equation (9.49) implies that $\eta = \eta(y)$. Equation (9.48) thus shows that $\eta_y = const.$, and equation (9.50) shows that $\xi_x = const.$ The functions ξ and η must therefore be of the form

$$\xi = c_1 x + c_2, \quad \eta = c_3 y + c_4. \tag{9.51}$$

Substituting expressions (9.51) into equation (9.48) gives

$$2x \left(c_1 x + c_2 \right) + 2x^2 c_3 - x^2 c_1 = 0,$$

which must be satisfied for all $x \in [x_0, x_1]$; consequently,

$$c_2 = 0, \quad 2c_3 + c_1 = 0. \tag{9.52}$$

Substituting expressions (9.51) into equation (9.50) gives

$$4y^3 (c_3 y + c_4) + c_1 y^4 = 0,$$

which must be satisfied for all y; hence,

$$c_4 = 0, \quad 4c_3 + c_1 = 0. \tag{9.53}$$

The only choice of constants that satisfies both (9.52) and (9.53) is $c_1 = c_3 = 0$, so that $\xi = 0, \eta = 0$ is the only solution to (9.38) for all $x \in [x_0, x_1]$ and all y. In this case there are no variational symmetries for the functional.

The condition for finding a variational symmetry for a functional of the form

$$J(\mathbf{q}) = \int_{t_0}^{t_1} L(t, \mathbf{q}, \dot{\mathbf{q}}) \, dt, \tag{9.54}$$

where $\mathbf{q} = (q_1, \ldots, q_n)$ and $n > 1$, is somewhat more complicated than that given in Theorem 9.4.1. Let ξ and $\eta_1, \ldots, \eta_n$ be the infinitesimal generators for the transformation (9.19), and let

$$pr^{(1)}\mathbf{v}(L) = \xi \frac{\partial L}{\partial t} + \sum_{k=1}^{n} \left(\eta_k \frac{\partial L}{\partial q_k} + p_k \left(\dot{\eta}_k - \dot{\xi} \dot{q}_k \right) \right),$$

where

$$p_k = \frac{\partial L}{\partial \dot{q}_k}$$

$$\dot{\xi} = \frac{\partial \xi}{\partial t} + \sum_{j=1}^{n} \dot{q}_j \frac{\partial \xi}{\partial q_j},$$

$$\dot{\eta}_k = \frac{\partial \eta_k}{\partial t} + \sum_{j=1}^{n} \dot{q}_j \frac{\partial \eta_k}{\partial q_j}.$$

Theorem 9.4.2 *The transformation (9.51) is a variational symmetry for the functional defined by (9.54) if and only if*

$$pr^{(1)}\mathbf{v}(L) + L\dot{\xi} = 0 \tag{9.55}$$

for all smooth $\mathbf{q}$ on $[t_0, t_1]$.

Example 9.4.3:   Kepler Problem

Let

$$L(t, \mathbf{q}, \dot{\mathbf{q}}) = \frac{m}{2} \left(\dot{q}_1^2 + \dot{q}_2^2 \right) + \frac{K}{\sqrt{q_1^2 + q_2^2}},$$

where $m > 0$ and K are constants. Here $n = 2$, and

$$\frac{\partial L}{\partial t} = 0, \quad p_k = m\dot{q}_k, \quad \frac{\partial L}{\partial q_k} = -\frac{q_k K}{(q_1^2 + q_2^2)^{3/2}}.$$

Now,

$$\begin{aligned}
pr^{(1)}\mathbf{v}(L) &= \eta_1 \frac{\partial L}{\partial q_1} + p_1 \left(\dot{\eta}_1 - \dot{\xi}\dot{q}_1 \right) + \eta_2 \frac{\partial L}{\partial q_2} + p_2 \left(\dot{\eta}_2 - \dot{\xi}\dot{q}_2 \right) \\
&= -K \frac{\eta_1 q_1 + \eta_2 q_2}{(q_1^2 + q_2^2)^{3/2}} + m \left(\dot{\eta}_1 \dot{q}_1 - \dot{\xi}\dot{q}_1^2 + \dot{\eta}_2 \dot{q}_2 - \dot{\xi}\dot{q}_2^2 \right) \\
&= -K \frac{\eta_1 q_1 + \eta_2 q_2}{(q_1^2 + q_2^2)^{3/2}} \\
&\quad + m \left(-\dot{q}_1^3 \xi_1 - \dot{q}_2^3 \xi_2 - \dot{q}_1^2 \dot{q}_2 \xi_2 - \dot{q}_1 \dot{q}_2^2 \xi_1 \right. \\
&\quad + \dot{q}_1^2 \left(\eta_{1,1} - \xi_t \right) + \dot{q}_2^2 \left(\eta_{2,2} - \xi_t \right) \\
&\quad \left. + \dot{q}_1 \dot{q}_2 \left(\eta_{1,2} + \eta_{2,1} \right) + \dot{q}_1 \eta_{1,t} + \dot{q}_2 \eta_{2,t} \right),
\end{aligned}$$

and

$$\begin{aligned}
L\dot{\xi} &= \left(\frac{m}{2} \left(\dot{q}_1^2 + \dot{q}_2^2 \right) + \frac{K}{\sqrt{q_1^2 + q_2^2}} \right) \left(\xi_t + \dot{q}_1 \xi_1 + \dot{q}_1 \xi_1 \right) \\
&= \frac{m}{2} \left(\dot{q}_1^3 \xi_1 + \dot{q}_2^3 \xi_2 + \dot{q}_1^2 \dot{q}_2 \xi_2 + \dot{q}_1 \dot{q}_2^2 \xi_1 + \dot{q}_1^2 \xi_t + \dot{q}_2^2 \xi_t \right) \\
&\quad + \frac{K}{\sqrt{q_1^2 + q_2^2}} \left(\xi_t + \dot{q}_1 \xi_1 + \dot{q}_2 \xi_2 \right),
\end{aligned}$$

where, for succinctness, we use the notation

$$\xi_k = \frac{\partial \xi}{\partial q_k}, \quad \xi_t = \frac{\partial \xi}{\partial t}, \quad \eta_{j,k} = \frac{\partial \eta_j}{\partial q_k}, \quad \eta_{j,t} = \frac{\partial \eta_j}{\partial t}.$$

Equation (9.55) is thus

$$\begin{aligned}
pr^{(1)}\mathbf{v}(L) + L\dot{\xi} &= -\frac{m}{2} \left(\dot{q}_1^3 \xi_1 + \dot{q}_2^3 \xi_2 + \dot{q}_1^2 \dot{q}_2 \xi_2 + \dot{q}_1 \dot{q}_2^2 \xi_1 \right) \\
&\quad + m\dot{q}_1^2 \left(\eta_{1,1} - \frac{1}{2}\xi_t \right) + m\dot{q}_2^2 \left(\eta_{2,2} - \frac{1}{2}\xi_t \right) \\
&\quad + m\dot{q}_1 \dot{q}_2 \left(\eta_{1,2} + \eta_{2,1} \right) + m\dot{q}_1 \left(\eta_{1,t} + \frac{K\xi_1}{\sqrt{q_1^2 + q_2^2}} \right)
\end{aligned}$$

$$+ m\dot{q}_2 \left(\eta_{2,t} + \frac{K\xi_2}{\sqrt{q_1^2 + q_2^2}} \right)$$

$$+ K \left(\frac{\xi_t}{\sqrt{q_1^2 + q_2^2}} - \frac{\eta_1 q_1 + \eta_2 q_2}{(q_1^2 + q_2^2)^{3/2}} \right)$$

$$= 0.$$

The same arguments used in the $n = 1$ case can be leveled at the above equation, which is an identity for all $\mathbf{q}$. The coefficients of $\dot{q}_1^3$ and $\dot{q}_2^3$ must vanish, and this means that ξ depends only on t. Since $\xi_1 = \xi_2 = 0$, the coefficients of the other cubic terms in the derivatives also vanish. In a similar way we argue that the coefficients of $\dot{q}_k^2$ etc. vanish, and this leads to the following system of equations for ξ, η_1 and η_2.

$$\eta_{1,1} - \frac{1}{2}\xi_t = 0, \tag{9.56}$$

$$\eta_{2,2} - \frac{1}{2}\xi_t = 0, \tag{9.57}$$

$$\eta_{1,2} + \eta_{2,1} = 0, \tag{9.58}$$

$$\eta_{1,t} = 0, \tag{9.59}$$

$$\eta_{2,t} = 0, \tag{9.60}$$

$$\left(q_1^2 + q_2^2 \right) \xi_t - (\eta_1 q_1 + \eta_2 q_2) = 0. \tag{9.61}$$

Equations (9.59) and (9.60) show that the η_k do not depend on t. Since ξ depends only on t, equations (9.56) and (9.57) indicate that there is a constant c_1 such that

$$\xi_t = 2c_1, \tag{9.62}$$

$$\eta_{1,1} = c_1,$$

$$\eta_{2,2} = c_1;$$

hence, η_1 and η_2 are of the form

$$\eta_1 = c_1 q_1 + g(q_2),$$

$$\eta_2 = c_1 q_2 + h(q_1),$$

where g and h are functions to be determined. Substituting the above expressions for the η_k into equation (9.58) gives

$$\frac{\partial g}{\partial q_2} + \frac{\partial h}{\partial q_1} = 0,$$

which implies that there is a constant c_2 such that

$$\frac{\partial g}{\partial q_2} = -\frac{\partial h}{\partial q_1} = c_2.$$

We thus see that η_1 and η_2 are of the form

$$\eta_1 = c_1 q_1 + c_2 q_2 + c_3,$$
$$\eta_2 = -c_2 q_1 + c_1 q_2 + c_4,$$

where c_3 and c_4 are constants of integration. Equation (9.61) can now be written

$$\left(q_1^2 + q_2^2\right) \xi_t - \left((c_1 q_1 + c_2 q_2 + c_3) q_1 + (-c_2 q_1 + c_1 q_2 + c_4) q_2\right) = 0,$$

which, using equation (9.62), reduces to

$$c_1 \left(q_1^2 + q_2^2\right) - c_3 q_1 - c_4 q_2 = 0. \tag{9.63}$$

Now, equation (9.63) is an identity that must be satisfied for all $\mathbf{q}$. We thus conclude that $c_1 = c_3 = c_4 = 0$.

Equation (9.55) thus shows that the infinitesimal generators of a variational symmetry of J must be of the form

$$\eta_1 = c_2 q_2,$$
$$\eta_2 = -c_2 q_1,$$
$$\xi = c_5,$$

where c_5 is another constant of integration. We thus have a two-parameter family of generators that lead to variational symmetries. If $c_2 = 0$ and $c_5 \neq 0$, then the transformation is a time translation. If $c_2 \neq 0$ and $c_5 = 0$, then the transformation is a rotation (cf. case C, Example 9.3.4). Theorem 9.4.2 shows that the only variational symmetries of J are combinations of rotations and time translations.

Noether's theorem as given in this chapter along with conditions for variational symmetries can be further generalized to accommodate functionals involving higher-order derivatives and/or several independent variables. The quantity $pr^{(1)}\mathbf{v}(L)$ is called the **first prolongation** of the vector field $\mathbf{v}$ (defined by the infinitesimal generators) acting on L. If a functional has an integrand that involves derivatives of order n then a higher prolongation $pr^{(n)}\mathbf{v}(L)$ is needed. The expression for $pr^{(n)}\mathbf{v}(L)$ escalates in complexity as n increases, but the condition for a variational symmetry remains deceptively simple, viz.,

$$pr^{(n)}\mathbf{v}(L) + L\dot{\xi} = 0.$$

The reader is directed to Bluman and Kumei [11] and Olver [57] for the general expression of $pr^{(n)}\mathbf{v}(L)$ in terms of the generators and their derivatives. Here, we have given only a basic "no frills" version of Noether's theorem and the reader is encouraged to consult the above references for deeper insights into this result.

Exercises 9.3:

1. Consider the functional of Example 9.4.1. Show that the transformation $Y = y$, $X = (1+\epsilon)x$ is a variational symmetry and find the corresponding conservation law.

2. The **Emden-Fowler equation** of astrophysics is

$$y'' + \frac{2}{x}y' + y^5 = 0,$$

which arises as the Euler-Lagrange equation for the functional

$$J(y) = \int_{x_0}^{x_1} \frac{x^2}{2}\left(y'^2 - \frac{1}{3}y^6\right)\,dx.$$

Find the infinitesimal generators that lead to a variational symmetry for this functional and establish the conservation law

$$x^2\left(y'y + 2x\left(y'^2 + y^5\right)\right) = const.$$

3. The **Thomas-Fermi equation**,

$$y'' - \frac{y^{3/2}}{\sqrt{x}} = 0,$$

corresponds to the Euler-Lagrange equation for the functional

$$J(y) = \int_{x_0}^{x_1} \left(\frac{1}{2}y'^2 + \frac{2}{5}\frac{y^{5/2}}{\sqrt{x}}\right)\,dx.$$

Show that this functional does not have a variational symmetry. (Exact solutions to the Thomas-Fermi and Emden-Fowler equations are discussed in detail in [8].)

10

The Second Variation

The Euler-Lagrange equation forms the centrepiece of the necessary condition for a functional to have an extremum. The Euler-Lagrange equation is analogous to the first derivative (or gradient) test for optimization problems in finite dimensions, and we know from elementary calculus that a vanishing first derivative is not sufficient for a local extremum. Likewise, satisfaction of the Euler-Lagrange equation is not a sufficient condition for a local extremum for a functional. In essence, we need a result analogous to the second derivative test in order to assert that a solution to the Euler-Lagrange equation produces a local extremum. In this chapter we investigate the next term in the expansion of $J(\hat{y}) - J(y)$, the second variation, and develop more refined necessary conditions for local extrema. We also develop sufficient conditions for a function y to produce a local extremum for a functional J. We restrict our attention almost exclusively to the basic fixed endpoint problem in the plane.

10.1 The Finite-Dimensional Case

The reader is doubtless aware of the second derivative test for determining whether a stationary point is a local extremum for a function of one variable. In this section we review a few concepts from the finite-dimensional case, primarily to motivate our study of conditions for functionals to have extrema. We begin with the familiar case of two independent variables.

Let $f : \Omega \to \mathbb{R}$ be a smooth function on the region $\Omega \subset \mathbb{R}^2$. Let $\mathbf{x} = (x_1, x_2) \in \Omega$ and let $\hat{x} = \mathbf{x} + \epsilon\eta$, where $\epsilon > 0$ and $\eta = (\eta_1, \eta_2) \in \mathbb{R}^2$. If ϵ is small, Taylor's theorem implies that

$$f(\hat{x}) = f(\mathbf{x}) + \epsilon \left\{ \eta_1 \frac{\partial f(\mathbf{x})}{\partial x_1} + \eta_2 \frac{\partial f(\mathbf{x})}{\partial x_2} \right\}$$

$$+ \frac{\epsilon^2}{2!} \left\{ \eta_1^2 \frac{\partial^2 f(\mathbf{x})}{\partial x_1^2} + 2\eta_1\eta_2 \frac{\partial^2 f(\mathbf{x})}{\partial x_1 \partial x_2} + \eta_2^2 \frac{\partial^2 f(\mathbf{x})}{\partial x_2^2} \right\}$$

$$+O(\epsilon^3).$$

If f has a stationary point at $\mathbf{x}$ then $\nabla f(\mathbf{x}) = \mathbf{0}$ (cf. Section 2.1), and the Taylor expansion reduces to

$$f(\hat{\mathbf{x}}) = f(\mathbf{x}) + \frac{\epsilon^2}{2} Q(\eta) + O(\epsilon^2), \tag{10.1}$$

where

$$Q(\eta) = \eta_1^2 \frac{\partial^2 f(\mathbf{x})}{\partial x_1^2} + 2\eta_1 \eta_2 \frac{\partial^2 f(\mathbf{x})}{\partial x_1 \partial x_2} + \eta_2^2 \frac{\partial^2 f(\mathbf{x})}{\partial x_2^2}.$$

The nature of a stationary point is determined by the lowest-order derivatives that are nonzero at $\mathbf{x}$. If one of the second derivatives is nonzero, then the sign of $f(\hat{\mathbf{x}}) - f(\mathbf{x})$ is controlled by the sign of Q.

For nonzero $\eta \in \mathbb{R}^2$, the quadratic term may be always positive or always negative, but it may be that this term is positive for some η and negative for others. The character of the extremum revolves around what signs Q may have for various choices of η, and we can track the sign changes by examining when $Q(\eta) = 0$. If $\eta \neq \mathbf{0}$, then either η_1 or η_2 is nonzero. Without loss of generality suppose that $\eta_2 \neq 0$. Now, Q is a continuous function of η, and if Q changes sign, there must be some $\eta \neq \mathbf{0}$ such that $Q(\eta) = 0$. Hence, there must be a real solution to the quadratic equation

$$\left(\frac{\eta_1}{\eta_2} \right)^2 \frac{\partial^2 f(\mathbf{x})}{\partial x_1^2} + 2 \frac{\eta_1}{\eta_2} \frac{\partial^2 f(\mathbf{x})}{\partial x_1 \partial x_2} + \frac{\partial^2 f(\mathbf{x})}{\partial x_2^2} = 0.$$

The nature of the solutions to this equation is determined by the discriminant,

$$\Delta = \frac{\partial^2 f}{\partial x_1^2} \frac{\partial^2 f}{\partial x_2^2} - \left(\frac{\partial^2 f}{\partial x_1 \partial x_2} \right)^2,$$

of the quadratic form Q at $\mathbf{x}$. There can be at most two solutions to this quadratic equation. If real solutions exist, then Q may change sign. If there are no real solutions to the quadratic equation then Q, being a continuous function, will never change sign. Whether $Q = 0$ has a nontrivial solution depends on the sign of Δ: if $\Delta(\mathbf{x}) < 0$ and $\partial^2 f / \partial x_1^2 \neq 0$ (or $\partial^2 f / \partial x_2^2 \neq 0$) at $\mathbf{x}$ then Q is indefinite and vanishes along two distinct lines. For this case, a small neighbourhood of $\mathbf{x}$, $B(\mathbf{x}; \epsilon)$, can be divided into four sets, two in which $Q > 0$ and two in which $Q < 0$. In this case Q is called **indefinite**. Evidently, $\mathbf{x}$ cannot produce a local extremum because the sign of $f(\hat{\mathbf{x}}) - f(\mathbf{x})$ depends on the choice of η. Stationary points $\mathbf{x}$ for which $\Delta(\mathbf{x}) < 0$ are called **saddle points**.

In contrast, if $\Delta(\mathbf{x}) > 0$, then there are no real solutions to the quadratic equation; consequently, Q cannot change sign. In this case Q is called **definite**, and $\mathbf{x}$ corresponds to a local extremum. The type of extremum can be deduced from the examination of any particular curve through $\mathbf{x}$. The simplest such curves correspond to $\gamma_1(\eta_1) = (\eta_1, 0)$ and $\gamma_2(\eta_2) = (0, \eta_2)$. If $\mathbf{x}$ is a local maximum/minimum, then $\eta_1 = 0$ corresponds to a local maximum/minimum for $\gamma_1(\eta_1)$ (and $\eta_2 = 0$ corresponds to a local maximum/minimum for $\gamma_2(\eta_2)$). Thus $f(\mathbf{x})$ is a local maximum if $\partial^2 f(\mathbf{x})/\partial x_1^2 < 0$ (or $\partial^2 f(\mathbf{x})/\partial x_2^2 < 0$), and a local minimum if $\partial^2 f(\mathbf{x})/\partial x_1^2 > 0$ (or $\partial^2 f(\mathbf{x})/\partial x_2^2 > 0$).

It may be that $\Delta(\mathbf{x}) = 0$ even though the second derivatives of f at $\mathbf{x}$ are not all zero. In this case there is a line $(x_1(t), x_2(t))$ in $B(\mathbf{x}; \epsilon)$ through $\mathbf{x} = (x_1(0), x_2(0))$ where Q vanishes. The nature of this point is determined by the third-order (or higher) derivatives. If $\Delta(\mathbf{x}) = 0$, then $\mathbf{x}$ is called a **degenerate stationary point** (or parabolic point). We must examine the cubic terms (or higher-order terms) in the expansion to discern the nature of the stationary point.

Note that if all the second derivatives of f vanish at $\mathbf{x}$ then it is clear that the sign of $f(\mathbf{x}) - f(\hat{\mathbf{x}})$ is determined by the third-order derivatives. These must also vanish at $\mathbf{x}$ for a local extremum, and if this is the case, the quartic terms control the sign.

The above approach can be adapted to functions of three or more independent variables, although the increase in variables escalates the number of possibilities and the complexity of the computations. Let $f : \Omega \to \mathbb{R}$ be a smooth function on the region $\Omega \subset \mathbb{R}^n$ and suppose $\mathbf{x} = (x_1, x_2, \ldots, x_n)$ is a stationary point. Then, as in the two-variable case, we have $\nabla f(\mathbf{x}) = 0$, and the sign of $f(\mathbf{x}) - f(\hat{\mathbf{x}})$ is controlled by the quadratic terms in the Taylor expansion. Let $\hat{\mathbf{x}} = \mathbf{x} + \epsilon\eta$, where $\epsilon > 0$ and $\eta = (\eta_1, \eta_2, \ldots, \eta_n)$. The quadratic terms in the Taylor expansion may be written in the form

$$Q(\eta) = \eta^T \mathbf{H}(\mathbf{x}) \eta,$$

where $\mathbf{H}(\mathbf{x})$ is the **Hessian matrix** for f at $\mathbf{x}$; i.e.,

$$\mathbf{H}(\mathbf{x}) = \begin{pmatrix} \frac{\partial^2 f(\mathbf{x})}{\partial x_1^2} & \frac{\partial^2 f(\mathbf{x})}{\partial x_1 \partial x_2} & \cdots & \frac{\partial^2 f(\mathbf{x})}{\partial x_1 \partial x_n} \\ \frac{\partial^2 f(\mathbf{x})}{\partial x_2 \partial x_1} & \frac{\partial^2 f(\mathbf{x})}{\partial x_2^2} & \cdots & \frac{\partial^2 f(\mathbf{x})}{\partial x_2 \partial x_n} \\ \vdots & & & \\ \frac{\partial^2 f(\mathbf{x})}{\partial x_n \partial x_1} & \frac{\partial^2 f(\mathbf{x})}{\partial x_n \partial x_2} & \cdots & \frac{\partial^2 f(\mathbf{x})}{\partial x_n^2} \end{pmatrix}.$$

The nature of a stationary point depends on whether $\mathbf{H}$ is definite. If $\mathbf{H}$ is definite, then f has a local extremum at $\mathbf{x}$; if $\mathbf{H}$ is indefinite, then $\mathbf{x}$ corresponds to some type of saddle point. The Morse lemma can be used to classify the types of stationary points provided the Hessian matrix at the stationary point has the same rank as the number of independent variables (i.e., $\mathbf{H}$ is nondegenerate). Stationary points satisfying this condition are called **nondegenerate**.

Lemma 10.1.1 (Morse Lemma) *Let $\mathbf{x}_0$ be a nondegenerate stationary point for the smooth function f. Then there exists a smooth invertible coordinate transformation $x_j \to x_j(\mathbf{v})$, where $\mathbf{v} = (v_1, v_2, \ldots, v_n)$ defined in a neighbourhood $N(\mathbf{x}_0)$ of $\mathbf{x}_0$ such that the identity*

$$f(\mathbf{x}) = \hat{f}(\mathbf{v}) = f(\mathbf{x}_0) - v_1^2 - v_2^2 - \cdots - v_\lambda^2$$
$$+ v_{\lambda+1}^2 + \cdots + v_n^2$$

holds throughout $N(\mathbf{x}_0)$. The integer λ is called the **index** of f at $\mathbf{x}_0$.

A proof of this lemma can be found in [53]. The function $v_1^2 + v_2^2 + \cdots + v_\lambda^2 - v_{\lambda+1}^2 - \cdots - v_n^2$ is called a **Morse λ-saddle**. The index is an invariant under smooth invertible coördinate transformations; therefore, it can be used to classify nondegenerate stationary points. A Morse n-saddle is a local maximum; a Morse 0-saddle is a local minimum. If λ is not 0 or n, then the Morse λ-saddle indicates that the difference $f(\mathbf{x}) - f(\hat{\mathbf{x}})$ can be positive or negative depending on the choice of $\mathbf{x}$. The Morse lemma has another consequence: the stationary points of the saddle are isolated, and since smooth invertible coördinate transformations leave isolated stationary points isolated, all nondegenerate stationary points must be isolated.

There is a wealth of results regarding conditions under which a quadratic form is definite. An example is provided by the following theorem.

Theorem 10.1.2 (Sylvester Criterion) Let $\mathbf{X} = (X_1, X_2, \ldots, X_n)$ and let $\mathbf{A}$ denote an $n \times n$ symmetric matrix with entries a_{ij}. A necessary and sufficient condition for a quadratic form $\mathbf{X}^T \mathbf{A} \mathbf{X}$ to be positive definite, is that every principal minor determinant of $\mathbf{A}$ is positive. In particular, $\det \mathbf{A} > 0$ and every diagonal element a_{jj} is positive.

Suppose that $\mathbf{x}$ is a stationary point for f and let

$$\mathbf{H} = \begin{pmatrix} h_{11} & h_{12} & \cdots & h_{1n} \\ h_{21} & h_{22} & \cdots & h_{2n} \\ \vdots & & & \\ h_{n1} & h_{n2} & \cdots & h_{nn} \end{pmatrix},$$

denote the Hessian matrix at $\mathbf{x}$. The above theorem indicates that the quadratic form is positive definite if $h_{11} > 0$ and the determinants of the matrices

$$\begin{pmatrix} h_{11} & h_{12} \\ h_{21} & h_{22} \end{pmatrix}, \begin{pmatrix} h_{11} & h_{12} & h_{13} \\ h_{21} & h_{22} & h_{23} \\ h_{31} & h_{32} & h_{33} \end{pmatrix}, \begin{pmatrix} h_{11} & h_{12} & \cdots & h_{1k} \\ h_{21} & h_{22} & \cdots & h_{2k} \\ \vdots & & & \\ h_{k1} & h_{k2} & \cdots & h_{kk} \end{pmatrix},$$

are all positive for $k \in \mathbb{N}$, $k \leq n$. The quadratic form is negative definite if $\mathbf{X}^T(-\mathbf{H})\mathbf{X}$ is positive definite, where $-\mathbf{H}$ is the matrix with elements $-h_{ij}$.

10.2 The Second Variation

Let us return to the basic fixed endpoint variational problem. Recall that we seek a smooth function $y : [x_0, x_1] \to \mathbb{R}$ such that $y(x_0) = y_0$, $y(x_1) = y_1$, and the functional

$$J(y) = \int_{x_0}^{x_1} f(x, y, y') \, dx \tag{10.2}$$

is an extremum. We make the blanket assumption throughout this chapter that f is smooth in the indicated arguments. More specifically, we need f to have derivatives of at least third order for some of our arguments. We assume that for any given extremal y, $f(x, y(x), y'(x))$ is smooth in a neighbourhood of x_0 and in a neighbourhood of x_1.

Suppose that J has an extremum at y, and let $\hat{y}$ be a "nearby" function of the form $\hat{y} = y + \epsilon \eta$, where $\epsilon > 0$ is a small and η is a smooth function on $[x_0, x_1]$ such that $\eta(x_0) = \eta(x_1) = 0$. Our brief foray into the finite-dimensional case indicates that we need to consider the $O(\epsilon^2)$ terms of $J(\hat{y}) - J(y)$ in order to glean information regarding the nature of the extremal. Taylor's theorem implies

$$f(x, \hat{y}, \hat{y}') = f(x, y, y') \epsilon \left(\eta \frac{\partial f}{\partial y} + \eta' \frac{\partial f}{\partial y'} \right)$$
$$+ \frac{\epsilon^2}{2} \left(\eta^2 \frac{\partial^2 f}{\partial y^2} + 2\eta\eta' \frac{\partial^2 f}{\partial y \partial y'} + \eta'^2 \frac{\partial^2 f}{\partial y'^2} \right) + O(\epsilon^3),$$

where the partial derivatives are evaluated at (x, y, y'). In the interest of simplicity, we use the following notation,

$$f_{yy} = \frac{\partial^2 f}{\partial y^2}, \quad f_{yy'} = \frac{\partial^2 f}{\partial y \partial y'}, \quad f_{y'y'} = \frac{\partial^2 f}{\partial y'^2},$$

where, unless otherwise noted, the partial derivatives are evaluated at (x, y, y'). Thus,

$$J(\hat{y}) - J(y) = \epsilon \delta J(\eta, y) + \frac{\epsilon^2}{2} \delta^2 J(\eta, y) + O(\epsilon^3),$$

where $\delta J(\eta, y)$ is the first variation (cf. Section 2.2) and

$$\delta^2 J(\eta, y) = \int_{x_0}^{x_1} \left(\eta^2 f_{yy} + 2\eta\eta' f_{yy'} + \eta'^2 f_{y'y'} \right) \, dx.$$

The term $\delta^2 J(\eta, y)$ is called the **second variation** of J. The second variation plays a rôle analogous to the $Q(\eta)$ term in the finite-dimensional case. Since y is an extremal for J we have that $\delta J(\eta, y) = 0$, and hence

$$J(\hat{y}) - J(y) = \frac{\epsilon^2}{2} \delta^2 J(\eta, y) + O(\epsilon^3).$$

The sign of $J(\hat{y}) - J(y)$ thus depends on the sign of $\delta^2 J(\eta, y)$, and consequently the nature of the extremal thus depends on whether the second variation changes sign for different choices of η. Note that, at this stage, we have already solved the Euler-Lagrange equations so that y is known and hence the functions f_{yy}, $f_{yy'}$, and $f_{y'y'}$ are known in terms of x. This situation parallels that for the finite-dimensional case, where we know the numerical values of the entries in the Hessian matrix.

Using the notation of Section 2.2, let S denote the set of functions y smooth on $[x_0, x_1]$ such that $y(x_0) = y_0$ and $y(x_1) = y_1$. Let H denote the set of functions η smooth on $[x_0, x_1]$ such that $\eta(x_0) = \eta(x_1) = 0$. The above arguments yield the following necessary condition.

Theorem 10.2.1 *Suppose that J has a local extremum in S at y. If y is a local minimum, then*

$$\delta^2 J(\eta, y) \geq 0 \qquad (10.3)$$

for all $\eta \in H$; if y is a local maximum, then

$$\delta^2 J(\eta, y) \leq 0 \qquad (10.4)$$

for all $\eta \in H$.

The above result is of limited value at present, because we have no method to test the second variation for sign changes. Inequality (10.3) is analogous to a positive semidefinite condition on a Hessian matrix, and we have seen how the addition of independent variables escalates the number of possibilities (types of saddles) and the complexity of verifying whether the matrix is definite. For the infinite-dimensional case we thus expect tests for establishing sign changes in the second variation to be complicated. It is thus a pleasant surprise that certain conditions can be derived that are tractable and simple to implement.

Exercises 10.2:

1. Suppose that J is a functional of the form

$$J(y) = \int_{x_0}^{x_1} f(x, y, y', y'') \, dx,$$

    where f is a smooth function of all its arguments. Derive an expression for the second variation of J.

2. Assuming f has the requisite number of derivatives, show that for ϵ small, $\eta \in H$, and $\hat{y} = y + \epsilon \eta$,

$$J(\hat{y}) - J(y) = \epsilon \delta J(\eta, y) + \frac{\epsilon^2}{2!} \delta^2 J(\eta, y)$$

$$+ \frac{\epsilon^3}{3!} \delta^3 J(\eta, y) + O(\epsilon^4),$$

    where

$$\delta^3 J(\eta, y) = \int_{x_0}^{x_1} (\eta'^3 \frac{\partial^3 f}{\partial y'^3} + 3\eta'^2 \eta \frac{\partial^3 f}{\partial y'^2 \partial y}$$

$$+ 3\eta' \eta^2 \frac{\partial^3 f}{\partial y' \partial y^2} + \eta^3 \frac{\partial^3 f}{\partial y^3}) \, dx.$$

    The functional $\delta^3 J(\eta, y)$ is called the **third variation** of J.

3. Suppose that J has a local extremum at y and that $\delta^2 J(\eta, y) = 0$. Show that $\delta^3 J(\eta, y) = 0$.

10.3 The Legendre Condition

In this section we develop a necessary condition for a functional to have a local extremum. This result is called the Legendre condition. Unlike Theorem 10.2.1, it is straightforward to apply and hence useful for filtering out extremals that do not produce a local minimum or maximum.

The second variation can be recast in a more convenient form that separates the η terms from the η' terms in the integrand. Note that

$$2\eta\eta' f_{yy'} = \left(\eta^2\right)' f_{yy'},$$

and hence

$$\int_{x_0}^{x_1} 2\eta\eta' f_{yy'}\, dx = \eta^2 f_{yy'} \Big|_{x_0}^{x_1} - \int_{x_0}^{x_1} \eta^2 \frac{d}{dx}(f_{yy'})\, dx$$

$$= - \int_{x_0}^{x_1} \eta^2 \frac{d}{dx}(f_{yy'})\, dx,$$

where we have used the conditions that $\eta(x_0) = \eta(x_1) = 0$. The second variation can thus be written

$$\delta^2 J(\eta, y) = \int_{x_0}^{x_1} \left(\eta^2 \left(f_{yy} - \frac{d}{dx}(f_{yy'}) \right) + \eta'^2 f_{y'y'} \right)\, dx.$$

The essence of the Legendre condition is that the second variation must change sign for certain choices of $\eta \in H$ if $f_{y'y'}$ changes sign. Before we launch into a statement and proof of the Legendre condition, a few comments to motivate the proof are in order. We reiterate that the coefficients of η^2 and η'^2 are known functions of x. Let A and B be the functions defined by

$$A(x) = f_{y'y'}(x, y(x), y'(x)), \tag{10.5}$$

$$B(x) = f_{yy}(x, y(x), y'(x)) - \frac{d}{dx}(f_{yy'}(x, y(x), y'(x))), \tag{10.6}$$

for $x \in [x_0, x_1]$. Since f and y are smooth, both A and B are continuous functions on the interval $[x_0, x_1]$. The reason that the sign of A plays a pivotal rôle in this theory is that there are functions $\eta \in H$ for which $|\eta(x)|$ is small for all $x \in [x_0, x_1]$, but $|\eta'(x)|$ is not. In contrast, if $|\eta'(x)|$ is small for all $x \in [x_0, x_1]$, then, since η must be smooth and satisfy the conditions $\eta(x_0) = \eta(x_1) = 0$ for membership in H, we have that $|\eta(x)|$ is also small for all $x \in [x_0, x_1]$. The simple mollifier

$$\eta(x) = \begin{cases} \exp\left(-\frac{\gamma}{\gamma - (x-c)^2}\right), & \text{if } x \in [c - \gamma, c + \gamma] \\ 0, & \text{if } x \notin [c - \gamma, c + \gamma], \end{cases}$$

where $c \in (x_0, x_1)$, and $\gamma > 0$ is some suitably small number, illustrates this scenario. It can be shown that this function is smooth [1] and vanishes at x_0 and x_1 so that it is in the set H. Now the maximum value of η is $1/e$, but the mean value theorem shows that there is a least one value of $x \in (c - \gamma, c)$ such that $\eta'(x) = 1/(\gamma e)$, and the continuity of the derivative thus implies that there is a subinterval $I \subset (c - \gamma, c)$ for which $\eta'(x) > 1/(2\gamma e)$ for all $x \in I$.

The issue at stake, of course, is not the pointwise behaviour of $|\eta'(x)|$ but rather the influence of this function on the integral defining the second variation. The subinterval I, after all, might be rather small, and although $\eta'^2(x)$ is large for $x \in I$ compared to $\eta^2(x)$ the overall effect on the value of the integral might be small owing to the small length of I. We must keep in mind, however, that we can construct a function $\eta \in H$ that is essentially a superposition of any number of mollifiers, each with a different value for $c \in (x_0, x_1)$ with $\gamma > 0$ chosen sufficiently small so that their supports (i.e., intervals in which they are nonzero) do not intersect. The net effect is that functions in H can always be found such that the derivative terms dominate the second variation. A superposition of mollifiers makes the importance of the sign of A transparent, at least conceptually. Rather than chase the above mollifiers any further, however, we opt (in the interests of computational simplicity) for a simpler function that captures the same behaviour for the proof of the Legendre condition.

Theorem 10.3.1 (Legendre Condition) *Let J be a functional of the form* (10.2), *where f is a smooth function of x, y, and y', and suppose that J has a local minimum in S at y. Then,*

$$f_{y'y'} \geq 0 \qquad\qquad (10.7)$$

for all $x \in [x_0, x_1]$.

Proof: Using the notation introduced above, suppose that there is a $c \in [x_0, x_1]$ such that $A(c) < 0$. Since J has a local minimum at y, Theorem 10.2.1 implies that $\delta^2 J(\eta, y) \geq 0$ for all $\eta \in H$. The theorem is thus established if it can be shown that there is an $\nu \in H$ such that $\delta^2 J(\nu, y) < 0$.

Since A is continuous on $[x_0, x_1]$ there is an $\gamma > 0$ such that $A(x) < A(c)/2$ for all $x \in (c - \gamma, c + \gamma)$. We construct a function in H that effectively filters out the influence of A and B for all x not in $(c - \gamma, c + \gamma)$ and magnifies the contribution of the derivative. Let

$$\nu(x) = \begin{cases} \sin^4\left(\frac{\pi(x-c)}{\gamma}\right), & \text{if } x \in [c - \gamma, c + \gamma] \\ 0, & \text{if } x \notin [c - \gamma, c + \gamma]. \end{cases}$$

Now, it can be shown that $\nu \in H$ and that

---

[1] In fact it has derivatives of all orders for all $x \in \mathbb{R}$. See also the comments in Appendix A.1.

$$\nu'(x) = \begin{cases} \frac{4\pi}{\gamma} \sin^3 \left(\frac{\pi(x-c)}{\gamma} \right) \cos \left(\frac{\pi(x-c)}{\gamma} \right), & \text{if } x \in [c - \gamma, c + \gamma] \\ 0, & \text{if } x \notin [c - \gamma, c + \gamma]. \end{cases}$$

We can get a rough upper bound on the ν' contribution to the second variation as follows.

$$\begin{aligned} \int_{x_0}^{x_1} A(x)\nu'^2(x)\, dx &= \int_{c-\gamma}^{c+\gamma} A(x)\nu'^2(x)\, dx \\ &= \int_{c-\gamma}^{c+\gamma} \frac{16\pi^2}{\gamma^2} \sin^6 \left(\frac{\pi(x-c)}{\gamma} \right) \cos^2 \left(\frac{\pi(x-c)}{\gamma} \right) dx \\ &< \frac{A(c)}{2} \frac{4^2 \pi^2}{\gamma^2} 2\gamma \\ &= \frac{16A(c)\pi^2}{\gamma}. \end{aligned}$$

Since B is continuous on $[c - \gamma, c + \gamma]$ there is an $N > 0$ such that $|B(x)| < N$ for all $x \in [c - \gamma, c + \gamma]$; hence, a rough upper bound for the ν contribution to the second variation is given by

$$\begin{aligned} \int_{x_0}^{x_1} B(x)\nu^2(x)\, dx &= \int_{c-\gamma}^{c+\gamma} B(x)\nu^2(x)\, dx \\ &= \int_{c-\gamma}^{c+\gamma} B(x) \sin^8 \left(\frac{\pi(x-c)}{\gamma} \right) dx \\ &< 2N\gamma. \end{aligned}$$

Now,

$$\delta^2 J(\nu, y) < \frac{16A(c)\pi^2}{\gamma} + 2N\gamma,$$

so that the second variation is negative for ν, if γ is chosen arbitrarily small. We have the freedom to choose γ small, so that there are functions in H that make the second variation negative. This contradicts Theorem 10.2.1 and we conclude that $A(x) \geq 0$ for all $x \in [x_0, x_1]$. □

Evidently, the above result can be readily modified for the case where J has a local maximum at y. Inequality (10.7) is called the **Legendre condition**.

Aside from the theoretical benefits (which we reap later) the Legendre condition is a practical tool for deciding whether a solution to the Euler-Lagrange equation is even in the running for a solution.

Example 10.3.1: Let

$$J(y) = \int_{-1}^{1} x\sqrt{1 + y'^2}\, dx.$$

For a given set of boundary conditions at $x = -1$ and $x = 1$ we can find the corresponding extremal explicitly for this functional (see Exercises 2.3-1).

However, we need not even do this to deduce that the extremals cannot give a local extremum for J. We can use elementary arguments to show that J can be made arbitrarily small, but the Legendre condition conveniently answers the question. In particular, we have for any smooth y,

$$f_{y'y'}(x, y, y') = \frac{x}{(1 + y'^2)^{3/2}},$$

so that the sign of this derivative changes in the interval $[-1, 1]$.

Example 10.3.2:   Catenary

Consider the catenary problem of Example 4.2.1. We consider the problem here as an unconstrained fixed endpoint problem with the appropriate Lagrange multiplier, so that in the present notation

$$f(x, y, y') = (y - \lambda)\sqrt{1 + y'^2},$$

where

$$\lambda = h - \frac{1}{2\hat{\xi}} \cosh(\hat{\xi}).$$

Here, h is the height of the poles supporting the cable and $\hat{\xi}$ is one of two possible nonzero solutions to

$$L\xi = \sinh(\xi), \tag{10.8}$$

where L is the arclength of the cable. As discussed in Example 4.2.1, we can distinguish which solution is relevant from simple physical considerations. The Legendre condition also makes a distinction. Now, for any y,

$$f_{y'y'}(x, y, y') = \frac{y - \lambda}{(1 + y'^2)^{3/2}},$$

so that the sign of this derivative is the same as the sign of $y - \lambda$. The solution to the Euler-Lagrange equations is given by equation (4.32) and therefore

$$y - \lambda = \frac{1}{2\hat{\xi}} \cosh(\hat{\xi}(2x - 1)).$$

Recall that there is precisely one positive solution and one negative solution $\hat{\xi}$ to equation (10.8). Only the positive solution satisfies the Legendre condition for a local minimum.

The Legendre condition cannot be converted into a sufficient condition even if we replace inequality (10.7) by the **strengthened Legendre condition**

$$f_{y'y'} > 0. \tag{10.9}$$

The following well-worn but simple example illustrates this comment.

Example 10.3.3:   Consider the fixed-endpoint problem with the functional

$$J(y) = \int_0^\ell \left(y'^2 - y^2 \right) dx,$$

and the boundary conditions $y(0) = 0$, $y(\ell) = 0$. We have seen in Chapter 5 that problems of this sort lead to Sturm-Liouville type problems, and that the existence of a nontrivial solution depends on the choice of ℓ. (Modify Example 5.1.1 by fixing $\lambda = 1$ and replace the upper limit π with ℓ.) Suppose that we choose $\ell > \pi$.

For this simple problem the strengthened Legendre condition is evidently satisfied for any y, and the second variation is given by

$$\delta^2 J(\eta, y) = 2 \int_0^\ell \left(\eta'^2(x) - \eta^2(x) \right) dx.$$

Let

$$\eta = \sin(\frac{\pi x}{\ell}).$$

Clearly $\eta \in H$, and

$$\delta^2 J(\eta, y) = \int_0^\ell \left(\frac{\pi^2}{\ell^2} \cos^2(\frac{\pi x}{\ell}) - \sin^2(\frac{\pi x}{\ell}) \right) dx$$

$$= \frac{1}{4} \left(\pi^2 - \ell^2 \right) < 0.$$

The trivial solution $y = 0$ is an extremal for the problem, but the above calculation shows that it cannot give a local minimum for J.

Legendre himself tried to frame a sufficient condition for a local minimum around the inequality (10.9) but ran into various snags such as the above example, and it became apparent that more information was needed. The essence of the problem is that the strengthened Legendre condition and Euler-Lagrange equation place only *pointwise* restrictions on the functions. The great circles on a sphere (i.e., the geodesics) give an intuitive example of why global information is needed. Consider three points on the earth (which we assume is a perfect sphere) all on the same great circle, say the North Pole, London, and the South Pole. The shortest distance from London to the North Pole is along the meridian connecting these points. But there are two choices: one can proceed directly north, or one can go initially south, through the South Pole, and then turn northwards to eventually arrive at the North Pole. Evidently, the latter option produces a longer path, but pointwise the Euler-Lagrange equation is satisfied on the meridian as is the Legendre condition. It is only when we look at "the big picture" that we realize the latter option cannot be even a local minimum: there are paths *near* the South Pole route that are shorter (they avoid the South Pole at the "last minute" and jump onto a suitably close line of longitude and head north).

Exercises 10.3:

1. Geodesics on a sphere of radius $R > 0$ correspond to the extremals of the functional

$$J(\phi) = \int_{\theta_0}^{\theta_1} \sqrt{1 + \sin^2 \theta \phi'^2} \, d\theta,$$

   where ϕ is the polar angle, θ is the azimuth angle, and ϕ' denotes $d\phi/d\theta$. Show that J satisfies the Legendre condition (10.7).

2. Let

$$J(y) = \int_{x_0}^{x_1} \frac{1 + y^2}{y'^2} \, dx.$$

   Suppose that J has a local extremum at y. Use the Legendre condition to determine the nature of the extremum.

10.4 The Jacobi Necessary Condition

The major shortcoming of the Legendre condition is that it is a pointwise restriction. Example 10.3.3 makes it clear that other considerations are needed before a sufficient condition can be formulated. In this section we present a necessary condition that builds on the Legendre condition, but is distinctly global in character. The key concept of a conjugate point is introduced, and it turns out that this necessary condition paves the way for the formulation of a sufficient condition. We focus on local minima trusting the reader to make the necessary adjustments to get analogous results for local maxima.

10.4.1 A Reformulation of the Second Variation

For a finite-dimensional optimization problem we can appeal to the Morse Lemma 10.1.1 to argue that the relevant quadratic form can be written as a sum/difference of squares. This special transformation of the quadratic form allows us to classify critical points as described in Section 10.1. The infinite-dimensional analogue of this process for a Morse 0-saddle is to convert the second variation into a functional of the form

$$\delta^2 J(\eta, y) = \int_{x_0}^{x_1} f_{y'y'} \Upsilon^2 \, dx, \tag{10.10}$$

where Υ is a function of x, η, and (indirectly) the extremal.[2] Ideally, we seek a function Υ such that Υ is identically zero on $[x_0, x_1]$ only if η is identically zero on $[x_0, x_1]$. In this case, the sign of the second variation would depend on that of $f_{y'y'}$.

---

[2] The analogy is made more precise in Gelfand and Fomin [31], pp. 125–129. See also [22], pp. 571–572.

The idea of transforming the second variation into the form (10.10) dates back to Legendre, who tried to establish the existence of Υ by completing the square of the quadratic form. Although he failed to achieve a sufficient condition, his idea of completing the square proved fruitful.

We know that the second variation can be written in the form

$$\delta^2 J(\eta, y) = \int_{x_0}^{x_1} \left(f_{y'y'} \eta'^2 + B\eta^2 \right) \, dx,$$

where B is as defined by equation (10.6). Now, for any smooth function w the conditions $\eta(x_0) = \eta(x_1) = 0$ give

$$\int_{x_0}^{x_1} \left(w\eta^2 \right)' \, dx = 0;$$

consequently, we can always add a term of the form $(w\eta^2)'$ to the integrand of the second variation without changing the value of the functional. The strategy is to select a function w such that

$$f_{y'y'} \eta'^2 + B\eta^2 + \left(w\eta^2 \right)' = f_{y'y'} \Upsilon^2,$$

for some Υ. We know from the Legendre condition that $f_{y'y'}$ cannot change sign in $[x_0, x_1]$ if y produces a local extremum for J. We assume that the strengthened Legendre condition (10.9) is satisfied. Thus,

$$f_{y'y'} \eta'^2 + B\eta^2 + \left(w\eta^2 \right)' = f_{y'y'} \left(\eta'^2 + 2\frac{w}{f_{y'y'}}\eta\eta' + \frac{B + w'}{f_{y'y'}}\eta^2 \right).$$

Suppose that w satisfies the differential equation

$$w^2 = f_{y'y'} \left(B + w' \right) \tag{10.11}$$

for all $x \in [x_0, x_1]$. Then

$$f_{y'y'} \eta'^2 + B\eta^2 + \left(w\eta^2 \right)' = f_{y'y'} \left(\eta' + \frac{w}{f_{y'y'}}\eta \right)^2,$$

so that the second variation could be recast in the form (10.10).

Following the analogy with the finite-dimensional case, the second variation is called **positive definite** if $J(\eta, y) > 0$ for all $\eta \in H - \{0\}$. Given a solution to (10.11), we have

$$\Upsilon = \eta' + \frac{w}{f_{y'y'}}\eta,$$

which is zero for all $x \in [x_0, x_1]$ only if η satisfies the first-order differential equation

$$\eta' + \frac{w}{f_{y'y'}}\eta = 0. \tag{10.12}$$

Since $\eta \in H$, however, the above differential equation is accompanied by the conditions $\eta(x_0) = \eta(x_1) = 0$. Now, f is assumed smooth in x, y, and y' and y is a smooth extremal; hence, $f_{y'y'}$ is a smooth function. Picard's theorem shows that there exists a unique solution to equation (10.12) that satisfies the initial condition $\eta(x_0) = 0$, and a quick inspection shows that this must be the trivial solution $\eta(x) = 0$ for all $x \in [x_0, x_1]$. We thus see that Υ is identically zero on $[x_0, x_1]$ only if η is identically zero on this interval. Otherwise, under the strengthened Legendre condition, the integral defining the second variation must be positive. In summary, we have the following lemma.

Lemma 10.4.1 *Let f be a smooth function of x, y, and y', and let y be a smooth extremal for the functional J defined by (10.2) such that $f_{y'y'} > 0$ for all $x \in [x_0, x_1]$. If there is a solution w to the differential equation (10.11) valid on the interval $[x_0, x_1]$, then the second variation is positive definite.*

The "fly in the ointment" is the existence of the solution w. We can appeal to Picard's theorem to assert the existence of *local* solutions to equation (10.11), but this is not good enough. We need solutions that are defined over the entire interval $[x_0, x_1]$ rather than in some small subinterval. The reformulation of $\delta^2 J(\eta, y)$ thus hinges on whether a *global* solution w exists to equation (10.11).

10.4.2 The Jacobi Accessory Equation

Relation (10.11) is an example of a well-known class of equations called **Riccati equations**. A standard solution technique for such equations entails converting the nonlinear first-order equation to a second-order linear equation by use of the transformation

$$w = \frac{u'}{u} f_{y'y'} \tag{10.13}$$

(cf. [41], [61]). Under this transformation, the Riccati equation becomes

$$\frac{d}{dx}(f_{y'y'}u') - Bu = 0;$$

i.e.,

$$\frac{d}{dx}(f_{y'y'}u') - \left(f_{yy} - \frac{d}{dx}f_{yy'}\right)u = 0. \tag{10.14}$$

Equation (10.14) is called the **Jacobi accessory equation**. If there is a solution u to this equation that is valid on $[x_0, x_1]$ and such that $u(x) \neq 0$ for all $x \in [x_0, x_1]$, then transformation (10.13) implies that the Riccati equation (10.11) has a solution valid for $x \in [x_0, x_1]$.

Certainly one advantage of working with a second-order linear ordinary differential equation as opposed to a first-order nonlinear equation is that the theory underlying the linear equation is well developed and perhaps more tractable. It is beyond the scope of this book to recount the theory in any

detail. The reader is directed to textbooks such as Birkhoff and Rota [9] for
a full discussion. It suffices here to mention a few results concerning the existence of solutions to the Jacobi accessory equation. Note that the smoothness
assumptions on f and y mean that the coefficients $f_{y'y'}$ and B are smooth
functions of x on $[x_0, x_1]$, and the strengthened Legendre condition ensures
that the problem is not singular.[3] We can thus use Picard's theorem to deduce
that, given initial values $u(x_0) = u_0$ and $u'(x_0) = u'_0$, there exists a unique solution $u = u(x, u_0, u'_0)$ to equation (10.14) that satisfies the initial conditions.
Picard's theorem guarantees only a local solution near x_0, but we can now
appeal to standard results concerning the extension of such solutions to the
interval $[x_0, x_1]$. In fact, it can be shown that there exist linearly independent
solutions u_1 and u_2 to equation (10.14) such that any solution to (10.14) can
be represented in the form

$$u(x) = \alpha u_1(x) + \beta u_2(x), \tag{10.15}$$

where α and β are constants. Finally, another result from the general theory
shows that the solution to the initial-value problem depends continuously on
the initial data; i.e., $u(x, u_0, u'_0)$ is continuous with respect to the parameters
u_0 and u'_0.

We need more than a global existence result for solutions to equation
(10.14): in order to assert the existence of a solution to the Riccati equation,
we need to show that there are solutions u that do not vanish on the interval
$[x_0, x_1]$. This problem leads us to the important concept of conjugate points.
Let $\kappa \in \mathbb{R} - \{x_0\}$. If there exists a *nontrivial* solution u to equation (10.14)
that satisfies $u(x_0) = u(\kappa) = 0$, then κ is called a **conjugate point** to x_0.

Lemma 10.4.2 *Let f satisfy the conditions of Lemma 10.4.1, and suppose
that there are no conjugate points to x_0 in $(x_0, x_1]$. Then, there exists a solution u to equation (10.14) such that $u(x) \neq 0$ for all $x \in [x_0, x_1]$.*

Proof: (Sketch) Given that, for any initial conditions $u(x_0) = u_0, u'(x_0) = u'_0$,
there exists a solution to equation (10.14) valid in $[x_0, x_1]$, we need to show
that the absence of a conjugate point to x_0 in $(x_0, x_1]$ implies the existence of
a solution u that does not vanish on $[x_0, x_1]$.

Consider a family of initial conditions of the form $u(x_0) = \epsilon, u'(x_0) = 1$, where ϵ is a small parameter. For each ϵ there is a solution $u(x, \epsilon)$ to
equation (10.14) valid in $[x_0, x_1]$, and u is a continuous function of ϵ near
$\epsilon = 0$. Moreover, the initial condition $u'(x_0) = 1$ precludes the possibility of
$u(x, \epsilon)$ being a trivial solution. Now, $u(x_0, 0) = 0$, and the absence of conjugate
points to x_0 in $(x_0, x_1]$ implies that $u(x, 0) \neq 0$ for all $x \in (x_0, x_1]$. Thus either
$u(x, 0) > 0$ for all $x \in (x_0, x_1]$ or $u(x, 0) < 0$ for all $x \in (x_0, x_1]$, because u is
a continuous function of x. Suppose that $u(x, 0) > 0$. We know that $u(x, \epsilon)$
is continuous with respect to ϵ and hence for ϵ sufficiently small we have
$u(x, \epsilon) > 0$ except perhaps in a small neighbourhood of x_0. To construct a

---

[3] In particular, the coefficient of u'' is not zero.

solution that is nonzero near x_0 we choose $\epsilon > 0$ so that $u(x_0, \epsilon) = \epsilon > 0$, and the condition $u'(x_0) = 1$ ensures that u is nonzero near x_0. A similar argument can be used for the case $u(x, 0) < 0$. Some technical details need to be tightened up, but the essence of the argument is that the zeros (hence conjugate points) must change continuously with the parameter ϵ. If we are careful with our choice of sign for ϵ, the initial conditions imply that the zero at x_0 will shift *out* of the interval $[x_0, x_1]$. $\qquad\qquad\square$

In summary, Lemmas 10.4.1 and 10.4.2 combine to give the following result.

Theorem 10.4.3 *Let f be a smooth function of $x, y,$ and y', and let y be a smooth extremal for the functional J defined by (10.2) such that $f_{y'y'} > 0$ for all $x \in [x_0, x_1]$. If there are no points in the interval $(x_0, x_1]$ conjugate to x_0, then the second variation is positive definite.*

Example 10.4.1:   Let
$$J(y) = \int_{x_0}^{x_1} y'^2\, dx.$$
Then the corresponding Jacobi accessory equation is
$$u''(x) = 0;$$
hence the general solution is $u(x) = \alpha + \beta x$, where α and β are constants. Clearly, only the trivial solution can satisfy the conditions $u(x_0) = 0$ and $u(\kappa) = 0$ for any $\kappa \in \mathbb{R} - \{x_0\}$. Hence, there are no points conjugate to x_0.

Example 10.4.2:   Consider the functional of Example 10.3.3. The Jacobi accessory equation is
$$u''(x) + u(x) = 0.$$
Now, $u(x) = \sin(x)$ is a nontrivial solution to this equation, and $u(0) = u(\pi) = 0$. Hence, π is a point conjugate to 0. We thus see that there is a point conjugate to 0 in the interval $(0, \ell]$.

Finally, we note that if we consider the second variation as a functional (of η) in its own right, the Jacobi accessory equation is the Euler-Lagrange equation for this functional. There is, however, a distinction to be made concerning solutions. Specifically, the functions η that solve the Euler-Lagrange equation must vanish at the endpoints. In contrast, we are actively seeking solutions to the Jacobi accessory equation that do *not* vanish in (x_0, x_1).

10.4.3 The Jacobi Necessary Condition

Theorem 10.4.3 gives a sufficient condition for the second variation to be positive definite. We show that the absence of conjugate points is also necessary for positive definiteness. Before we launch into a statement of this result, however, we establish two small lemmas.

Lemma 10.4.4 *Let u be solution to the Jacobi accessory equation (10.14) in $[x_0, x_1]$. If there is a point $c \in [x_0, x_1]$ such that $u(c) = 0$ and $u'(c) = 0$, then u must be the trivial solution.*

Proof: Suppose there is a point $c \in [x_0, x_1]$ such that $u(c) = 0$ and $u'(c) = 0$. Consider the initial-value problem formed by the Jacobi accessory equation and these conditions at $x = c$. Picard's theorem implies that there is a unique solution to this problem in a neighbourhood of c; hence, this solution must be the trivial solution. From the theory of linear differential equations we know that this solution has a unique extension into the interval $[x_0, x_1]$, and consequently $u(x) = 0$ for all $x \in [x_0, x_1]$.   □

Lemma 10.4.5 *Let u be a solution to the Jacobi accessory equation (10.14) in $[x_0, x_1]$ such that $u(x_0) = u(x_1) = 0$. Then*

$$\int_{x_0}^{x_1} \left(f_{y'y'} u'^2 + B u^2 \right) dx = 0. \tag{10.16}$$

Proof: Integration by parts gives this result immediately. Specifically, since u is a solution to (10.14),

$$\int_{x_0}^{x_1} \left(\frac{d}{dx} \left(f_{y'y'} u' \right) - Bu \right) u \, dx = 0. \tag{10.17}$$

Now,

$$\int_{x_0}^{x_1} \frac{d}{dx} \left(f_{y'y'} u' \right) u \, dx = u u' f_{y'y'} \Big|_{x_0}^{x_1} - \int_{x_0}^{x_1} f_{y'y'} u'^2 \, dx$$

$$= -\int_{x_0}^{x_1} f_{y'y'} u'^2 \, dx;$$

therefore, equation (10.17) implies equation (10.16).   □

Theorem 10.4.6 *Let f be a smooth function of x, y, and y', and let y be a smooth extremal for the functional J defined by (10.2) such that $f_{y'y'} > 0$ for all $x \in [x_0, x_1]$.*

1. *If $\delta^2 J(\eta, y) > 0$ for all $\eta \neq 0$, then there is no point conjugate to x_0 in (x_0, x_1).*

2. *If $\delta^2 J(\eta, y) \geq 0$ for all $\eta \in H$, then there is no point conjugate to x_0 in* (x_0, x_1).

Proof: We begin with a proof of the first statement. Suppose that the second variation is positive definite. We can quickly eliminate the possibility of a conjugate point at x_1 using Lemma 10.4.5. If there exists a nontrivial solution to the Jacobi accessory equation u such that $u(x_0) = u(x_1) = 0$, then we may take $\eta = u$. Lemma 10.4.5 shows that there is a nontrivial $\eta \in H$ such that the second variation vanishes contradicting the assumption that the second variation is positive definite. We thus conclude that x_1 cannot be conjugate to x_0.

To show that there is no point conjugate to x_0 in (x_0, x_1) we follow the proof given by Gelfand and Fomin ([31], p. 109). The strategy is to construct a family of positive-definite functionals $K(\mu)$, that depends on the parameter $\mu \in [0, 1]$, such that $K(1)$ is the second variation and $K(0)$ is free from conjugate points to x_0. Any solution to the Jacobi accessory equation associated with K will thus be a continuous function of μ. We exploit this continuity to show that the absence of a conjugate point for $K(0)$ implies that for $K(\mu)$, and in particular $K(1)$.

Let K be the functional defined by

$$K(\mu) = \mu \delta^2 J(\eta, y) + (1 - \mu) P(\eta),$$

where $\mu \in [0, 1]$ and P is the functional defined by

$$P(\eta) = \int_{x_0}^{x_1} \eta'^2 \, dx.$$

We know from Example 10.4.1 that P has no points conjugate to x_0, and it is clear that P is positive definite. Since $\delta^2 J(\eta, y)$ is also positive definite we see that K is positive definite for all $\mu \in [0, 1]$. The Jacobi accessory equation associated with K is

$$\frac{d}{dx} \left\{ (\mu f_{y'y'} + (1 - \mu)) u' \right\} - \mu B u = 0. \tag{10.18}$$

Now, any solution $u(x, \mu)$ to (10.18) is continuous with respect to $\mu \in [0, 1]$. Indeed, we can assert that $u(x, \mu)$ has a continuous derivative with respect to μ for all μ in an open interval containing $[0, 1]$, because $\mu f_{y'y'} + (1 - \mu) > 0$ for all $\mu \in [0, 1]$, and this coefficient (along with that for u) is smooth with respect to the parameter μ. Thus, u has continuous partial derivatives with respect to both x and μ. Let u_x and u_μ denote these derivatives.

Let U denote the family of nontrivial solutions to (10.18) such that $u(x_0, \mu) = 0$ for all $\mu \in [0, 1]$. Let $\bar{R} = [x_0, x_1] \times [0, 1]$ and $R = (x_0, x_1) \times (0, 1)$. Suppose that $K(1)$ has a conjugate point $\kappa \in (x_0, x_1)$. Then there is a $u \in U$ such that $u(\kappa, 1) = 0$. Now, u has continuous derivatives in $\bar{R}$, and by Lemma 10.4.4 (applied to $K(\mu)$) we have that $u_x(\kappa, 1) \neq 0$. We can thus invoke the implicit function theorem in a neighbourhood of $(\kappa, 1)$ to assert that there is a unique function $x = x(\mu)$ such that $u(x(\mu), \mu) = 0$ and $x(1) = \kappa$. In fact, since u_μ is continuous, we have that x_μ has a continuous derivative, and

$$x'(\mu) = -\frac{u_\mu}{u_x}. \tag{10.19}$$

The function $(x(\mu), \mu)$ thus describes parametrically a curve γ in some neighbourhood of $(\kappa, 1)$ with a continuous tangent that is nowhere horizontal; hence, the intersection of γ with the line $\mu = 1$ must be transverse. Consequently, $\gamma \cap R \neq \emptyset$; i.e., γ must have points in the interior of $\bar{R}$. Although the implicit function theorem gives only the existence of a curve near $(\kappa, 1)$, it is straightforward to see that γ cannot simply terminate in R. Specifically, suppose that γ did terminate at some point $(a, b) \in R$. The conditions of the implicit function theorem are still satisfied and we thus conclude that there is a unique nontrivial continuation of γ in a neighbourhood of (a, b) contradicting our assumption that (a, b) is a terminus for γ. We conclude that γ must continue to the boundary of R.

Relation (10.19) makes it clear that any continuation of γ cannot include a point where the tangent is horizontal, since the conditions of the implicit function theorem are satisfied at any point on this curve and this theorem guarantees that x' is finite and continuous. We thus see γ cannot "double back" and intersect the line $\mu = 1$. Since $K(0)$ does not have any conjugate points it is also clear that γ cannot intersect the line $\mu = 0$ except perhaps at the point $(x_0, 0)$. In any event, the only possible boundary curves that γ might intersect are the lines $x = x_1$ and $x = x_0$ (figure 10.1). Suppose that γ intersects the line $x = x_1$. Then, there is a μ_1 and a $u \in U$ such that $u(x_0, \mu_1) = u(x_1, \mu_1) = 0$. But the arguments used to prove Lemma 10.4.5 can be applied to $K(\mu_1)$ to show that $K(\mu_1) = 0$ for the choice $\eta = u$ and this contradicts the fact that $K(\mu)$ is positive definite. Hence γ cannot intersect the line $x = x_1$.

Consider now the line $x = x_0$. By construction we have $u(x_0, \mu) = 0$ for all $\mu \in [0, 1]$ and hence the function $x(\mu) = x_0$ is a solution of $u(x, \mu) = 0$ for all $\mu \in [0, 1]$. The conditions of the implicit function theorem are satisfied at any point on this line and therefore $x(\mu) = x_0$ is the *unique* solution that intersects this line. Evidently, γ is distinct from this line because $\kappa > x_0$. Hence, γ cannot intersect the line $x = x_0$. We thus conclude that no such curve γ can exist and hence that $K(1)$ has no points conjugate to x_0 in the interval (x_0, x_1).

To prove the second statement, note that even if $\delta^2 J \geq 0$, the functional $K(\mu)$ is still positive definite for all $\mu \in [0, 1)$. The proof that there is no point conjugate to x_0 in the interval (x_0, x_1) given above is thus valid. Lemma 10.4.5, however, does not preclude x_1 from being a conjugate point since $K(1)$ can be zero. $\square$

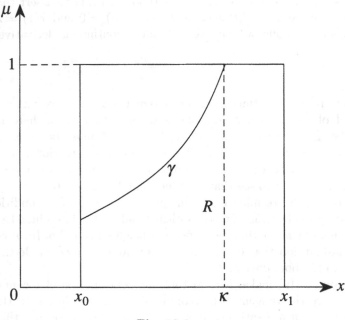

Fig. 10.1.

Theorem 10.2.1 and the second part of Theorem 10.4.6 combine to form a more refined necessary condition for a local minimum known as **Jacobi's necessary condition**.

Theorem 10.4.7 (Jacobi) *Let y be a smooth extremal for the functional*

$$J(y) = \int_{x_0}^{x_1} f(x, y, y') \, dx$$

such that for all $x \in [x_0, x_1]$

$$f_{y'y'} > 0$$

along y. If y produces a local minimum for J then there are no points conjugate to x_0 in the interval (x_0, x_1).

Note that Jacobi's necessary condition does not preclude the possibility that x_1 is conjugate to x_0.

Exercises 10.4:

1. Derive the Riccati equation (10.11) associated with the functional of Example 10.3.3. Solve the Riccati equation directly and show that there are no solutions w defined for *all* $x \in [0, \ell]$ if $\ell > \pi$.

2. Let
$$f(x, y, y') = y'^2 - y'y + y^2.$$
Show, using elementary arguments, that $\delta^2 J(\eta, y) \geq 0$ for all $\eta \in H$. Derive the Jacobi accessory equation and show by solving this equation that any nontrivial solution u can have at most one zero.

3. Suppose that f does not depend on y explicitly and that f satisfies the strengthened Legendre condition along an extremal $y(x)$. Prove that there are no points conjugate to x_0.

4. The proof of the first statement of Theorem 10.4.6 is considerably simpler if we "open up" the space H to a more general one $\hat{H}$ that includes piecewise smooth functions on $[x_0, x_1]$ that vanish at x_0 and x_1. If $\delta^2 J(\eta, y) > 0$ for all $\eta \in \hat{H}$, $\eta \neq 0$, prove that there are no points conjugate to x_0 in (x_0, x_1) (cf. [71], p. 91).

10.5 A Sufficient Condition

In Section 10.2 we derived the expression

$$J(\hat{y}) - J(y) = \frac{\epsilon^2}{2} \delta^2 J(\eta, y) + O(\epsilon^3), \tag{10.20}$$

which an extremal y for J satisfies for any "neighbouring curve" $\hat{y} = y + \epsilon\eta$. The necessity of the condition $\delta^2 J(\eta, y) \geq 0$ for a local minimum is clear, but the sufficiency of this condition is suspect. Indeed, we know from our brief tour of finite-dimensional optimization that semidefinite quadratic forms do not necessarily lead to local extrema. The problem is that if there is a nontrivial η such that $\delta^2 J(\eta, y) = 0$, then the $O(\epsilon^3)$ terms in the above expansion control the sign of $J(\hat{y}) - J(y)$. Certainly, we can avoid this problem by requiring that $\delta^2 J(\eta, y)$ be positive definite, but even this strengthened requirement has snags because there may be nontrivial η such that $\delta^2 J(\eta, y) > 0$ but of order ϵ, in which case the sign of $J(\hat{y}) - J(y)$ depends on the higher order terms for ϵ small. Harsher restrictions are needed to control the magnitude of $\delta^2 J(\eta, y)$ relative to the remainder terms.

Let $\| \cdot \|_1$ be the norm on the space $C^2([x_0, x_1])$ defined by

$$\|y\|_1 = \sup_{x \in [x_0, x_1]} |y(x)| + \sup_{x \in [x_0, x_1]} |y'(x)|.$$

We say that a functional $J : C^2([x_0, x_1]) \to \mathbb{R}$ has a **weak local minimum** at $y \in S$ if there is a $\Delta > 0$ such that $J(\hat{y}) - J(y) \geq 0$ for all $\hat{y} \in S$ such that $\|\hat{y} - y\|_1 < \Delta$. Similarly, J is said to have a **weak local maximum** at $y \in S$ if $-J$ has a weak local minimum at y. The adjective "weak" creeps into the definition to distinguish such extrema from **strong extrema**, the definition of which is identical to the weak extrema except that the norm $\| \cdot \|_0$, defined by

$$\|y\|_0 = \sup_{x \in [x_0, x_1]} |y(x)|,$$

is used. Clearly, $\|\hat{y} - y\|_1 < \Delta$ implies that $\|\hat{y} - y\|_0 < \Delta$, but the converse is not true. Hence, "weak" signifies that we are restricting the competition to a *subset* of the set of functions that are candidates for a strong extremum.

We need to get an expression for the remainder term in the Taylor expansion (10.20) in order to develop a sufficient condition. Now, it can be shown that there is a function $m(\eta)$ such that

$$J(y + \epsilon \eta) = J(y) + \epsilon \delta J(\eta, y) + \frac{\epsilon^2}{2} \delta^2 J(\eta, y) + \epsilon^2 m(\eta) \|\eta_1\|_1^2,$$

where $m(\eta) \to 0$ as $\|\eta\|_1 \to 0,$[4] so that, for an extremal,

$$J(\hat{y}) - J(y) = \epsilon^2 \left(\frac{1}{2} \delta^2 J(\eta, y) + m(\eta) \|\eta_1\|_1^2 \right).$$

In fact, it can be shown[5] that there are functions ν and ρ such that

$$J(\hat{y}) - J(y) = \epsilon^2 \left(\frac{1}{2} \delta^2 J(\eta, y) + \int_{x_0}^{x_1} \left(\nu \eta'^2 + \rho \eta^2 \right) dx \right), \tag{10.21}$$

and $\nu, \rho \to 0$ as $\|\eta\|_1 \to 0$.

The main result of this section is the following theorem, which applies to the basic fixed endpoint problem.

Theorem 10.5.1 *Let $y \in S$ be an extremal for the functional*

$$J(y) = \int_{x_0}^{x_1} f(x, y, y') \, dx,$$

and suppose that along this extremal the strengthened Legendre condition

$$f_{y'y'}(x, y(x), y'(x)) > 0$$

is satisfied for all $x \in [x_0, x_1]$. Suppose further that there are no points conjugate to x_0 in (x_0, x_1). Then J has a weak local minimum in S at y.

Proof: Let $\mu \in \mathbb{R}$ be a small parameter and consider the family of functionals K defined by

$$K(\mu) = \int_{x_0}^{x_1} \left(f_{y'y'} \eta'^2 + B \eta^2 \right) dx - \mu^2 \int_{x_0}^{x_1} \eta'^2 \, dx$$

$$= \int_{x_0}^{x_1} \left(\left(f_{y'y'} - \mu^2 \right) \eta'^2 + B \eta^2 \right) dx.$$

---

[4] Taylor's theorem has an extension to operators on general Banach spaces (cf. [22]).

[5] See [31], pp. 101–102.

The Jacobi accessory equation for K is

$$\frac{d}{dx}\left(\left(f_{y'y'} - \mu^2\right) u'\right) - Bu = 0. \tag{10.22}$$

The smoothness conditions on f and the strengthened Legendre condition imply that there is a positive number σ such that $f_{y'y'} \geq \sigma$ for all $x \in [x_0, x_1]$. Thus, for all $\mu^2 < \sigma$, we have

$$f_{y'y'} - \mu^2 > 0, \tag{10.23}$$

for all $x \in [x_0, x_1]$, and this in turn means that the solutions $u(x, \mu)$ to equation (10.22) are continuous functions of both x and μ, provided $|\mu|$ is small. By hypothesis we have that there are no points conjugate to x_0 in (x_0, x_1) for the case $\mu = 0$, and the continuity of u with respect to μ implies that there are no points conjugate to x_0 in $(x_0, x_1]$ for all $|\mu|$ sufficiently small. Therefore, there is a $\mu_1 > 0$ such that for all μ, with $|\mu| < \mu_1$, the functional $K(\mu)$ satisfies the strengthened Legendre condition (10.23) and has no points conjugate to x_0 in $(x_0, x_1]$. Theorem 10.4.3 therefore implies that $K(\mu)$ is positive definite for all $|\mu| < \mu_1$. We thus conclude that there is a number $p > 0$ such that

$$\frac{1}{2}\delta^2 J(\eta, y) > p \int_{x_0}^{x_1} \eta'^2 \, dx. \tag{10.24}$$

We now consider the remainder term

$$R = \int_{x_0}^{x_1} \left(\nu\eta'^2 + \rho\eta^2\right) \, dx.$$

Since $\nu, \rho \to 0$ as $\|\eta\|_1 \to 0$, there is a q such that $q \to 0$ as $\|\eta\|_1 \to 0$, and

$$|R| \leq q \left\{ \int_{x_0}^{x_1} \eta'^2 \, dx + \int_{x_0}^{x_1} \eta^2 \, dx \right\}.$$

For any continuous functions g, h on $[x_0, x_1]$ and any $x \in [x_0, x_1]$, the Schwarz inequality is

$$\langle g, h \rangle^2 = \left(\int_{x_0}^{x} g(\xi) h(\xi) \, d\xi \right)^2$$
$$\leq \langle g, g \rangle \langle h, h \rangle$$
$$= \int_{x_0}^{x} g^2(\xi) \, d\xi \int_{x_0}^{x} h^2(\xi) \, d\xi.$$

Now,

$$\eta^2 = \left(\int_{x_0}^{x} \eta'(\xi) \, d\xi \right)^2,$$

so that for $g = 1$, $h = \eta$, the Schwarz inequality gives

$$\eta^2 \le (x - x_0) \int_{x_0}^{x} \eta'^2 \, d\xi \le (x - x_0) \int_{x_0}^{x_1} \eta'^2 \, dx.$$

Thus,

$$\int_{x_0}^{x_1} \eta^2 \, dx \le \frac{(x_1 - x_0)^2}{2} \int_{x_0}^{x_1} \eta'^2 \, dx,$$

and hence

$$|R| \le |q| \left(1 + \frac{(x_1 - x_0)^2}{2} \right) \int_{x_0}^{x_1} \eta'^2 \, dx.$$

For $\|\eta\|_1$ sufficiently small, we have

$$|q| \left(1 + \frac{(x_1 - x_0)^2}{2} \right) < \frac{p}{2};$$

consequently,

$$J(\hat{y}) - J(y) > \epsilon^2 \left(p \int_{x_0}^{x_1} \eta'^2 \, dx - \frac{p}{2} \int_{x_0}^{x_1} \eta'^2 \, dx \right)$$

$$> \frac{\epsilon^2}{2} p \int_{x_0}^{x_1} \eta'^2 \, dx$$

$$> 0.$$

Thus, $J(\hat{y}) - J(y) > 0$ for all nontrivial η, provided $\|\eta\|_1$ sufficiently small, and therefore y is a weak local minimum for J.   □

Exercises 10.5:

1. The second variation $\delta^2 J(\eta, y)$ is called **positive and nondegenerate** (or strongly positive) if there is a constant $\Lambda > 0$ such that

$$\delta^2 J(\eta, y) \ge \Lambda \|\eta\|_1^2.$$

Suppose that y is an extremal for J in S and that $\delta^2 J(\eta, y)$ is positive and nondegenerate. Show that y is a weak local minimum.

10.6 More on Conjugate Points

The Jacobi necessary condition and the sufficient condition of Section 10.5 both require verification that there are no points conjugate to x_0 in the interval (x_0, x_1). In this section we discuss a simple method for finding conjugate points and a geometrical interpretation of these points.

10.6.1 Finding Conjugate Points

Suppose that y is an extremal for the functional J. Recall that a point $\kappa \in \mathbb{R} - \{x_0\}$ is conjugate to x_0 if there is a nontrivial solution u to the Jacobi accessory equation (10.14) such that $u(x_0) = u(\kappa) = 0$. In order to test whether an extremal has a conjugate point in the interval $(x_0, x_1]$ we are thus obliged to somehow procure a general solution u to equation (10.14) and check whether there is a zero of u in the interval $(x_0, x_1]$. Although the Jacobi accessory equation is linear, finding a general solution to such equations can prove a formidable task. It is thus a relief to discover that solutions to equation (10.14) can be *derived* from the general solution to the Euler-Lagrange equation.

Suppose that y is a general solution to the Euler-Lagrange equation

$$\frac{d}{dx}\frac{\partial f}{\partial y'} - \frac{\partial f}{\partial y} = 0, \tag{10.25}$$

associated with the functional

$$J(y) = \int_{x_0}^{x_1} f(x, y, y')\, dx.$$

The general solution to a second-order ordinary differential equation contains two parameters c_1, c_2 (constants of integration) and it can be shown that y depends smoothly on these parameters. Since y depends on c_1 and c_2, so does f in the Euler-Lagrange equation, and the smoothness of f with respect to y and y' implies that f also depends smoothly on c_1 and c_2. Differentiating equation (10.25) with respect to c_1, noting that the smoothness assumptions on f allow the orders of differentiation to be changed, gives

$$\frac{\partial}{\partial c_1}\left(\frac{d}{dx}\frac{\partial f}{\partial y'} - \frac{\partial f}{\partial y}\right) = \frac{d}{dx}\left(\frac{\partial}{\partial y'}\left(\frac{\partial f}{\partial y}\frac{\partial y}{\partial c_1} + \frac{\partial f}{\partial y'}\frac{\partial y'}{\partial c_1}\right)\right)$$
$$- \frac{\partial}{\partial y}\left(\frac{\partial f}{\partial y}\frac{\partial y}{\partial c_1} + \frac{\partial f}{\partial y'}\frac{\partial y'}{\partial c_1}\right)$$
$$= \frac{d}{dx}\left(\frac{\partial^2 f}{\partial y \partial y'}\frac{\partial y}{\partial c_1} + \frac{\partial^2 f}{\partial y' \partial y'}\frac{\partial y'}{\partial c_1} + \frac{\partial f}{\partial y'}\right)$$
$$- \left(\frac{\partial^2 f}{\partial y \partial y}\frac{\partial y}{\partial c_1} + \frac{\partial^2 f}{\partial y \partial y'}\frac{\partial y'}{\partial c_1} + \frac{\partial f}{\partial y}\right)$$
$$= 0.$$

Let $u_1 = \partial y/\partial c_1$. Then,

$$u_1' = \frac{d}{dx}\frac{\partial y}{\partial c_1} = \frac{\partial y'}{\partial c_1}.$$

Using the notation of Section 10.2 and noting that y is a solution to the Euler-Lagrange equation, the above calculation yields

$$\frac{d}{dx}(f_{yy'}u_1 + f_{y'y'}u_1' + f_{y'}) - (f_{yy}u_1 + f_{yy'}u_1' + f_y)$$

$$= \frac{d}{dx}(f_{y'y'}u_1') + \frac{d}{dx}(f_{yy'}u_1) - f_{yy}u_1 - f_{yy'}u_1'$$

$$= \frac{d}{dx}(f_{y'y'}u_1') - \left(f_{yy} - \frac{d}{dx}f_{yy'}\right)u_1$$

$$= 0.$$

In this manner we see that u_1 must be a solution to the Jacobi accessory equation (10.14). A similar argument shows that $u_2 = \partial y/\partial c_2$ is also a solution to equation (10.14). *We can thus obtain solutions to the Jacobi accessory equation by simply differentiating the general solution to the Euler-Lagrange equation with respect to c_1 and c_2.* In fact, it can be shown ([15], pp.68–72) that u_1 and u_2 form a basis for the solution space.

Let $\mathbf{c} = (c_1, c_2)$, $\mathbf{k} = (k_1, k_2)$, and suppose that $y(x, \mathbf{k})$ is a solution to equation (10.25) that satisfies the boundary conditions of the fixed endpoint problem. Let

$$u_1(x, \mathbf{k}) = \frac{\partial y}{\partial c_1}\bigg|_{\mathbf{c}=\mathbf{k}}, \quad u_2(x, \mathbf{k}) = \frac{\partial y}{\partial c_2}\bigg|_{\mathbf{c}=\mathbf{k}}.$$

Then, the general solution $u(x, \mathbf{k})$ to equation (10.14) is given by

$$u(x, \mathbf{k}) = \alpha u_1(x, \mathbf{k}) + \beta u_2(x, \mathbf{k}),$$

where α and β are constants. We are interested in nontrivial solutions u, so that α and β are not both zero. If κ is conjugate to x_0, then there are values of α and β such that

$$u(x_0, \mathbf{k}) = \alpha u_1(x_0, \mathbf{k}) + \beta u_2(x_0, \mathbf{k}) = 0,$$

and

$$u(\kappa, \mathbf{k}) = \alpha u_1(\kappa, \mathbf{k}) + \beta u_2(\kappa, \mathbf{k}) = 0;$$

hence,

$$u_2(\kappa, \mathbf{k})u_1(x_0, \mathbf{k}) = u_2(x_0, \mathbf{k})u_1(\kappa, \mathbf{k}). \tag{10.26}$$

Note that u_1 and u_2 cannot both vanish at the same value of x, because this would imply that the Wronskian $W(x) = u_1(x)u_2'(x) - u_1'(x)u_2(x) = 0$ for all values of x and hence that u_1 and u_2 would be linearly dependent.[6] Thus, relation (10.26) is an equation for a conjugate point κ. If $u_1(x_0, \mathbf{k}) \neq 0$ and $u_1(\kappa, \mathbf{k}) \neq 0$, the above relation is usually written in the form

$$\frac{u_2(x_0, \mathbf{k})}{u_1(x_0, \mathbf{k})} = \frac{u_2(\kappa, \mathbf{k})}{u_1(\kappa, \mathbf{k})}.$$

---

[6] See [9], pp. 42–45 for more details.

Example 10.6.1: Let J be the functional of Example 10.3.3. Here, $f(x, y, y') = y'^2 - y^2$, so that the Euler-Lagrange equation has the general solution

$$y(x, c_1, c_2) = c_1 \cos x + c_2 \sin x.$$

The above arguments show that

$$u_1 = \frac{\partial y}{\partial c_1} = \cos x, \quad u_2 = \frac{\partial y}{\partial c_2} = \sin x$$

are solutions to the Jacobi accessory equation. (This can be verified directly, since the Euler-Lagrange equation is equivalent to the Jacobi accessory equation for this integrand.) Hence, any conjugate point κ to 0 must satisfy

$$u_2(\kappa)u_1(0) = u_1(\kappa)u_2(0);$$

i.e.,

$$\sin \kappa = 0.$$

The points conjugate to 0 are therefore of the form $\kappa = \pm n\pi$, where $n = 1, 2, \ldots$. In Example 10.3.3 we chose $\ell > \pi$ and hence the interval $(0, \ell)$ included the point π, which is conjugate to 0.

Example 10.6.2: Geodesics in the Plane
Consider the arclength functional of Example 2.2.1, where $f(x, y, y') = \sqrt{1 + y'^2}$. The general solution of the Euler-Lagrange equation is

$$y(x, c_1, c_2) = c_1 x + c_2.$$

The corresponding solutions to the Jacobi accessory equation are $u_1 = x$ and $u_2 = 1$. Any point κ that is conjugate to x_0 must satisfy equation (10.26). The only solution to this equation, however, is $\kappa = x_0$, and therefore there are no points conjugate to x_0. Since $f_{y'y'} > 0$, Theorem 10.5.1 implies J has a weak local minimum at y. In fact, we can do better than this (cf. Example 10.7.1).

Example 10.6.3: Catenary
Consider the catenary problem of Section 1.2, where $f(x, y, y') = y\sqrt{1 + y'^2}$ and $x \in [0, 1]$. We showed in Example 2.3.3 that a general solution to the Euler-Lagrange equation is of the form

$$y(x, c_1, c_2) = c_1 \cosh\left(\frac{x}{c_1} + c_2\right),$$

where c_1 and c_2 are constants. Solutions u_1 and u_2 to the Jacobi accessory equation for this problem are thus

$$u_1 = \frac{\partial y}{\partial c_1} = \cosh\left(\frac{x}{c_1} + c_2\right) - \left(\frac{x}{c_1} + c_2\right)\sinh\left(\frac{x}{c_1} + c_2\right),$$

$$u_2 = \frac{\partial y}{\partial c_2} = -\sinh\left(\frac{x}{c_1} + c_2\right).$$

Consider now the extremals associated with the boundary values

$$y(0) = 1, \quad y(1) = 1.$$

In Example 2.6.1 we studied the general problem of determining the c_1 and c_2 to satisfy such boundary conditions. In that example we argued that for any value of $y(1) > 0.6$ there are precisely two solutions for the integration constants. For this example, we thus know that there are two sets of parameters that satisfy the boundary conditions. Equation (2.42) implies

$$\cosh c_2 = \cosh(\cosh c_2 + c_2);$$

i.e.,

$$\pm c_2 = \cosh c_2 + c_2.$$

Since $\cosh c_2 > 0$ for all $c_2 \in \mathbb{R}$, we must have

$$c_2 < 0,$$

and

$$\cosh c_2 = -2c_2. \tag{10.27}$$

The boundary condition $y(0) = 1$ and equation (10.27) give

$$\frac{1}{c_1} = \cosh c_2 = -2c_2.$$

The solutions u_1 and u_2 can thus be written

$$u_1 = \cosh(c_2(1 - 2x)) - c_2(1 - 2x)\sinh(c_2(1 - 2x)),$$
$$u_2 = -\sinh(c_2(1 - 2x)),$$

where c_2 is a solution to equation (10.27). Let $\xi = c_2(1-2x)$. Equation (10.26) shows that a conjugate point to 0 must satisfy the relation

$$(\cosh c_2 - c_2\sinh c_2)\sinh \xi = (\cosh \xi - \xi \sinh \xi)\sinh c_2;$$

i.e.,

$$\coth c_2 - c_2 = \coth \xi - \xi. \tag{10.28}$$

Now, equation (10.27) has two solutions $r_1 \approx -0.6$ and $r_2 \approx -2.1$, and for any fixed c_2 equation (10.28) also has precisely two solutions, one of which is simply $\xi = c_2$; i.e., $x = 0$ (see figure 10.2). For the choice $c_2 = r_1$, the second solution to (10.28) corresponds to $x \approx 2.4 \notin (0, 1]$. The strengthened Legendre

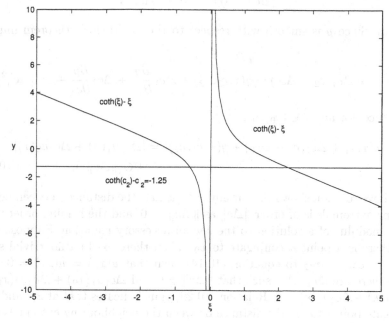

Fig. 10.2.

condition is satisfied for any choice of c_2, and since there are no points in $(0, 1]$ conjugate to 0, Theorem 10.5.1 implies that the extremal

$$y_1 = -\frac{1}{2r_1} \cosh(r_1(1 - 2x))$$

corresponds to a weak local minimum for J.

In contrast, if $c_2 = r_2$, then the second solution to equation (10.28) corresponds to $x \approx 0.6 \in (0, 1)$; hence, there is a point conjugate to 0 in the interval $(0, 1)$. Theorem 10.4.7 thus shows that the extremal

$$y_2 = -\frac{1}{2r_2} \cosh(r_2(1 - 2x))$$

does not correspond to a local extremum for J. A fuller discussion of conjugate points for catenaries (and catenoids) can be found in Forsyth [27], pp. 98–104.

10.6.2 A Geometrical Interpretation

The general solution $y(x, c_1, c_2)$ to equation (10.25) describes a two-parameter family of extremals for J. Suppose that we consider "neighbouring" extremals $y(x, c_1, c_2)$ and $y(x, c_1 + \Delta c_1, c_2 + \Delta c_2)$, where

$$\|\Delta \mathbf{c}\| = \sqrt{(\Delta c_1)^2 + (\Delta c_1)^2}$$

is small. Since y is smooth with respect to the c_k, Taylor's theorem implies that

$$y(x, c_1 + \Delta c_1, c_2 + \Delta c_2) = y(x, c_1, c_2) + \Delta c_1 \frac{\partial y}{\partial c_1} + \Delta c_2 \frac{\partial y}{\partial c_2} + O(\|\Delta \mathbf{c}\|^2),$$

and hence, for any $x \in [x_0, x_1]$,

$$|y(x, c_1 + \Delta c_1, c_2 + \Delta c_2) - y(x, c_1, c_2)| = |\Delta c_1 u_1(x) + \Delta c_2 u_2(x)|$$
$$+ O(\|\Delta \mathbf{c}\|^2). \qquad (10.29)$$

The above relation shows that at any $x \in [x_0, x_1]$ the distance between neighbouring extremals is of order $\|\Delta \mathbf{c}\|$ as $\|\Delta \mathbf{c}\| \to 0$, and the leading order term is the modulus of a solution to the Jacobi accessory equation. Suppose now that there is a point κ conjugate to x_0. Then there exists a nontrivial solution $u = \alpha u_1 + \beta u_2$ to equation (10.14) such that $u(x_0) = u(x_1) = 0$, and hence there are $\Delta c_1, \Delta c_2$ such that $\|\Delta \mathbf{c}\| \neq 0$ and $\Delta c_1 u_1(x_0) + \Delta c_2 u_2(x_0) = \Delta c_1 u_1(\kappa) + \Delta c_2 u_2(\kappa) = 0$. Relation (10.29) thus indicates that at x_0, and any conjugate points to x_0, the distance between the neighbouring extremals is of order $\|\Delta \mathbf{c}\|^2$ as $\|\Delta \mathbf{c}\| \to 0$, since we can always scale our choice of Δc_1 and Δc_2. Roughly speaking, equation (10.29) indicates that the neighbouring extremals "nearly intersect" at conjugate points. At any rate, conjugate points bear the distinctive hallmark of an envelope for the family of extremals.

Let $h(x, y, c) = 0$ describe a one-parameter family $\mathcal{F}$ of curves in the xy-plane parametrized by c. A curve ν is called an **envelope** of the family $\mathcal{F}$ if:

(a) at each point $(\hat{x}, \hat{y}) \in \nu$ there is a $\gamma \in \mathcal{F}$ that is tangent to ν; and
(b) there are infinitely many curves in $\mathcal{F}$ tangent to each arc of ν.

Suppose that $\mathcal{F}$ has an envelope ν. Then at each point $(\hat{x}, \hat{y}) \in \nu$ there is a $c = c(\hat{x}, \hat{y})$ such that the curve described by $h(x, y, c(\hat{x}, \hat{y})) = 0$ is tangent to ν at $(\hat{x}, \hat{y})$. For simplicity, we assume here that h is a smooth function of x, y, and c, and that the arc under consideration is such that c can be regarded as a smooth function of $\hat{x}$. (The latter assumption is true, for instance, if the arc of ν can be described parametrically in the form $(\hat{x}, \hat{y}(\hat{x}))$ for $\hat{x}$ in some interval I, where $\hat{y}$ is differentiable with respect to $\hat{x}$, and $\hat{y}'(\hat{x}) \neq 0$ for all $\hat{x} \in I$.) On the envelope ν, the function h must therefore satisfy

$$h(\hat{x}, \hat{y}, c(\hat{x})) = 0. \qquad (10.30)$$

Now, the crucial property of ν is that each point on ν is tangent to some curve in $\mathcal{F}$. The derivative $\hat{y}'$ can be determined from the relation

$$\frac{d}{d\hat{x}} h(\hat{x}, \hat{y}, c(\hat{x})) = \frac{\partial h}{\partial \hat{x}} + \frac{\partial h}{\partial \hat{y}} y'(\hat{x}) + \frac{\partial h}{\partial c} c'(\hat{x}). \qquad (10.31)$$

For any member of $\mathcal{F}$ the derivative can be found from the relation

$$\frac{d}{dx} h(x, y, c) = \frac{\partial h}{\partial x} + \frac{\partial h}{\partial y} y'(x).$$

In particular, at $(x, y, c) = (\hat{x}, \hat{y}, c(\hat{x}))$ for $\gamma \in \mathcal{F}$ we have

$$\frac{\partial h}{\partial \hat{x}} + \frac{\partial h}{\partial \hat{y}} \hat{y}'(\hat{x}) = 0. \tag{10.32}$$

Equations (10.31) and (10.32) thus give the condition

$$\frac{\partial h}{\partial c} c'(\hat{x}) = 0,$$

and, assuming $c'(\hat{x}) \neq 0$ for all $\hat{x} \in I$, we have the relation

$$\frac{\partial h}{\partial c} = 0. \tag{10.33}$$

Equations (10.30) and (10.33) can be regarded as a pair of implicit equations involving the three variables $\hat{x}$, $\hat{y}$, and c. Under the assumption that

$$\frac{\partial h}{\partial \hat{y}} \neq 0,$$

we can invoke the implicit function theorem to solve equation (10.30) for $\hat{y}$, and regard equation (10.33) as an implicit equation for c as a function of $\hat{x}$ (or vice versa). Once c is eliminated, equation (10.30) can then be used to determine $\hat{y}$ as a function of $\hat{x}$ and hence the curve ν.

Example 10.6.4: Let

$$h(x, y, c) = y - (x + c)^3 = 0. \tag{10.34}$$

The family of curves described by h consists of the cubic curve $y = x^3$ shifted parallel to the x-axis by the value c. Now,

$$\frac{\partial h}{\partial c} = -3(x + c)^2,$$

so that equation (10.33) gives $c = -x$. Equation (10.34) thus implies that $y(x) = 0$, so that if there is an envelope ν, it must be the x-axis. In this case the x-axis is an envelope for the family of curves.

Example 10.6.5: Let

$$h(x, y, c) = y^2 + (x + c)^2 - 1 = 0. \tag{10.35}$$

The family $\mathcal{F}$ corresponds to circles centred at $(-c, 0)$ of radius 1. Here,

$$\frac{\partial h}{\partial c} = 2(x + c),$$

so that equation (10.33) gives $c = -x$. Equation (10.35) thus implies $y(x) = \pm 1$. These lines form envelopes for $\mathcal{F}$.

Note that satisfaction of equations (10.30) and (10.33) is a necessary but not sufficient condition for an envelope. A simple counterexample is given by the family of lines defined by $h(x, y, c) = y - cx = 0$. Equations (10.30) and (10.33) are satisfied only if $x = y = 0$. The family has a "focus" at $(0, 0)$, but this singularity is not an envelope.

Returning to the calculus of variations, if κ is conjugate to x_0 and $u = \alpha u_1 + \beta u_2$ is a nontrivial solution to the Jacobi accessory equation, then α and β cannot both be zero, and it is clear that we can choose Δc_2 such that $\Delta c_2 = \Delta c_1 \beta / \alpha$ (or Δc_1 such that $\Delta c_1 = \Delta c_2 \alpha / \beta$) and get a nontrivial solution $\hat{u}$ to equation (10.14) that vanishes at x_0 and κ. In other words, for the purposes of studying conjugate points we can let $c_2 = c_1 \beta / \alpha$ and thus regard $y(x, c_1, c_2)$ as a one-parameter family $y(x, c)$. Now,

$$\hat{u} = \frac{\partial y}{\partial c} = \frac{c_1}{\alpha} u,$$

so that at any conjugate point κ we have

$$\frac{\partial y}{\partial c} = 0;$$

consequently, the necessary condition for an envelope is satisfied at conjugate points.

Envelopes abound in nature, and many arise through variational principles. For example, caustics are formed when light rays form an envelope. The extremals are the solutions to the Euler-Lagrange equations that arise from Fermat's Principle. A convenient example is the bright curve formed in a partially full tea cup on a sunny day as a result of the sun's rays reflecting on the inside of the cup.[7]

Example 10.6.6: Parabola of Safety

Another prominent example of an envelope is the so-called "parabola of safety" familiar to artillery gunners (and combat pilots). The path of the projectile is governed by Hamilton's Principle. Suppose that the cannon is fixed at the origin, but that it may be elevated at any angle ϕ, $0 < \phi < \pi/2$. The resulting trajectories for projectiles leaving the cannon are the parabolas given by

---

[7] True, these extremals are certainly not smooth, but if needed, we can restrict our attention to the family of light rays after the reflection. The curve that forms the caustic is called a **nephroid**.

$$y(x, \phi) = x \tan \phi - \frac{gx^2}{2v_0^2 \cos^2 \phi}, \tag{10.36}$$

where g is the gravitation constant and v_0 is the initial speed (muzzle velocity), which assumed to be the same for all projectiles. For this example,

$$h(x, y, \phi) = y - x \tan \phi + \frac{gx^2}{2v_0^2 \cos^2 \phi},$$

so that

$$\frac{\partial h}{\partial \phi} = x \sec \phi \left(\frac{gx}{v_0^2} \tan \phi - 1 \right).$$

The solution to equation (10.33) is

$$\frac{gx}{v_0^2} \tan \phi = 1. \tag{10.37}$$

Equations (10.36) and (10.37) imply that

$$y(x) = \frac{v_0^2}{2g} - \frac{gx^2}{2v_0^2}. \tag{10.38}$$

Equation (10.38) defines the **parabola of safety**. Each extremal in the family defined by relation (10.36) lies below this parabola except at one point where the extremal and the parabola intersect and have a common tangent. The "firing zone" is the space between the parabola of safety and the x-axis. The projectiles never exit this zone, so a pilot can fly safely above the parabola.

An introduction to envelopes and applications is given by Boltyanskii [13]. A more rigorous and advanced (but still quite accessible) account of envelopes and other singularities is given by Bruce and Giblin [18].

Conjugate points need not always yield envelopes for a family of extremals. The family of geodesics on a sphere through the North Pole, for instance, defines a family of extremals (lines of longitude) that intersect at a common point (the South Pole), which is a conjugate point that is certainly not an envelope. There is, in fact, an optical device that mimics geodesics on a sphere. The lens is called the **Maxwell fisheye**. The refractive index for this lens is

$$n(r) = \frac{1}{1 + (r/a)^2} n_0,$$

where r denotes the distance from a fixed point, and a and n_0 are constants. Born and Wolf [14], pp. 147–149 and Luneberg [50] pp. 197–214 discuss this remarkable lens and certain generalizations.

10.6.3 Saddle Points*

Theorem 10.4.7 shows that if an extremal produces a local minimum for a functional J, then it cannot have conjugate points in the interval (x_0, x_1). Theorem 10.5.1 shows that extremals that do not have conjugate points in the interval $(x_0, x_1]$ produce weak local minima for J. Geodesics on the sphere from the North Pole to the South Pole show that if x_1 is conjugate to x_0 then the extremal (albeit not uniquely determined) may still produce a local minimum. These results can be easily adapted to the case of local maxima of a functional J by simply applying the results to the functional $K = -J$ instead of J. We can thus conclude that an extremal y corresponds to neither a local minimum nor a local maximum if there is a point conjugate to x_0 in the interval (x_0, x_1). The question arises whether it is possible to classify extremals with conjugate points in a manner analogous to that used in finite dimensions for saddle points. Although such a classification may be of limited interest in many physical applications, it turns out that it is certainly a fruitful line of enquiry in topology and differential geometry.

The classification of extremals with conjugate points is the starting point for a broad subject called "The Calculus of Variations in the Large" pioneered by M. Morse. It is well beyond the scope of this book to give even a rudimentary account of this topic, but we do give a few comments (no proofs), which we hope will whet the appetite of the reader to look at a serious study of this subject. A standard reference is the book by Morse [55]. Milnor [53] and Spivak [65] also give accounts of the theory as it applies to geodesics. As with the previous material, we focus exclusively on candidates for local minima and the strengthened Legendre condition (10.9) is assumed to be satisfied.

The key to obtaining a classification of extremals lies in extending the Morse index to infinite-dimensional spaces. In finite-dimensional spaces, the Morse index counts the number of minus signs in the canonical representation of f near $\mathbf{x}_0$ (Lemma 10.1.1). On a slightly deeper level, the Morse index corresponds to the maximum dimension of a subspace of $\mathbb{R}^n$ wherein the Hessian matrix is negative definite. This idea can be transferred to infinite dimensional spaces. If the function space is a Hilbert space, then a decomposition of quadratic functionals such as $\delta^2 J$ analogous to that given in Lemma 10.1.1 is possible (cf. [22], pp. 571–572). Here we take the direct approach. Let y be an extremal for the functional J and let $\delta^2 J : H \times H \to \mathbb{R}$ be the corresponding second variation. The **Morse index** λ of y is defined to be the maximal dimension of the subspace of H on which $\delta^2 J$ is negative definite.

The problem with the above definition is that it is not clear how one might determine λ, or for that matter, if λ is even finite. It turns out that there is a tractable way to calculate λ thanks to the Morse index theorem. The general statement of this result concerns functionals that involve several dependent variables, and the notion of multiplicity for conjugate points is needed. The **multiplicity** of a point κ conjugate to x_0 is defined to be the number of linearly independent solutions $\mathbf{u}$ to the Jacobi accessory equation that satisfy

$\mathbf{u}(x_0) = \mathbf{u}(\kappa) = 0$. For our case,[8] the general solution to the Jacobi accessory equation (10.14) is of the form $u = \alpha u_1 + \beta u_2$, where u_1 and u_2 are linearly independent solutions to (10.14) and α, β are constants. It is thus clear that the multiplicity cannot exceed two. The problem of finding nontrivial solutions to this boundary-value problem is a thinly disguised Sturm-Liouville problem, and we know that all the eigenvalues associated with such problems are simple (see Section 5.1). In short, the multiplicity of conjugate points is one, for functionals of the type considered here.

A general statement of the Morse index theorem along with a proof is given in Milnor [53]. Here, we give a simple "no frills" version for extremals to functionals of the form

$$J(y) = \int_{x_0}^{x_1} f(x, y, y') \, dx$$

that satisfy the strengthened Legendre condition.

Theorem 10.6.1 (Morse Index Theorem) *Let y be an extremal for J. The index λ of $\delta^2 J$ is equal to the number of points in (x_0, x_1) conjugate to x_0. This index is always finite.*

The above result allows us to classify extremals in a spirit similar to that used to classify critical points in finite-dimensional spaces. For instance, if $\lambda = 0$ for an extremal, and x_1 is not conjugate to x_0, then Theorem 10.5.1 indicates that J has a weak local minimum at y. For the functional of Example 10.3.3, the index λ is at least 1 since $\pi \in (0, \ell)$; if, say $\ell = 7\pi/2$, then $\lambda = 3$, since the conjugate points π, 2π, and 3π are all in $(0, \ell)$. For this example the coefficients of the Jacobi accessory equation do not depend on y, so that all extremals with the same endpoints have the same index. Geodesics on the sphere can have an index of 0 or 1 depending on whether they contain antipodal points. Similarly, extremals for the catenary can have a Morse index of 0 or 1 depending on the choice of solutions for the integration constants (Example 10.6.3).

Exercises 10.6:

1. Derive the Jacobi accessory equation for the catenary and verify directly that the functions u_1 and u_2 in Example 10.6.3 are solutions to this equation.

2. In Example 10.6.3 suppose that the boundary values are $y(0) = 1$ and $y(1) = \cosh(1)$. Find a solution (c_1, c_2) such that the corresponding extremal produces a weak local minimum.

---

[8] The Jacobi accessory equation is a vector differential equation when the functional involves several dependent variables. For our case, the Jacobi accessory equation is scalar.

3. Let

$$J(y) = \int_0^{\pi/4} \left(y^2 - y'^2 - 2y \cosh x \right) \, dx.$$

Find the extremals for J and show that for the fixed endpoint problem these extremals produce weak local maxima.

4. Let

$$J(y) = \int_{x_0}^{x_1} y' \left(1 + x^2 y' \right) \, dx,$$

where $0 < x_0 < x_1$. Find the extremals for J and the general solution to the Jacobi accessory equation. Find any conjugate points to x_0 and determine the nature of the extremals for the fixed endpoint problem.

5. Let J be the functional of Exercises 10.3-2.
   (a) Derive a two-parameter family of functions $y(x, c_1, c_2)$ that are extremals for J.
   (b) Find the general solution to the Jacobi accessory equation and show that there are no points conjugate to x_0 for any choice of x_0 and $y(x_0)$. Determine the nature of the extremals for this problem.

6. Let J be the functional of Exercises 10.3-1 (geodesics on the sphere). The extremals for J satisfy the implicit equation

$$\tan \theta \cos(\phi + c_2) = \tan c_1,$$

where c_1 and c_2 are constants. Find the general solution to the Jacobi accessory equation. If A is the point with spherical coordinates (R, ϕ_0, θ_0) show that the points conjugate to A have coordinates $(R, \phi_0, \theta_0 \pm \pi)$.

7. Let

$$h(x, y, c) = y^5 - (x + c)^3.$$

Find the curve along with the points (x, y) that satisfy $h(x, y, c) = 0$ and equation (10.33). Does this curve form an envelope?

8. Let c and ℓ be constants such that $c > \ell > 0$, and consider the one-parameter family of circles given by

$$h(x, y, \alpha) = \alpha^2 \left(1 - \frac{\ell^2}{c^2} \right) - 2\alpha x + (x^2 + y^2 + \ell^2) = 0.$$

Solve the equations $h(x, y, \alpha) = 0$ and (10.33), and show that this family of curves forms an envelope corresponding to the hyperbola

$$\frac{x^2}{c^2 - \ell^2} - \frac{y^2}{\ell^2} = 1.$$

This family of circles arises in the study of the sound made by a supersonic aircraft. In the model, ℓ is the height of the aircraft and $c = \ell v/u$, where v is the speed of the aircraft and u is the speed of sound in air (hence $c > \ell$ for supersonic aircraft). The right branch of the hyperbola encloses a region known as the **zone of audibility**. See [13] for more details and other applications.

10.7 Convex Integrands

In this section we present a sufficient condition for a minimum that does not involve conjugate points. The condition exploits the case where the integrand is a convex function of y and y', and uses a basic result about convex functions to establish the requisite inequalities. The requirement of convexity, however, is harsh: many functionals such as that for the catenary do not have convex integrands. Nonetheless, the test for convexity is straightforward and the result is simple to use.

Recall that a set $\Omega \subseteq \mathbb{R}^2$ is **convex** if the line segments connecting any two points $\mathbf{z}_1, \mathbf{z}_2 \in \Omega$ lie in Ω. In other words, if $\mathbf{z}_1, \mathbf{z}_2 \in \Omega$ then

$$\mathbf{w}(t) = (1 - t)\mathbf{z}_1 + t\mathbf{z}_2 \in \Omega$$

for all $t \in [0, 1]$. The sets $\mathbb{R}^2$, $\{(y, w) \in \mathbb{R}^2 : y^2 + w^2 < 1\}$ and $\{(y, w) \in \mathbb{R}^2 : |y| < 1$ and $|w| < 1\}$ are examples of convex sets.

Let $\Omega \subseteq \mathbb{R}^2$ be a convex set. A function $f : \Omega \to \mathbb{R}$ is said to be **convex** if

$$f(\mathbf{w}(t)) = f((1 - t)\mathbf{z}_1 + t\mathbf{z}_2) \leq (1 - t)f(\mathbf{z}_1) + tf(\mathbf{z}_2) \tag{10.39}$$

for all $\mathbf{z}_1, \mathbf{z}_2 \in \Omega$ and all $t \in [0, 1]$. Geometrically, inequality (10.39) implies that the set $M = \{(\mathbf{z}, x) \in \Omega \times \mathbb{R} : x \geq f(\mathbf{z})\}$ is a convex set. Roughly speaking, M is the set of points that "lies above" the graph of f.

In finite-dimensional optimization, convexity is a desirable property because one can proceed directly to the classification of a critical point without resorting to the Hessian matrix. More importantly, the minimum thus found is global in the sense that it is a minimum of f for all $\mathbf{z} \in \Omega$. The crucial inequality that leads to this result comes directly from the mean value theorem.

Let $\Omega \subseteq \mathbb{R}^2$ be a convex set and let $f : \Omega \to \mathbb{R}$ be a convex function that has continuous partial derivatives on Ω. Let $\mathbf{z}_1, \mathbf{z}_2 \in \Omega$. Then

$$\mathbf{w}(t) = (1 - t)\mathbf{z}_1 + t\mathbf{z}_2 = \mathbf{z}_1 + t(\mathbf{z}_2 - \mathbf{z}_1) \in \Omega$$

for all $t \in [0, 1]$, and the mean value theorem implies that there is a $\tau \in (0, t)$ such that

$$f(\mathbf{w}(t)) = f(\mathbf{z}_1) + t(\mathbf{z}_2 - \mathbf{z}_1) \cdot \nabla f(\mathbf{w}(\tau)). \tag{10.40}$$

Equation (10.40) and inequality (10.39) imply

$$f(\mathbf{z}_1) + t(\mathbf{z}_2 - \mathbf{z}_1) \cdot \nabla f(\mathbf{w}(\tau)) \leq (1 - t)f(\mathbf{z}_1) + tf(\mathbf{z}_2);$$

i.e.,

$$(\mathbf{z}_2 - \mathbf{z}_1) \cdot \nabla f(\mathbf{w}(\tau)) \leq f(\mathbf{z}_2) - f(\mathbf{z}_1),$$

for all $t \in (0, 1)$. Now, $\mathbf{z}_1$ and $\mathbf{z}_2$ are fixed points in Ω, but τ depends on t and $0 < \tau < t$. Since the partial derivatives of f are continuous, $\lim_{t \to 0} \nabla f(\mathbf{w}(\tau)) = \nabla f(\mathbf{z}_1)$; hence, for any $\mathbf{z}_1, \mathbf{z}_2 \in \Omega$,

$$(\mathbf{z}_2 - \mathbf{z}_1) \cdot \nabla f(\mathbf{z}_1) \le f(\mathbf{z}_2) - f(\mathbf{z}_1). \tag{10.41}$$

Suppose now that z_1 is a stationary point for f so that $\nabla f(\mathbf{z}_1) = 0$. The above inequality shows that

$$f(\mathbf{z}_1) \le f(\mathbf{z}_2)$$

for all $\mathbf{z}_2 \in \Omega$. In this manner we see that stationary points for convex functions lead to a minimum for f in Ω.

We can exploit the above result to develop a sufficient condition for a functional to have a minimum at an extremal. Let

$$J(y) = \int_{x_0}^{x_1} f(x, y, y') \, dx,$$

and let $D_f \subset \mathbb{R}^3$ denote the domain of definition for f. Suppose that for each $x \in [x_0, x_1]$ the set

$$\Omega_x = \{(y, y') \in \mathbb{R}^2 : (x, y, y') \in D_f\}$$

is convex, and that f as a function on Ω_x is convex. Then for any points $(y, y'), (\hat{y}, \hat{y}') \in \Omega_x$ inequality (10.41) implies

$$f(x, \hat{y}, \hat{y}') - f(x, y, y') \ge (\hat{y} - y)f_y(x, y, y') + (\hat{y}' - y')f_{y'}(x, y, y').$$

Suppose now that y is a smooth extremal for J and that $\hat{y} \in S$, so that $\hat{y}(x_0) = y(x_0)$ and $\hat{y}(x_1) = y(x_1)$. Then,

$$J(\hat{y}) - J(y) = \int_{x_0}^{x_1} \left(f(x, \hat{y}, \hat{y}') - f(x, y, y') \right) dx$$

$$\ge \int_{x_0}^{x_1} \left((\hat{y} - y)f_y(x, y, y') + (\hat{y}' - y')f_{y'}(x, y, y') \right) dx.$$

Since $\hat{y} \in S$, integration by parts shows that

$$\int_{x_0}^{x_1} (\hat{y}' - y')f_{y'}(x, y, y') \, dx = -\int_{x_0}^{x_1} \frac{d}{dx} \left(f_{y'}(x, y, y') \right) (\hat{y} - y) \, dx,$$

and hence

$$J(\hat{y}) - J(y) \ge \int_{x_0}^{x_1} (\hat{y} - y) \left(f_y(x, y, y') - \frac{d}{dx} \left(f_{y'}(x, y, y') \right) \right) dx.$$

Now, y satisfies the Euler-Lagrange equation, so that the integrand in the above inequality is zero; therefore,

$$J(\hat{y}) - J(y) \ge 0,$$

and consequently J has a minimum at y. In summary we have the following result.

Theorem 10.7.1 *Suppose that for each $x \in [x_0, x_1]$ the set Ω_x is convex, and that f is a convex function of the variables $(y, y') \in \Omega_x$. If y is a smooth extremal for J, then J has a minimum at y for the fixed endpoint problem.*

In order to apply the above theorem we need a method for discerning whether a given function of two variables is convex. Fortunately, there is a tractable characterization when f is a smooth function. We omit the proof of the next result.[9]

Theorem 10.7.2 *Let $\Omega \subseteq \mathbb{R}^2$ be a convex set and let $f : \Omega \to \mathbb{R}$ be a function with continuous first- and second-order partial derivatives. The function f is convex if and only if for each $(y, w) \in \Omega$:*

$$f_{yy}(y, w) \geq 0;$$
$$f_{ww}(y, w) \geq 0;$$
$$f_{yy}(y, w) f_{ww}(y, w) - f_{yw}^2(y, w) = \Delta \geq 0.$$

The final inequality in the above theorem is simply the requirement that the quadratic form Q introduced in Section 10.1 be positive semidefinite. Geometrically, this inequality ensures that the Gaussian curvature is nonnegative and hence each point on the surface described by $(y, w, f(y, w))$, $(y, w) \in \Omega$ is either elliptic or parabolic. Elliptic and parabolic points are characterized by the property that the tangent plane at $(y, w, f(y, w))$ does not intersect the surface in a neighbourhood of $(y, w, f(y, w))$. The other inequalities ensure that the surface always "lies above" the tangent plane. A paradigm for a convex function is the paraboloid described by $f(y, w) = y^2 + w^2$ for $(y, w) \in \mathbb{R}^2$. A convexity condition for functions of three or more independent variables can be derived using the Hessian matrix of Section 10.1. For example, if f is a smooth function on the convex set $\Omega \subseteq \mathbb{R}^n$ and the Hessian matrix for f is positive definite, then f is a convex function (cf. Theorem 10.1.2 for conditions on the Hessian matrix).

Example 10.7.1:   Geodesics in the Plane

Let J be the arclength functional

$$J(y) = \int_{x_0}^{x_1} \sqrt{1 + y'^2} \, dx$$

(Example 2.2.1). Here, $f(x, y, y') = \sqrt{1 + y'^2}$, and hence $\Omega_x = \mathbb{R}^2$ is a convex set. Moreover, for all $(y, y') \in \Omega_x$, we have $f_{yy} = f_{yy'} = 0$, and

$$f_{y'y'} = \frac{1}{(1 + y'^2)^{3/2}} > 0;$$

hence, f is convex. Theorem 10.7.1 thus implies that (among smooth curves) line segments are the curves of shortest arclength between two points in the plane.

---

[9] A proof can be found in [15], pp. 41–43.

Example 10.7.2:   Catenary

The integrand of the catenary problem (Example 2.3.3) is $f(x, y, y') = y\sqrt{1 + y'}$. Now, $f_{yy} = 0$, but

$$f_{yy'} = \frac{y'}{\sqrt{1 + y'^2}},$$

so that $\Delta < 0$ for all $y' \neq 0$. The integrand is thus not convex. We could have deduced this from Example 10.6.3, because there are conjugate points for certain solutions of the boundary value equations.

Example 10.7.3:   Consider an integrand of the form

$$f(x, y, y') = (c_1 y - y' - c_2)^2,$$

where c_1 and c_2 are constants. (The Ramsey growth model of Exercises 7.1-2 has an integrand of this form.) Here $\Omega_x = \mathbb{R}^2$, $f_{yy} = 2c_1^2 > 0$, $f_{y'y'} = 2 > 0$, and $f_{yy'} = -2c_1 > 0$. Hence, $\Delta = 0$, and the integrand is convex. Extremals to the fixed endpoint problem thus correspond to minima.

Exercises 10.7:

1. Let Ω be a convex set and suppose that f and g are convex functions on Ω. Show that the function $f + g$ is also convex.
2. Determine whether the integrand for the brachystochrone functional (Example 2.3.4) is convex.
3. Show that the functions $\sqrt{y^2 + y'^2}$ and $e^y \sqrt{1 + y'^2}$ are convex.
4. Is the integrand convex for the functional of Exercises 10.3-2?
5. Develop a result analogous to Theorem 10.7.1 for functionals of the form

$$J(y) = \int_{x_0}^{x_1} f(x, y, y'') \, dx,$$

and apply it to the fixed endpoint problem for the beam of Example 7.1.3.

A

Analysis and Differential Equations

In this appendix we review some elementary analytical concepts that are used frequently in the book. The review is intended to be simply a brief summary of a few key results from analysis and differential equations that are relevant to material presented in the text. It is not intended as a "quick introduction" to these topics: it is merely a budget of handy results collected for the convenience of the reader. The first two sections concern Taylor polynomials and the implicit function theorem. A full account of these topics resplendent with proofs can be found in any book on real analysis or advanced calculus (e.g., [19], [56], [29]). The third section deals with the theory of ordinary differential equations. Here, one can consult Birkhoff and Rota [9], Coddington and Levinson [24], or Petrovski [60] for detailed presentations.

A.1 Taylor's Theorem

A good deal of the mathematics in this book relies on an exceedingly useful result known as Taylor's theorem. We commonly encounter transcendental functions such as e^x or algebraic functions such as $\sqrt{1+x^2}$, that need to be approximated near a given point in terms of a polynomial. Taylor's theorem provides the analytical framework to do such approximations. Let us first warm up with the mean value theorem.

Theorem A.1.1 (Mean Value Theorem) *Let x_0 and x_1 be real numbers such that $x_0 < x_1$. Let f be a function continuous in $[x_0, x_1]$ and differentiable in (x_0, x_1). Then there is a number ξ such that $x_0 < \xi < x_1$ and*

$$f(x_1) = f(x_0) + (x_1 - x_0)f'(\xi). \tag{A.1}$$

The mean value theorem is easy to explain geometrically. The slope of the line segment that connects the points $(x_0, f(x_0))$ and $(x_1, f(x_1))$ is

$$m = \frac{f(x_1) - f(x_0)}{x_1 - x_0}.$$

The mean value theorem asserts that somewhere in the interval (x_0, x_1) the graph of the function f has a tangent parallel to the line segment; i.e., $m = f'(\xi)$ (figure A.1). It is clear that there may be several values of $\xi \in (x_0, x_1)$

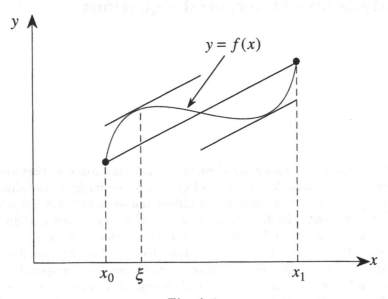

Fig. A.1.

for which equation (A.1) is valid. The "catch" is that the mean value theorem does not give us any value for ξ. We know only that it is in the open interval (x_0, x_1), so that all the uncertainty of the representation lies in the derivative term. We can nonetheless use this result to approximate f near x_0 with some control over the error through the derivative.

Note that the mean value theorem can easily be interpreted to be a representation of f at x_0 in terms of $f(x_1)$ and the derivative term. In other words, the relation is symmetric and it does not matter whether $x_0 < x_1$ or $x_0 > x_1$. The point is that there is a number ξ between these numbers such that (A.1) is satisfied provided f satisfies the continuity and differentiability conditions in the relevant interval.

The mean value theorem can be extended to provide representations of f in terms of a nonlinear polynomial. This extension goes by various names such as the "generalized mean value theorem," the "higher mean value theorem," "Taylor's theorem with remainder," or simply Taylor's theorem.

Theorem A.1.2 (Taylor's Theorem) *Let f be a function such that its first n derivatives are continuous in the interval $[x_0, x_1]$, and $f^{(n+1)}(x)$ exists for all $x \in (x_0, x_1)$. Then, there is a number $\xi \in (x_0, x_1)$ such that*

$$f(x_1) = f(x_0) + (x_1 - x_0)f'(x_0)\frac{(x_1 - x_0)^2}{2}f''(x_0) + \cdots$$

$$+ \frac{(x_1 - x_0)^n}{n!} f^{(n)}(x_0) + \frac{(x_1 - x_0)^{n+1}}{(n+1)!} f^{(n+1)}(\xi).$$

The polynomial

$$P_n(x_1) = f(x_0) + (x_1 - x_0)f'(x_0)\frac{(x_1 - x_0)^2}{2} f''(x_0) + \cdots + \frac{(x_1 - x_0)^n}{n!} f^{(n)}(x_0)$$

is called the nth degree **Taylor polynomial** of f at x_0. The term

$$R_{n+1} = \frac{(x_1 - x_0)^{n+1}}{(n+1)!} f^{(n+1)}(\xi)$$

is called the **remainder**. If there is a number M such that $M \geq |f^{n+1}(x)|$ for all $x \in (x_0, x_1)$, then we can approximate f by the Taylor polynomial P_n with an error bound of the form

$$|f(x_1) - P_n(x_1)| \leq \frac{(x_1 - x_0)^{n+1}}{(n+1)!} M.$$

Note that if $f^{n+1}(x)$ is continuous in the interval $[x_0, x_1]$, then this function must be bounded. Hence, for this case there is always a number M such that $M \geq |f^{n+1}(x)|$ for all $x \in [x_0, x_1]$.

Taylor's theorem can be generalized to functions of several independent variables. To keep things simple we give a version for two variables and restrict the geometry to discs in the plane. First, however, to avoid swimming in notation we introduce the operator

$$(x_1 - x_0, y_1 - y_0) \cdot \nabla = (x_1 - x_0)\frac{\partial}{\partial x} + (y_1 - y_0)\frac{\partial}{\partial y},$$

and the notation $[(x_1 - x_0, y_1 - y_0) \cdot \nabla] = (x_1 - x_0, y_1 - y_0) \cdot \nabla$. For $n \geq 1$, let

$$[(x_1 - x_0, y_1 - y_0) \cdot \nabla]^{n+1} = (x_1 - x_0, y_1 - y_0) \cdot \nabla \left[(x_1 - x_0, y_1 - y_0) \cdot \nabla\right]^n.$$

For example,

$$\begin{aligned}
[(x_1 - x_0, y_1 - y_0) \cdot \nabla]^2 &= \left((x_1 - x_0)\frac{\partial}{\partial x} + (y_1 - y_0)\frac{\partial}{\partial y}\right) \\
&\quad \left((x_1 - x_0)\frac{\partial}{\partial x} + (y_1 - y_0)\frac{\partial}{\partial y}\right) \\
&= (x_1 - x_0)^2 \frac{\partial^2}{\partial x^2} + 2(x_1 - x_0)(y_1 - y_0)\frac{\partial^2}{\partial x \partial y} \\
&\quad + (y_1 - y_0)^2 \frac{\partial^2}{\partial y^2}.
\end{aligned}$$

We also use the notation

$$\left[(x_1 - x_0, y_1 - y_0) \cdot \nabla\right] f \Big|_{(c,d)}$$

to indicate that the operator acts on f and the derivatives are evaluated at the point $(x, y) = (c, d)$.

Theorem A.1.3 Let $D_R = \{(x, y) \in \mathbb{R}^2 : (x - x_0)^2 + (y - y_0)^2 < R^2\}$ and suppose that $f : D_R \to \mathbb{R}^2$ has continuous partial derivatives up through order $n + 1$ in D_R. Then for any point $(x_1, y_1) \in D_R$, there is a point (a, b) on the line segment connecting (x_1, y_1) to (x_0, y_0) such that

$$f(x_1, y_1) = f(x_0, y_0) + \left[(x_1 - x_0, y_1 - y_0) \cdot \nabla\right] f \Big|_{(x_0, y_0)}$$

$$+ \frac{1}{2!} \left[(x_1 - x_0, y_1 - y_0) \cdot \nabla\right]^2 f \Big|_{(x_0, y_0)} + \cdots$$

$$+ \frac{1}{n!} \left[(x_1 - x_0, y_1 - y_0) \cdot \nabla\right]^n f \Big|_{(x_0, y_0)}$$

$$+ \frac{1}{(n+1)!} \left[(x_1 - x_0, y_1 - y_0) \cdot \nabla\right]^{n+1} f \Big|_{(a,b)}.$$

As in the one-variable case we can use Taylor's theorem to approximate f and control the error by finding suitable bounds for the $n + 1$th-order partial derivatives. We use this version of Taylor's theorem several times in the book. One might draw comfort, however, from the fact that we seldom need terms beyond the second order.

The reader is doubtless familiar with Taylor series or at least the special case of Maclaurin series. For example, we are familiar with the series representations

$$e^x = 1 + x + \frac{x^2}{2!} + \frac{x^3}{3!} + \cdots, \quad x \in \mathbb{R},$$

$$\sin(x) = x - \frac{x^3}{3!} + \frac{x^5}{5!} - \cdots, \quad x \in \mathbb{R},$$

$$\frac{1}{1 - x} = 1 + x + x^2 + x^3 + \cdots, \quad |x| < 1,$$

and it is natural to enquire if the Taylor polynomial tends to f as $n \to \infty$, assuming f has derivatives of all orders. In other words, if f has derivatives of all orders in a neighbourhood of x_0 does $P_n(x) \to f(x)$ as $n \to \infty$ for all x sufficiently close to x_0? The answer is *no*. The Cauchy function

$$f(x) = \begin{cases} e^{-1/x^2} & \text{if } x \neq 0, \\ 0 & \text{if } x = 0, \end{cases}$$

provides a counterexample. It can be shown using the definition of a derivative that f has derivatives of all orders at $x = 0$ and that $f^{(n)}(0) = 0$ for all $n = 1, 2, \ldots$. The Taylor polynomial is thus

$$P_n(x) = 0,$$

for all n. Hence $\lim_{n \to \infty} P_n(x) = 0$. It is clear that $f(x) > 0$ for all $x \neq 0$; consequently, $\lim_{n \to \infty} P_n(x) \neq f(x)$ except at $x = 0$. Functions that can be represented by a convergent power series with a nonzero radius of convergence are thus special. If there exists a representation of f of the form

$$f(x) = \sum_{n=0}^{\infty} a_n (x - x_0)^n,$$

valid for all $|x - x_0| < \rho$, where ρ is some positive number, then f is said to be (real) **analytic** at x_0. Such functions always have Taylor series representations at x_0 and

$$a_n = \frac{f^{(n)}(x_0)}{n!}.$$

Similar statements can be made concerning functions of several independent variables.

Finally, we note that the Cauchy function can also be used to construct a **mollifier**. Roughly speaking, a mollifier is a smooth function that is zero outside a bounded interval I and nonzero within I. Specifically, choose any $a \in \mathbb{R}$, $a \neq 0$, and consider the function

$$m(x) = \begin{cases} e^{-1/(a^2 - x^2)} & \text{if } |x| \leq a, \\ 0 & \text{if } |x| > a. \end{cases}$$

It can be shown that m has derivatives of all orders for all $x \in \mathbb{R}$, and that $m(x) > 0$, if $x \in (-a, a)$; otherwise, $m(x) = 0$. Evidently, we can modify this function so that given any two points a, b with $a < b$ we get a mollifier that is zero outside (a, b) and positive inside (a, b). Such functions are always useful. In this book, however, we tend to use simpler functions that have similar properties, but are not as smooth. Such functions have the merit of simplicity for our calculations.

A.2 The Implicit Function Theorem

Frequently, we are confronted with equations of the form

$$g(x, y) = 0, \tag{A.2}$$

which we need to either solve for x or y, or at least discern whether such an equation defines y as a function of x (or vice versa). Often, we cannot solve implicit equations, but it is important to know qualitative details such as whether a solution exists and is unique. We also usually need to know certain analytical properties such as continuity of solutions. When we cannot find an explicit solution (or need only qualitative properties), the implicit function theorem comes to our rescue.

Theorem A.2.1 (Implicit Function Theorem) *Let $g : \Omega \to \mathbb{R}$ be a function defined in a neighbourhood $\Omega \subseteq \mathbb{R}^2$ of the point (x_0, y_0) such that*

$$g(x_0, y_0) = 0, \tag{A.3}$$

and suppose that g is differentiable with respect to y and that $\partial g/\partial y$ is continuous in Ω. If

$$\left.\frac{\partial g}{\partial y}\right|_{(x_0, y_0)} \neq 0, \tag{A.4}$$

then there exist neighbourhoods $I_{x_0} \subset \mathbb{R}$ of x_0 and $I_{y_0} \subset \mathbb{R}$ of y_0, and a function $\phi : I_{x_0} \to \mathbb{R}$ such that:

1. *$\phi(x_0) = y_0$;*
2. *for all $x \in I_{x_0}$ we have $(x, \phi(x)) \in \Omega$, and*

$$g(x, \phi(x)) = 0; \tag{A.5}$$

3. *the function ϕ with the above properties is unique; and*
4. *ϕ is continuous in I_{x_0}*

Moreover, if $\partial g/\partial x$ exists and is continuous in Ω, then the function ϕ is differentiable for all $x \in I_{x_0}$, and

$$\phi'(x) = -\frac{\frac{\partial g}{\partial x}}{\frac{\partial g}{\partial y}}. \tag{A.6}$$

Loosely speaking, given a point (x_0, y_0) at which $g(x_0, y_0) = 0$, the implicit function theorem guarantees that implicit relations such as (A.2) are solvable for y, provided g satisfies the requisite conditions. The solution ϕ is local in character: we do not know I_{x_0} explicitly. Also, it is worth noting that the above theorem does not preclude the existence of another solution $\theta(x)$ to (A.2), but it does preclude two distinct solutions θ, ϕ such that $\theta(x_0) = \phi(x_0) = y_0$. For example, let $g(x, y) = x - y^2$ and $x_0 = y_0 = 1$. Evidently, g satisfies the conditions of the implicit function theorem and $\phi(x) = \sqrt{x}$ is the unique solution with the properties listed in 1 and 2 of the implicit function theorem. The function $\theta(x) = -\sqrt{x}$ is also a solution to $g(x, y) = 0$, but $\theta(x_0) \neq y_0$.

The implicit function theorem can be extended to systems of implicit equations such as

$$f(x, y, u, v) = 0,$$
$$\tag{A.7}$$
$$g(x, y, u, v) = 0.$$

Suppose that we wish to solve the above system for, say, u and v in terms of x and y in a neighbourhood of a point (x_0, y_0, u_0, v_0) that satisfies the equations. In this case, condition (A.4) generalizes to the Jacobian condition

$$J(x_0, y_0, u_0, v_0) = \frac{\partial(f, g)}{\partial(u, v)}$$

$$= \det \begin{pmatrix} \frac{\partial f}{\partial u} & \frac{\partial f}{\partial v} \\ \frac{\partial g}{\partial u} & \frac{\partial g}{\partial v} \end{pmatrix}$$

$$= \frac{\partial f}{\partial u} \frac{\partial g}{\partial v} - \frac{\partial f}{\partial v} \frac{\partial g}{\partial u} \neq 0,$$

where the partial derivatives are evaluated at (x_0, y_0, u_0, v_0). Under conditions analogous to those given in Theorem A.2.1 it can be shown that there is a unique solution $\phi(x, y)$, $\psi(x, y)$ with properties analogous to those of $\phi(x)$ in Theorem A.2.1. An important special case of this result concerns coordinate transformations of the form

$$x = x(u, v),$$

$$y = y(u, v).$$

(A.8)

The central question is whether such transformations can be inverted to get u and v in terms of x and y.

Theorem A.2.2 (Inverse Transformations) *Let $x_0 = x(u_0, v_0)$ and $y_0 = y(u_0, v_0)$. Suppose that the functions $x(u, v)$ and $y(u, v)$ have continuous partial derivatives of order 1 in a neighbourhood $\Omega \subset \mathbb{R}^2$ of the point (u_0, v_0). Suppose further that the Jacobian condition*

$$J = \frac{\partial(x, y)}{\partial(u, v)} \neq 0$$

is satisfied at (u_0, v_0). Then there is a neighbourhood $N(x_0, y_0) \subset \mathbb{R}^2$ of (x_0, y_0) and functions $u(x, y)$, $v(x, y)$ defined in $N(x_0, y_0)$ such that:

1. *$u(x_0, y_0) = u_0$ and $v(x_0, y_0) = v_0$;*
2. *the identities*

$$x(u(x, y), v(x, y)) = x,$$
$$y(u(x, y), v(x, y)) = y,$$

are valid throughout $N(x_0, y_0)$;
3. *the functions $u(x, y)$ and $v(x, y)$ that satisfy the above properties are unique; and*
4. *$u(x, y)$ and $v(x, y)$ have continuous partial derivatives in $N(x_0, y_0)$, and*

$$\frac{\partial u}{\partial x} = \frac{1}{J} \frac{\partial y}{\partial v}, \quad \frac{\partial u}{\partial y} = -\frac{1}{J} \frac{\partial x}{\partial v},$$

$$\frac{\partial v}{\partial x} = -\frac{1}{J} \frac{\partial y}{\partial u}, \quad \frac{\partial v}{\partial y} = \frac{1}{J} \frac{\partial x}{\partial u}.$$

Consider, for example, the familiar transformation to polar coordinates

$$x = u \cos v,$$
$$y = u \sin v.$$

The Jacobian is given by

$$\frac{\partial(x,y)}{\partial(u,v)} = \frac{\partial x}{\partial u}\frac{\partial y}{\partial v} - \frac{\partial x}{\partial v}\frac{\partial y}{\partial u}$$
$$= u \cos^2 v + u \sin^2 v$$
$$= u;$$

hence, the Jacobian is nonzero provided $u \neq 0$. Clearly $x(u,v)$ and $y(u,v)$ satisfy the requisite smoothness conditions for any $(u,v) \in \mathbb{R}^2$. We thus conclude that given any point (u_0, v_0), with $u_0 \neq 0$, there is a neighbourhood of (u_0, v_0) wherein the transformation is invertible. We see that

$$u = (x^2 + y^2)^{1/2},$$
$$v = \arctan\left(\frac{y}{x}\right).$$

Although the above expressions for u and v involve multifunctions, note that the conditions $u(x_0, y_0) = u_0$ and $v(x_0, y_0) = v_0$ determine the branches of these functions. The exceptional point for this transformation is the pole $(u_0, v_0) = (0, v_0)$. Here, we know that the equations $x_0 = u_0\cos v_0 = 0$ and $y_0 = u_0\sin v_0 = 0$ place no restrictions on v_0, so that there cannot be a unique inverse.

A.3 Theory of Ordinary Differential Equations

Much of our study of the calculus of variations revolves around the Euler-Lagrange equation, which is a second-order nonlinear ordinary differential equation. We also need to study ordinary differential equations arising from constraints and sufficient conditions (the Jacobi accessory equation). Suffice it to say that ordinary differential equations loom large in the subject. Some of the theory underlying these equations is developed as needed in the context of its application. There are some results, however, that we use several times and it is perhaps best to collect them in a single section for reference.

Given an equation of the form

$$g(x, y, y') = 0, \tag{A.9}$$

where y' denotes dy/dx, we face a more formidable problem than that posed by implicit equations. Assuming g satisfies the conditions of the implicit function theorem with respect to y', and $g(x_0, y_0, y_0') = 0$, we can (at least in principle) solve equation (A.9) for y' and thus study an equation of the form

$$y'(x) = f(x, y) \tag{A.10}$$

along with the condition

$$y(x_0) = y_0. \tag{A.11}$$

Equation (A.10) along with the condition (A.11) is an example of an **initial-value problem**. There are no systematic solution techniques available for solving such problems explicitly in closed form. If f has some special properties (e.g., if f is separable) then there are special methods for solution, but for the general f we must concede defeat. As with the implicit function theorem, we often do not need to know the solution explicitly, but we do need to know whether a solution exists and perhaps some qualitative properties such as uniqueness and smoothness. The following result is basic to the theory of differential equations and plays a rôle analogous to the implicit function theorem in that it guarantees the existence of a unique local solution.

Theorem A.3.1 (Picard's Theorem) *Suppose that $f(x, y)$ is continuous with respect to x in a neighbourhood $N(x_0, y_0) \subset \mathbb{R}^2$ of (x_0, y_0), and there is a constant $K > 0$ such that for all $(x, y_1), (x, y_2) \in N(x_0, y_0)$*

$$|f(x, y_2) - f(x, y_1)| \le K|y_2 - y_1|. \tag{A.12}$$

Then there exists a neighbourhood $N(x_0) \subset \mathbb{R}$ of x_0 and a function $y(x)$ such that

$$y'(x) = f(x, y)$$

for all $x \in N(x_0)$, and

$$y(x_0) = y_0.$$

Moreover, the solution is unique.

Inequality (A.12) is called the **Lipschitz condition**, and if f satisfies this inequality for some K then f is called **Lipschitz continuous** in y. The requirement of Lipschitz continuity is stronger than that of continuity. If f is Lipschitz continuous in y then it is continuous in y, but the converse is not true. We note that if we loosen the requirement on f to continuity in y, then we still have the existence of a solution (Peano's existence theorem, [60], p. 29), but uniqueness is not guaranteed. For example, the simple problem

$$y'(x) = y^{1/3},$$
$$y(0) = 0$$

has the two distinct solutions

$$y(x) = \left(\frac{2}{3}x\right)^{3/2}, \quad y(x) = 0.$$

For our purposes we seldom need the generality afforded by the Lipschitz condition. Usually, f is differentiable in y, and this is a stronger condition than Lipschitz continuity. Suppose, for example, that $\partial f/\partial y$ is continuous in the disc $\bar{D}(x_0, y_0) = \{(x, y) : (x - x_0)^2 + (y - y_0)^2 \leq R^2\}$, where $R > 0$. For any choice of $(x, y_1), (x, y_2) \in \bar{D}(x_0, y_0)$ we can apply the mean value theorem to assert that there is a number ξ between y_1 and y_2 such that

$$f(x, y_2) = f(x, y_1) + (y_2 - y_1) \frac{\partial f}{\partial y}\bigg|_{(x, \xi)}.$$

Since $\partial f/\partial y$ is continuous in $\bar{D}(x_0, y_0)$ it is bounded in this disc and hence there is a $K > 0$ such that

$$|f(x, y_2) - f(x, y_1)| = |y_2 - y_1| \left|\frac{\partial f}{\partial y}\bigg|_{(x, \xi)}\right| \leq K|y_2 - y_1|.$$

Thus, if $f(x, y)$ has a continuous partial derivative with respect to y in a neighbourhood of (x_0, y_0) then we can find a suitable disc $\bar{D}(x_0, y_0)$ and conclude that in this disc f is Lipschitz continuous in y.

A general solution to equation (A.10) contains a parameter (the constant of integration) the value of which is determined by the condition (A.11). It is natural thus to enquire whether the general solution depends continuously on this parameter. The next result gives conditions under which differential equations containing parameters have solutions that are smooth with respect to the parameters. We show that the initial condition parameters are a special case.

Theorem A.3.2 (Dependence of Solutions on Parameters) *Let $\alpha = (\alpha_1, \ldots, \alpha_n)$ and define the set $B = \{\alpha \in \mathbb{R}^n : |\alpha_1| < \beta_1, \ldots, |\alpha_n| < \beta_n\}$, where the β_k are positive numbers. Let $\Omega \subset \mathbb{R}^2$ be an open set with closure $\bar{\Omega}$ and define the set $\Upsilon = \bar{\Omega} \times B \subset \mathbb{R}^{n+2}$. Suppose that $f : \Upsilon \to \mathbb{R}$ has continuous derivatives with respect to $y, \alpha_1, \ldots, \alpha_n$ of order $k \geq 0$ on Υ, and that f satisfies the Lipschitz condition*

$$|f(x, y_2, \alpha) - f(x, y_1, \alpha)| \leq K|y_2 - y_1|$$

for some $K > 0$ and all $(x, y_1, \alpha), (x, y_2, \alpha) \in \Upsilon$. Given any point $(x_0, y_0) \in \Omega$ there exists an interval $[a, b]$ with $a < x_0 < b$ such that the differential equation

$$y'(x) = f(x, y, \alpha) \tag{A.13}$$

has a unique solution $\phi(x, \alpha)$ that satisfies the condition

$$\phi(x_0, \alpha) = y_0.$$

Moreover, ϕ has continuous derivatives with respect to the α_j up to (and including) order k on $[a, b] \times B$.

Basically, the above theorem shows that the solutions to differential equations that contain parameters are smooth in these parameters, provided f is smooth in these parameters. Note that if $k \geq 1$ in the theorem then the Lipschitz condition will be satisfied automatically.

Returning to equation (A.10), consider the transformation

$$w = y - y_0, \quad z = x - x_0.$$

Under this transformation equation (A.10) is

$$w'(z) = f(z + x_0, w + y_0).$$

An immediate consequence of the above theorem is that the general solution to (A.10) depends smoothly on the initial data x_0, y_0, provided f is smooth in x and y near (x_0, y_0).

The results concerning the existence and uniqueness of solutions along with the continuous dependence on parameters can be extended to initial-value problems of the form

$$\mathbf{y}'(x) = \mathbf{f}(x, \mathbf{y}),$$
$$\mathbf{y}(x_0) = \mathbf{y}_0,$$

where $\mathbf{y}(x) = (y_1(x), \ldots, y_n(x))$, $\mathbf{f} = (f_1, \ldots, f_n)$, and $\mathbf{y}_0 \in \mathbb{R}^n$. The Lipschitz condition in this framework is

$$|\mathbf{f}(x, \mathbf{y}_2) - \mathbf{f}(x, \mathbf{y}_1)| \leq K|\mathbf{y}_2 - \mathbf{y}_1|,$$

where $|\cdot|$ is defined by

$$|\mathbf{y}| = \sqrt{|y_1|^2 + \cdots + |y_n|^2}$$

for all $\mathbf{y} = (y_1, \ldots, y_n)$.

A higher-order differential equation can be readily converted into a system of differential equations. For example, given the differential equation

$$y'' = f(x, y, y'),$$

let $y_1 = y$ and $y_2 = y'$. The above differential equation can then be recast as the system

$$y_1' = y_2,$$
$$y_2' = f(x, y_1, y_2).$$

In this manner we can tackle questions concerning existence and uniqueness for higher-order equations. The results, however, are local in character and concern the initial-value problem, where y and y' are specified at a point x_0. The calculus of variations is impregnated with second-order differential equations, but most of the problems are boundary-value not initial-value problems. Boundary-value problems consist of determining solutions to a differential that satisfy conditions of the form $y(x_0) = y_0$ and $y(x_1) = y_1$, where $x_0 < x_1$. These problems are global in character because we require a solution to be valid throughout the interval $[x_0, x_1]$ and satisfy the boundary conditions. The theory behind such problems is more complicated than that for initial-value problems. A sample of one existence result is given in Section 2.6.

B

Function Spaces

We give here a brief synopsis of some concepts from functional analysis. Although we do not rely heavily on this material, it is included because a deeper understanding of the calculus of variations requires at least a nodding familiarity with functional analysis. At a minimum we need a sensible definition of "neighbouring functions," and certain concepts from Hilbert space are helpful for topics such as eigenvalue problems. This said, the book has been written so that it is not essential that the reader know functional analysis. A person ignorant of the subject can nonetheless progress through the book and read virtually every section with profit. Complete accounts of this material can be found in any book on functional analysis such as Kreyszig [46] or Hutson and Pym [40]. A concentrated account from a physicist's standpoint can be found in Choquet-Bruhat et al. [22].

B.1 Normed Spaces

The calculus of variations is essentially optimization in spaces of functions. It is thus useful to introduce some concepts from functional analysis, and basic among these concepts is that of a normed vector space. The reader has probably encountered the concept of a finite-dimensional vector space in a course on linear algebra. These spaces are modelled after the set of vectors in $\mathbb{R}^n$. Vector spaces, however, can be defined more generally and need not be finite dimensional. In fact, most the vector spaces of interest in the calculus of variations are not finite dimensional. A **vector space** is a nonempty set X equipped with the operations of addition "+" and scalar multiplication. For any elements f, g, h in X and any scalars α, β these operations have the properties:

(i) $f + g \in X$;
(ii) $f + g = g + f$;
(iii) $f + (g + h) = (f + g) + h$;

(iv) there is a unique element $\mathbf{0}$ (called zero) in X such that $f + \mathbf{0} = f$ for all
$f \in X$;

(v) for each element $f \in X$ there is a unique element $(-f) \in X$ such that
$f + (-f) = \mathbf{0}$;

(vi) $\alpha f \in X$;

(vii) $\alpha(f + g) = \alpha f + \alpha g$;

(viii) $(\alpha + \beta)f = \alpha f + \beta f$;

(ix) $(\alpha\beta)f = \alpha(\beta f)$;

(x) $1 \cdot f = f$.

For our purposes, the scalars are the real numbers.

Example B.1.1:   The set of vectors $\{(x_1, x_2, \ldots, x_n) : x_k \in \mathbb{R}, k = 1, 2, \ldots, n\}$ is denoted by $\mathbb{R}^n$. Let $\mathbf{x} = (x_1, x_2, \ldots, x_n)$ and $\mathbf{y} = (y_1, y_2, \ldots, y_n)$ be vectors in $\mathbb{R}^n$. If addition is defined by

$$\mathbf{x} + \mathbf{y} = (x_1 + y_1, x_2 + y_2, \ldots, x_n + y_n),$$

and scalar multiplication by

$$\alpha \mathbf{x} = (\alpha x_1, \alpha x_2, \ldots, \alpha x_n),$$

for any $\alpha \in \mathbb{R}$, then $\mathbb{R}^n$ is a vector space. The vector spaces $\mathbb{R}^n$ and $\mathbb{C}^n$ are essentially the prototypes for more abstract vector spaces.

Example B.1.2: Let $C[x_0, x_1]$ denote the set of all functions $f : [x_0, x_1] \to \mathbb{R}$ that are continuous on the interval $[x_0, x_1]$. If, for any $f, g \in C[x_0, x_1]$, addition is defined by

$$(f + g)(x) = f(x) + g(x),$$

and scalar multiplication by

$$(\alpha f)(x) = \alpha f(x),$$

for $\alpha \in \mathbb{R}$, then it is not difficult to see that $C[x_0, x_1]$ is a vector space.

Example B.1.3:   Let ℓ^1 denote the set of sequences $\{a_n\}$ in $\mathbb{R}$ such that the series $\sum_{n=1}^{\infty} |a_n|$ is convergent, and define addition so that for any two elements $A = \{a_n\}$, $B = \{b_n\}$,

$$A + B = \{a_n + b_n\},$$

and scalar multiplication so that

$$\alpha A = \{\alpha a_n\}.$$

Then ℓ^1 is also a vector space.

The above examples show that the elements in different vector spaces can be quite different in nature. More important, however, there is a significant difference between a vector space such as $\mathbb{R}^n$ and one such as $C[x_0, x_1]$ having to do with "dimension." The space $\mathbb{R}^n$ has a basis: any set of n linearly independent vectors in $\mathbb{R}^n$ such as $\mathbf{e}_1 = (1, 0, \ldots, 0), \mathbf{e}_2 = (0, 1, \ldots, 0) \ldots \mathbf{e}_n = (0, 0, \ldots, 1)$ forms a basis. The concept of dimension is tied to the number of elements in a basis for spaces such as $\mathbb{R}^n$, but it is not clear what a basis would be for a space like $C[x_0, x_1]$. In order to make some progress on generalizing the concept of dimension we need first to define what is meant by a linearly independent set when the set itself might contain an infinite number of elements. We say that a set is **linearly independent** if every *finite* subset is linearly independent; otherwise it is called **linearly dependent**. If there exists a positive integer n such that a vector space X has n linearly independent vectors but any set of $n + 1$ vectors is linearly dependent, then X is called **finite dimensional**. If no such integer exists, then X is called **infinite dimensional**.

A **subspace** of a vector space X is a subset of X which is itself a vector space under the same operations of addition and scalar multiplication. For example, the set of functions $f : [x_0, x_1] \to \mathbb{R}$ such that f is differentiable on $[x_0, x_1]$ is a subspace of $C[x_0, x_1]$. Given any vectors $x_1, x_2, \ldots, x_n$ in a vector space X, a subspace can always be formed by generating all the linear combinations involving the x_k, i.e., all the vectors of the form $\alpha_1 x_1 + \alpha_2 x_2 + \cdots + \alpha_n x_n$, where the α_ks are scalars, is a subspace of X. Given any finite set $S \subset X$ the subspace of X formed in this manner is called the **span** of S and denoted by $[S]$. If $S \subset X$ has an infinite number of elements then the span of S is defined to be the set of all *finite* linear combinations of elements of S.

Vector spaces of functions such as $C[x_0, x_1]$ are called **function spaces**. We are concerned primarily with function spaces, and to avoid repetition we agree here that for any function space the operations of addition and scalar multiplication are defined pointwise as was done for the space $C[x_0, x_1]$ in Example B.1.2.

Vector spaces are purely algebraic objects. In order to do any analysis more structure is needed. In particular, basic concepts such as convergence require some means of measuring the "distance" between objects in the vector space. This leads us to the concept of a norm. A **norm** on a vector space X is a real-valued function on X whose value at $f \in X$ is denoted by $\|f\|$ and which has the properties:

(i) $\|f\| \geq 0$;
(ii) $\|f\| = 0$ if and only if $f = \mathbf{0}$;
(iii) $\|\alpha f\| = |\alpha| \|f\|$;
(iv) $\|f + g\| \leq \|f\| + \|g\|$ (the triangle inequality).

Here, f and g are arbitrary elements in X and α is any scalar. A vector space X equipped with a norm $\| \cdot \|$ is called a **normed vector space**.

Example B.1.4:   For any $\mathbf{x} \in \mathbb{R}^n$ let $\| \cdot \|_e$ be defined by

$$\|\mathbf{x}\|_e = \{(x_1^2 + (x_2)^2 + \cdots + (x_n)^2\}^{1/2}.$$

Then $\| \cdot \|_e$ is a norm on $\mathbb{R}^n$. This function is called the **Euclidean** norm on $\mathbb{R}^n$. Another norm on $\mathbb{R}^n$ is given by

$$\|\mathbf{x}\|_T = |x_1| + |x_2| + \cdots + |x_n|.$$

Example B.1.5:   The function $\| \cdot \|_\infty$, given by

$$\|f\|_\infty = \sup_{x \in [x_0, x_1]} |f(x)|,$$

is well defined for any $f \in C[x_0, x_1]$, and it can be shown that $\| \cdot \|_\infty$ is a norm for $C[x_0, x_1]$. Alternatively, since any function f in this vector space is continuous, the function $|f|$ is integrable and thus the function $\| \cdot \|_R$ given by

$$\|f\|_1 = \int_{x_0}^{x_1} |f(x)|\, dx$$

is well defined on $C[x_0, x_1]$. It can be shown that $\| \cdot \|_R$ is a norm on $C[x_0, x_1]$.

Example B.1.6:   Let n be a positive integer and let $C^n[x_0, x_1]$ denote the set of functions that have at least an nth order continuous derivative on the interval $[x_0, x_1]$. Since any function that is differentiable on the interval $[x_0, x_1]$ must also be continuous on this interval we have that $C^n[x_0, x_1] \subset C[x_0, x_1]$ for $n = 1, 2, \ldots$. In fact, we have the hierarchy $C^n[x_0, x_1] \subset C^{n-1}[x_0, x_1] \subset \cdots \subset C^1[x_0, x_1] \subset C[x_0, x_1]$. We leave it to the reader to show that for $n = 1, 2, \ldots$ $C^n[x_0, x_1]$ is a vector space and that the norms defined in Example B.1.5 are also norms for $C^n[x_0, x_1]$. Other norms, however, can be defined for the space $C^n[x_0, x_1]$ which take advantage of the extra property of differentiability. For example, suppose $n = 1$. Then the function $\| \cdot \|_{\infty,1}$ given by

$$\|f\|_{\infty,1} = \sup_{x \in [x_0, x_1]} |f(x)| + \sup_{x \in [x_0, x_1]} |f'(x)|$$

is a norm on $C^1[x_0, x_1]$. Note that this function is also a norm for the space $C^2[x_0, x_1]$. In general we can define a norm of the form

$$\|f\|_{\infty,n} = \sum_{k=0}^{k=n} \sup_{x \in [x_0, x_1]} |f^{(k)}(x)|,$$

for the space $C^n[x_0, x_1]$. Here, f^k denotes the kth derivative of f and $f^{(0)} = f$.

The above examples indicate that a given vector space may have several norms leading to different normed vector spaces. For this reason, the notation $(X, \|\cdot\|)$ is often used to denote the vector space X equipped with the norm $\|\cdot\|$.

Once a vector space is equipped with a norm $\|\cdot\|$, a generalized distance function (called the metric induced by the norm $\|\cdot\|$) can be readily defined. The **distance** $d(f, g)$ of an element $f \in X$ from another element $g \in X$ is defined to be

$$d(f, g) = \|f - g\|.$$

The distance function for the normed vector space $(\mathbb{R}^n, \|\cdot\|_e)$ corresponds to the ordinary notion of Euclidean distance. The distance function for the normed vector space $(C[x_0, x_1], \|\cdot\|_\infty)$ measures the maximum vertical separation of the graph of f from the graph of g.

Neighbourhoods of an element in a normed vector space $(X, \|\cdot\|)$ can be defined as in the familiar finite-dimensional case. Specifically, for $\epsilon > 0$ we define an ϵ-neighbourhood of an element $f \in X$ as

$$B(f, \epsilon, \|\cdot\|) = \{g \in X : \|f - g\| < \epsilon\}.$$

We suppress the $\|\cdot\|$ in the above notation.

Convergence can be defined for sequences in a normed vector space in a manner which mimics the familiar definition in real analysis. Let $(X, \|\cdot\|)$ be a normed vector space and let $\{f_n\}$ denote an infinite sequence in X. The sequence $\{f_n\}$ is said to **converge** in the norm if there exists an $f \in X$ such that for every $\epsilon > 0$ an integer N can be found with the property that $f_n \in B(f, \epsilon)$ whenever $n > N$. The element f is called the **limit** of the sequence $\{f_n\}$, and the relationship is denoted by $\lim_{n \to \infty} f_n = f$ or simply $f_n \to f$. Note that convergence depends on the choice of norm: a sequence may converge in one norm and diverge in another. Note also that the limit f must also be an element in X.

In a similar spirit, we can define Cauchy sequences for a normed vector space. A sequence $\{f_n\}$ in X is a **Cauchy sequence** (in the norm $\|\cdot\|$) if for any $\epsilon > 0$ there is an integer N such that

$$\|f_m - f_n\| < \epsilon,$$

whenever $m > N$ and $n > N$. Cauchy sequences play a vital rôle in the theory of normed vector spaces. As with convergence, a sequence $\{f_n\}$ in X may be a Cauchy sequence for one choice of norm but not a Cauchy sequence for another choice.

It may be possible to define any number of norms on a given vector space X. Two different norms, however, may yield exactly the same results concerning convergence and Cauchy sequences. Two norms $\|\cdot\|_a$ and $\|\cdot\|_b$ on a vector space X are said to be **equivalent** if there exist positive numbers α and β such that for all $f \in X$,

$$\alpha \|f\|_a \le \|f\|_b \le \beta \|f\|_a.$$

If the norms $\|\cdot\|_a$ and $\|\cdot\|_b$ are equivalent, then it is straightforward to show that convergence in one norm implies convergence in the other, and that the set of Cauchy sequences in $(X, \|\cdot\|_a)$ is the same as the set of Cauchy sequences in $(X, \|\cdot\|_b)$. Equivalent norms lead to the same analytical results.

Identifying norms as equivalent can be difficult. In finite-dimensional vector spaces, however, the situation is simple: all norms defined on a finite-dimensional vector space are equivalent. Thus the two norms defined in Example B.1.4 are equivalent. The situation is different for infinite-dimensional spaces. For example, it can be shown that the norms $\|\cdot\|_R$ and $\|\cdot\|_\infty$ defined on the space $C[x_0, x_1]$ in Example B.1.5 are not equivalent.

B.2 Banach and Hilbert Spaces

The definitions for convergence and Cauchy sequences for normed vector spaces are formally analogous to those given in elementary real analysis. Various results such as the uniqueness of the limit can be proved for general normed vector spaces by essentially the same techniques used to prove analogous results in real analysis. The space $(\mathbb{R}^n, \|\cdot\|_e)$, however, has a special property not inherent in the definition of a normed vector space. It is well known that a sequence in $(\mathbb{R}, \|\cdot\|_e)$ converges if and only if it is a Cauchy sequence. This result does not extend to the general normed vector space. Every convergent sequence in a normed vector space must be a Cauchy sequence, but the converse is not true.

A normed vector space is called **complete** if every Cauchy sequence in the vector space converges. Complete normed vector spaces are called **Banach spaces**. In finite-dimensional vector spaces, completeness in one norm implies completeness in any norm since all norms are equivalent. Thus, spaces such as $(\mathbb{R}^n, \|\cdot\|_e)$ and $(\mathbb{R}^n, \|\cdot\|_T)$ are Banach spaces. For finite-dimensional vector spaces, completeness depends entirely on the vector space; for infinite-dimensional vector spaces completeness depends also on the choice of norm. The space $(C[-1, 1], \|\cdot\|_\infty)$, for instance, is a Banach space, whereas the space $(C[-1, 1], \|\cdot\|_1)$ is not. If the norms $\|\cdot\|_a$ and $\|\cdot\|_b$ are equivalent, then the corresponding normed vector spaces are either both Banach or both incomplete since the set of Cauchy sequences is the same for each space, and convergence in one norm implies convergence in the other. The two norms $\|\cdot\|_1$ and $\|\cdot\|_\infty$ on $C[-1, 1]$ are evidently not equivalent.

In passing we note that if a normed vector space is not complete, it is possible to "enlarge" the vector space and redefine the norm so that the resulting space is complete, and the value of the norm in the original space is preserved. In finite dimensions, the paradigm is the completion of the set of rational numbers to form the set of real numbers. An example involving an infinite-dimensional space is given by the space $(C[x_0, x_1], \|\cdot\|_1)$. This normed space is not complete. If the vector space $C[x_0, x_1]$ is expanded to include all functions that are Lebesgue integrable over the interval $[x_0, x_1]$, and the norm is replaced by

$$\|f\|_{L1} = \int_{[x_0,x_1]} f(x)\, dx,$$

where the Lebesgue integral is now used, then it can be shown that the resulting space is complete.[1]

A special type of Banach space that plays a large rôle in analysis is called a **Hilbert space**. Hilbert spaces are simpler than the general Banach space owing to an additional structure called an inner product. Briefly, a (real) **inner product** on a vector space X is a function $\langle \cdot, \cdot \rangle$ on $X \times X$ such that for any $f, g, h \in X$ and any $\alpha \in \mathbb{C}$ the following conditions hold.

(i) $\langle f, f \rangle \geq 0$;
(ii) $\langle f, f \rangle = 0$ if and only if $f = \mathbf{0}$;
(iii) $\langle f + g, h \rangle = \langle f, h \rangle + \langle g, h \rangle$;
(iv) $\langle f, g \rangle = \langle g, f \rangle$;
(v) $\langle \alpha f, g \rangle = \alpha \langle f, g \rangle$.

A vector space X equipped with an inner product $\langle \cdot, \cdot \rangle$ is called an **inner product space** and denoted by $(X, \langle \cdot, \cdot \rangle)$. Note that condition (i) indicates that $\langle f, f \rangle$ is always a real nonnegative number. Note also that conditions (iii) and (iv) imply that

$$\langle f, g + h \rangle = \langle f, g \rangle + \langle f, h \rangle.$$

Example B.2.1:   Let $X = \mathbb{R}^n$ and for any $\mathbf{x} = (x_1, x_2, \ldots, x_n), \mathbf{y} = (y_1, y_2, \ldots, y_n) \in \mathbb{R}^n$ let the function $\langle \cdot, \cdot \rangle$ be defined by

$$\langle \mathbf{x}, \mathbf{y} \rangle = \sum_{j=1}^{n} x_j y_j.$$

Then $\langle \cdot, \cdot \rangle$ defines an inner product on $\mathbb{R}^n$. In fact, the definition of the inner product is modelled after the familiar inner product (dot product) defined for $\mathbb{R}^n$.

Example B.2.2:   Let ℓ^2 denote the set of sequences $\{a_n\}$ such that the series $\sum_{n=1}^{\infty} a_n^2$ is convergent. If addition and scalar multiplication are defined the same way as for the space ℓ^1 in Example B.1.3, then ℓ^2 is a vector space. Suppose that $\mathbf{a} = \{a_n\}, \mathbf{b} = \{b_n\} \in \ell^2$, and let $c_n = \max(a_n, b_n)$. Then the series $\sum_{n=1}^{\infty} c_n^2$ is convergent and hence the series $\sum_{n=1}^{\infty} a_n b_n$ is absolutely convergent. An inner product on this vector space is defined by

---

[1] Strictly speaking the function replacing the norm is not even a norm because $\|f\|_{L1} = 0$ does not imply that $f = 0$. This problem is easily remedied using equivalence classes; i.e., two functions f and g are equivalent if $f = g$ almost everywhere.

$$\langle \mathbf{a}, \mathbf{b} \rangle = \sum_{n=1}^{\infty} a_n b_n.$$

If $(X, \langle \cdot, \cdot \rangle)$ is an inner product space, then it can be shown that the function $\| \cdot \|$ defined by

$$\|f\| = \sqrt{\langle f, f \rangle},$$

is a norm on the vector space X. Thus, any inner product space leads to a normed vector space. The special norm defined above is called the norm induced by the inner product. The normed vector space formed by the induced norm may or may not be complete. If (X, N) is a Banach space then the inner product space $(X, \langle \cdot, \cdot \rangle)$ is called a **Hilbert space**. A Hilbert space is thus an inner product space that is complete in the norm induced by the inner product. The inner product space $(\mathbb{R}^n, \langle \cdot, \cdot \rangle)$ is an example of a finite-dimensional Hilbert space. It can be shown that the inner product space $(\ell^2, \langle \cdot, \cdot \rangle)$ of Example B.2.2 is also a Hilbert space. Another infinite-dimensional Hilbert space of importance in analysis is the space $L^2[x_0, x_1]$.

Example B.2.3: Let $L^2[x_0, x_1]$ denote the set of functions [2] $f : [x_0, x_1] \to \mathbb{R}$ such that the Lebesgue integral $\int_{[x_0, x_1]} f^2(x)\, dx$ exists (i.e., the set of "square integrable" functions), and let $\langle \cdot, \cdot \rangle$ be defined by

$$\langle f, g \rangle = \int_{[x_0, x_1]} f(x) g(x) \, dx.$$

It can be shown that for any $f, g \in L^2[x_0, x_1]$ the above function is well defined and satisfies the axioms of an inner product. The resulting inner product space is a Hilbert space.

Hilbert spaces have found widespread applications in pure and applied mathematics. The extra structure afforded by an inner product gives rise to a generalization of geometrical concepts in $\mathbb{R}^n$. In particular, there is a straightforward extension of the orthogonality based on the inner product. Recall that in $\mathbb{R}^n$ two nonzero vectors $\mathbf{u}, \mathbf{v}$ are orthogonal if and only if $\langle \mathbf{u}, \mathbf{v} \rangle = 0$. For the general Hilbert space, we say that two elements f, g are **orthogonal** if $\langle f, g \rangle = 0$. Thus, for example, in the space $L^2[x_0, x_1]$ two functions f, g are orthogonal if

$$\int_{[x_0, x_1]} f(x) g(x) \, dx = 0.$$

As in the finite-dimensional case, given a set of elements in a general Hilbert space it is possible to form an orthogonal set by an algorithm analogous to the

---

[2] Strictly speaking, the elements of this set are equivalence classes of functions modulo equality almost everywhere.

Gram-Schmidt process. It is also possible to construct orthogonal bases for Hilbert spaces. Although bases for general infinite-dimensional Banach spaces play a somewhat nominal rôle (in contrast with finite-dimensional spaces), bases play a significant rôle in the theory of Hilbert spaces.

References

1. Abraham, R. and Marsden, J., *Foundations of Mechanics*, 2nd ed., Benjamin/Cummings Publ. Co., 1985.
2. Anco, S.C. and Bluman, G.W., "Direct Construction of Conservation Laws from Field Equations," *Phys. Rev. Lett.*, **78** no. 15, pp. 2869-2873, 1997.
3. Anco, S.C. and Bluman, G.W., "Derivation of Conservation Laws from Nonlocal Symmetries of Differential Equations," *J. Math. Phys.*, **37** no. 5, pp. 2361-2375, 1996.
4. Anderson, I. and Thompson, G., "The Inverse Problem of the Calculus of Variations for Ordinary Differential Equations," *Memoirs of the Amer. Math. Soc.*, vol. **98**, No. 473, 1992.
5. Arfken, G., *Mathematical Methods for Physicists*, 2nd ed., Academic Press, 1970.
6. Arnold, V.I., *Mathematical Methods of Classical Mechanics*, Springer-Verlag, 1978.
7. Bernstein, S.N., "Sur les équations du calcul des variations," *Ann. Sci. École Norm. Sup.*, **29**, pp.431-485, 1912.
8. Bhutani, O.P. and Vijayakumar, K., "On Certain New and Exact Solutions of the Emden-Fowler Equation and Emden Equation via Invariant Variational Principles and Group Invariance," *J. Austral. Math. Soc. B*, **32**, pp. 457-468, 1991.
9. Birkhoff, G. and Rota, G., *Ordinary Differential Equations*, 4th ed., John Wiley and Sons, 1989.
10. Bliss, G.A., *Lectures on the Calculus of Variations*, University of Chicago Press, 1946.
11. Bluman, G.W. and Kumei, S., *Symmetries and Differential Equations*, Springer-Verlag, 1989.
12. Bolza, O., *Lectures on the Calculus of Variations*, G.E. Stechert and Co., 1931.
13. Boltyanskii, V.G., *Envelopes*, Pergamon Press, 1964.
14. Born, M. and Wolf, E., *Principles of Optics*, 6th ed., Pergamon Press, 1986.
15. Brechtken-Manderscheid, U., *Introduction to the Calculus of Variations*, Chapman & Hall, 1991.
16. Bromwich, T.A., *An Introduction to the Theory of Infinite Series*, Macmillan and Co., 1926.
17. Browder, F., ed. *Mathematical Developments Arising from Hilbert Problems*, Proceedings of the Symposium in Pure Mathematics of the American Mathematical Society, vol. **28**, 1976.

18. Bruce, J.W. and Giblin, P.J., *Curves and Singularities*, Cambridge University Press, 1984.
19. Buck, R. and Buck, E., *Advanced Calculus*, 2nd ed., McGraw-Hill, 1965.
20. Burke, W.L., *Applied Differential Geometry*, Cambridge University Press, 1985.
21. Carathéodory, C., *Calculus of Variations and Partial Differential Equations of the First Order*, Chelsea, 1982.
22. Choquet-Bruhat, Y., DeWitt-Morette, C., and Dillard-Bleick, M., *Analysis, Manifolds and Physics*, rev. ed., North-Holland, 1982.
23. Churchill, R.V., *Fourier Series and Boundary Value Problems*, 2nd ed., McGraw-Hill, 1963.
24. Coddington, E.A. and Levinson, N., *Theory of Ordinary Differential Equations*, McGraw-Hill, 1955.
25. Courant, R. and Hilbert, D., *Methods of Mathematical Physics*, vol. 1, John Wiley and Sons, 1953.
26. Ewing, G.M., *Calculus of Variations with Applications*, Dover, 1985.
27. Forsyth, A.R., *Calculus of Variations*, Cambridge University Press, 1927.
28. Fox, C., *An Introduction to the Calculus of Variations*, Dover, 1987.
29. Fulks, W., *Advanced Calculus*, 3rd ed., John Wiley, 1978.
30. Garabedian, P.R., *Partial Differential Equations*, 2nd ed., Chelsea, 1986.
31. Gelfand, I.M. and Fomin, S.V., *Calculus of Variations*, Prentice-Hall, 1963.
32. Giaquinta, M. and Hildebrandt, S., *Calculus of Variations I:The Lagrangian Formalism*, Springer-Verlag, 1996.
33. Giaquinta, M. and Hildebrandt, S., *Calculus of Variations II:The Hamiltonian Formalism*, Springer-Verlag, 1996.
34. Gilbarg D. and Trudinger, N.S., *Elliptic Partial Differential Equations of the Second Order*, Springer-Verlag, 1977.
35. Goldstein, H., *Classical Mechanics*, 2nd ed., Addison-Wesley Publ. Co., 1980.
36. Goldstine, H., *A History of the Calculus of Variations from the 17th through the 19th Century*, Springer-Verlag, 1980.
37. Goodrich, N., *The Medieval Myths*, Mentor, 1961.
38. Hilbert, D., *Mathematical Problems*, Lecture delivered before the International Congress of Mathematicians at Paris in 1900, *Bull. Amer. Math. Soc.*, vol. **8**, pp. 437-479, 1902.
39. Hochstadt, H., *Integral Equations*, Wiley, 1973.
40. Hutson, V. and Pym, J.S., *Applications of Functional Analysis and Operator Theory*, Academic Press, 1980.
41. Ince, E.L., *Ordinary Differential Equations*, Longmans, Green and Co., 1927.
42. John, F., *Partial Differential Equations*, 4th ed., Springer-Verlag, 1982.
43. Kalnins, E.G., *Separation of Variables for Riemannian Spaces of Constant Curvature*, Pitman Monograph, Longman, 1986.
44. Kaplan, W., *Advanced Calculus*, 2nd ed., Addison-Wesley, 1973.
45. Körner, T.W., *Fourier Analysis*, Cambridge University Press, 1988.
46. Kreyszig, E., *Introductory Functional Analysis with Applications*, John Wiley and Sons, 1978.
47. Kreyszig, E., *Advanced Engineering Mathematics*, 4th ed., John Wiley and Sons, 1979.
48. Lanczos, C., *The Variational Principles of Mechanics*, 4th ed., University of Toronto Press, 1970.
49. Landau, L.D. and Lifshitz, E.M., *Mechanics*, Course of Theoretical Physics, Vol. I, 3rd ed., Pergamon, 1976.

50. Luneberg, R.K., *Mathematical Theory of Optics*, mimeographed lecture notes, Brown University, Providence R.I., 1944.
51. Mach, E., *Science of Mechanics*, 2nd ed., Open Court Publ. Co., 1907.
52. McLachlan, N.W., *Theory and Application of Mathieu Functions*, Oxford University Press, 1947.
53. Milnor, J., *Morse Theory*, Princeton University Press, 1963.
54. Moiseiwitsch, B.L., *Variational Principles*, Interscience, 1966.
55. Morse, M., *The Calculus of Variations in the Large*, American Math. Soc. Colloquium Pub., Vol. **18**, 1932.
56. Olmsted, J., *Advanced Calculus*, Appleton-Century-Crofts, 1961.
57. Olver, P.J., *Applications of Lie Groups to Differential Equations*, Springer-Verlag, 1986.
58. Osserman, R., *A Survey of Minimal Surfaces*, Dover, 1986.
59. Pars, L.A., *A Treatise on Analytical Dynamics*, Heinemann, 1965.
60. Petrovski, I.G., *Ordinary Differential Equations*, Prentice-Hall, 1966.
61. Piaggio, H.T.H., *An Elementary Treatise on Differential Equations and Their Applications*, rev. ed., Bell, 1971.
62. Postnikov, M.M., *The Variational Theory of Geodesics*, Dover, 1983.
63. Rund, H. *The Hamilton-Jacobi Theory in the Calculus of Variations*, Van Nostrand, 1966.
64. Spivak, M., *A Comprehensive Introduction to Differential Geometry*, Vol. **III**, 2nd ed., Publish or Perish, 1979.
65. Spivak, M., *A Comprehensive Introduction to Differential Geometry*, Vol. **IV**, 2nd ed., Publish or Perish, 1979.
66. Stoker, J.J., *Differential Geometry*, Wiley-Interscience, 1969.
67. Tee, G., " Isochrones and Brachistochrones," *Neural, Parallel, & Scientific Computations*, **7**, pp. 311-342, 1999.
68. Tee, G., "Brachistochrones for Attractive Logarithmic Potential," *Auckland Department of Mathematics* Report Series No.410, 1999.
69. Tee, G., "Brachistochrones for Repulsive Inverse Square Force," *Auckland Department of Mathematics* Report Series No.409, 1999.
70. Titchmarsh, E.C., *Eigenfunction Expansions*, Part I, 2nd ed., Oxford University Press, 1962.
71. Wan, F.W., *Introduction to the Calculus of Variations and its Applications*, Chapman & Hall, 1995.
72. Webster, A.G., *The Dynamics of Particles and of Rigid, Elastic, and Fluid Bodies*, 3rd ed., Teubner, 1925.
73. Whittaker, E.T., *A Treatise on the Analytical Dynamics of Particles and Bodies*, 4th ed., Cambridge University Press, 1952.
74. Whittaker, E.T. and Watson, G.N., *A Course of Modern Analysis*, 4th ed., Cambridge University Press, 1952.
75. Willmore, T.J., *An Introduction to Differential Geometry*, Oxford University Press, 1959.

Index

Universitext   *(continued)*